Asit K. Biswas

Water Resources of North America

Springer
Berlin
Heidelberg
New York
Hong Kong
London
Milan
Paris
Tokyo

Asit K. Biswas (Ed.)

Water Resources of North America

with 66 Figures and 41 Tables

Springer

PROFESSOR ASIT K. BISWAS
Avenida Manantial Oriente 27
Los Clubes, Atizapan
Estado de México, 52958, Mexico

Email: akbiswas@att.net.mx

cover photo: © Bryan P. Bernart

ISBN 3-540-00284-7 Springer-Verlag Berlin Heidelberg New York

Cataloging-in-Publication Data

Biswas, Asit K.
Water resources of North America / Asit K. Biswas.
  p.cm.
  Includes bibliographical references (p.).
  ISBN 3-540-00284-7 (alk. paper)
  1.   Water resources development—North America. 2. Water-supply—North America—
  Management. I. Title.

Springer-Verlag Berlin Heidelberg New York
a member of BertelsmannSpringer Science+Business Media GmbH

http://www.springer.de

© Springer-Verlag Berlin Heidelberg 2003
Printed in Germany

Camera ready by authors
Cover design: E. Kirchner, Heidelberg
Printed on acid-free paper   32/3141/as 5 4 3 2 1 0

Preface

Leonardo da Vinci, the eminent Renaissance scholar and philosopher said, "water is the driver of nature". Many may have considered it to be an overstatement in the past, but at the beginning of the third millennium, no sane individual would disagree with Leonardo's view. Water is becoming an increasingly scarce resource for most of the world's citizens. The current trends indicate that the overall situation is likely to deteriorate further, at least for the next two decades, unless the water profession eschews its existing "business as usual" practices, which can only allow incremental changes to occur.

Somewhat surprisingly, the water profession as a whole neither realised nor appreciated the gravity of the global water situation as late as 1990, even though a few serious scholars have been pointing out the increasing criticality of the situation from around 1982. For example, the seriousness of the crisis was not a major issue, either at the International Conference on Water and the Environment, which was organised by the UN system in Dublin and also at the UN Conference on Environment and Development at Rio de Janeiro. Held in 1992, both are considered to be important events for the water sector of the past decade. It is now being increasingly recognised that the Dublin Conference was poorly planned and organised, and thus not surprisingly it produced very little, if any, worthwhile and lasting results. Furthermore, as the Dublin Conference was expected to make the necessary inputs to the Rio discussions, water did not receive proper emphasis at Rio. For all practical purposes, at Rio, water was basically ignored by all the Heads of States, whose primary interests were focused on issues like climate change, biodiversity, and deforestation. Water was at best a very minor issue during the plenary sessions at Rio. Equally, the chapter on water in Agenda 21 is not only the longest but also is the most poorly formulated.

Thus, in spite of the rhetorics of many international institutions, the impacts of the Dublin and the Rio Conferences on water management globally and regionally are now not discernible. In all probability, the developments in the water sector would not have been materially different even if these two events had never occurred. Much as though this simple fact was consistently denied by the institutions and the individuals who were associated with the organization of the Dublin and the Rio Conferences, the truth is that these two conferences only had a very marginal impacts on global developments during the ensuing 10 years. This fact was widely accepted during the UN World Summit on Sustainable Development that was held on Johannesburg in 2002. Thus, both the Dublin and the Rio constituted a lost opportunity for water, since they failed to put water in the international agenda. Only time will tell if the Johannesburg Summit will have any impact on the future water developments. However, the initial signs are not very encouraging.

By the second half of the 1990s, most of the water profession had accepted that the world is heading for a water crisis that is likely to be unprecedented in the human history. Just as at the beginning of the decade, the water crisis was not taken very seriously by most members of the profession, now nearly every one believes that such a crisis is inevitable. It is highly likely that just as the previous position was erroneous, so will be the current thinking.

While predicting the future is an extremely hazardous business, one item can be predicted with complete certainty: the world in the year 2025 will be vastly different from what it is today, in the same way that the world today is significantly different from what it was in 1975. Among the main driving forces which are likely to contribute to these changes are evolution of new development paradigms, demographic changes, technological advances in all fields, speed and extent of globalisation, improvements in human capital, and future national and inter-governmental policies.

The water sector is an integral component of the global system, and thus it will unquestionably undergo major changes during the next 25 years. In fact, *water development and management will change more during the next twenty years compared to what we have witnessed during the past 2000 years.* The water profession has generally ignored so far the global forces outside the water sector that are already shaping its future availability, use and management practices. These external impacts are likely to increase significantly in the next quarter of a century. And yet, the water profession has not yet started to consider seriously the implications of accelerating developments in areas like globalisation, public-private partnership, demand management practices, biotechnology, desalination, and information and communication revolution, which are likely to change water availability, use, and management practices dramatically in the coming decades.

It is now widely accepted that the world will face a major water crisis in the coming decades because of increasing physical scarcities in numerous countries. Many international organisations have published maps in recent years, all somewhat similar, which show that more and more countries of the world will become water-stressed because of increasing scarcities. It is high time to review critically the reliability of such forecasts for several reasons. First, the information base on which such forecasts and maps are based is highly unreliable. Extensive review by the Third World Centre for Water Management indicates that the national estimates of water availability and use, on which the current global figures are based, are often erroneous, in many cases very significantly. For many major countries, like India and China, some estimates of water availability and use are currently available, but no one has any clear idea about the reliability and usefulness of such national statistics. Thus, it is impossible to get any reasonably reliable picture of the global and regional water situations, which are based on the aggregation of such incomplete and unreliable national data sets.

Second, water abstraction is at present widely used as a proxy for water use. Methodologically, this of course is fundamentally wrong. Unlike oil, water is a reusable resource, which can be used and reused many times. For example, some scientists have pointed out that each drop of the Colorado River water is currently used 6-7 times before it reaches the sea. Globally water is being increasingly reused, both formally and informally, and all the indications are that the extent of reuse in all countries will accelerate further in the coming decades. Accordingly, the current practice of using water abstraction as a proxy for water use is already significantly erroneous. In 10-20 years, when reuse becomes even more extensive, this practice of using water abstraction data as the total water available would become virtually meaningless.

Currently, no reasonable estimates exist for reuse of water, even at the national levels, let alone for the world as a whole. Some data exist only for a very few developed countries like Japan. Since the water profession has not considered reuse as an important factor in global water availability and use considerations, all the existing forecasts are highly suspect.

Third, private sector, water pricing and cost recovery are likely to play increasingly important roles as the 21st century progresses. The net result of these developments is likely to be significant advances in demand management, which currently plays a minor role in most countries of the world. This would mean that within a short period of about a decade or so, all projections of future water requirements would have to significantly revised downwards because of increasing emphasis on demand management, other associated conservation practices, and better management processes.

The current estimates of the future global water requirements are likely to prove to be far too high, and thus would have to be revised significantly downwards during the next decade. Simultaneously, the amount of water that is available for use at present is seriously underestimated because reuse and recycling are ignored, estimates of groundwater availability would have to be revised upwards, and technological advances are making costs of desalination and new non-conventional ways of transportation of water (i.e. rubber bags to transport water over long distances over the sea) more and more attractive. Because of the upward adjustments in water availability and downward revisions in requirements, *one can now be cautiously optimistic of the global water future.*

This, of course, does not mean that it would be an easy process for countries to adjust to the new realities of a rapidly changing global water scene.

Unquestionably, many countries are likely to find it difficult to manage the transformation without discontinuities because of socio-political constraints, institutional inertia, increasing management complexities and current and past inefficient water management practices. However, since the "business as usual" will not be a feasible option for the future in nearly all countries, policy-makers, water professionals and public institutions, whether they like it or not, would be forced to embrace the new conditions, most probably within the next 10-15 years. All these and other associated developments are likely to make the present, "gloom and doom" forecasts of a global crisis due to water scarcities somewhat unlikely in the coming decades.

On the basis of the above analysis and other associated considerations, the threat of a global water crisis because of physical scarcities only, as is expected at present, is probably overstated. If there is likely to be a crisis in the water sector, it would probably be due to two reasons, none of which is receiving adequate attention at present.

The first reason which could contribute to a crisis could be due to continuous water quality deterioration. Globally, water quality is receiving inadequate attention, even though it has already become a critical issue. *While global data on water quantity is poor, it is virtually non-existent for water quality.* Even for major developed countries like the United States or Japan, a reliable picture of national water quality situation currently simply does not exist. For developing countries and for countries in transition, existing frameworks and networks for water quality monitoring are highly deficient, adequate expertise on water quality management simply does not exist, and laboratories for water quality assessments suffer very seriously from poor quality control and quality assurance practices. Furthermore, senior water policy-makers in most developing countries become interested in quality aspects primarily when there are major local crises due to political and /or media interventions. *Sadly, for all practical purposes, water quality is still receiving only lip service from the water professionals, most senior bureaucrats and politicians of developing countries and countries in transition.*

Not surprisingly, because of the above deficiencies, water quality problems are becoming increasingly serious in all developing countries. For example, nearly all surface water bodies within and near urban industrial centres are now highly polluted. While data on the existing groundwater quality are extremely poor, it is highly likely that groundwater is also getting increasingly contaminated near centres of population.

Because of poor and unreliable water quality data, we now have an incorrect picture of the existing water quality conditions. As a general rule, in developing countries, the official pictures of water quality situations are mostly rosier than the current conditions warrant. Such statements are often repeated by international institutions, which simply adds to a false sense of security.

Absence of good data and information means that many of the currently accepted wisdoms in the water sector need to be challenged. Furthermore, some of the current problem identifications may be wide of the mark because the data used may not be reliable, or data required are simply not available or accessible. If the problem identifications are not accurate, their solutions cannot be correct either.

The quality and extent of data available in most parts of the world at present do not allow us to assess, analyse and evaluate the degrees and extents of many global and/or regional water problems. Thus, when the World Commission on Water, of which I as a member, started its work, very soon it became clear that the current data availability is a serious constraint to conduct proper global water policy dialogues. In some instances, data may have been collected, but they may not be accessible.

Faced with these constraints, the Third World Centre for Water Management, with the support of the Nippon Foundation, initiated a project on the assessment of freshwater-related issues in certain major countries of the world. The study considered issues like water availability (quantity and quality), uses, demands, impacts on environment and health, etc.

Unlike earlier assessments, these studies were not carried out by bureaucrats in international organisations with limited knowledge of national water situations, but by knowledgeable water experts from the countries selected. This book is the first major publication from this project, which provides an objective and definitive assessment of the state of the water resources in North America, that is, Canada, the United States and Mexico. We believe that these studies will give the water professionals a good platform from which the current problems in these three countries can properly be assessed, and reliable forecasts can be made of the potential water and water-related problems of the future. It would also enable us to identify their possible solutions.

On behalf of the Third World Centre for Water Management, I would like to express our most sincere appreciation to the Nippon Foundation for supporting this overall study. We are specially grateful to Mr. Reizo Utagawa, Managing Director, and Mr. Masanori Tamazawa of the International Affairs Division, both of the Nippon Foundation, for their personal interest in this study. Without their support and encouragement, the present book could not have been prepared.

June 2003   Asit K. Biswas, President
                                             Third World Centre for Water
                                             Management, Mexico

Table of Contents

1. General Aspects

1.1 Physical Context

Located in the northern portion of the North American continent, Canada extends roughly 5,300 km east to west, from St. John's, Newfoundland, to the Queen Charlotte Islands in British Columbia, and nearly 4,600 km south to north, from Point Pelee in southern Ontario to Alert on Ellesmere Island in the high Arctic, and is the second largest country in the world after the Russian federation. At 9,970,610 km^2, it's surface area (land plus fresh water) represents 7% of the global land mass. Coniferous forest covers the largest area, almost 28%. Canada's coastline of 243,792 km is the longest in the world (Figure 1.1).

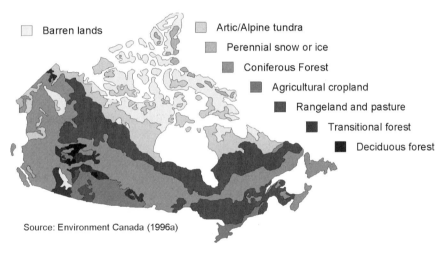

Barren lands

Artic/Alpine tundra

Perennial snow or ice

Coniferous Forest

Agricultural cropland

Rangeland and pasture

Transitional forest

Deciduous forest

Source: Environment Canada (1996a)

Fig. 1.1. Land cover in Canada

1.2 Socio-political Context

A self-governing *Dominion* in the British Empire since 1867, full independence for Canada (as for all British colonies) was established only in 1931 by the Statute of Westminster. Canada is a constitutional monarchy[1] and a federal state with a democratic Parliament. The Parliament in the capital, Ottawa, consists of the House of Commons, whose members are elected, and the Senate, whose members are appointed by the Governor General on the advice of the Prime Minister.

The federal structure, with the sharing of powers it entails, is the formula that has been chosen by Canadians to take into account the country's geographical realities, the diversity of its cultural communities and its dual legal and linguistic heritage (today English is spoken by 59.2% of the population and French by 23.3% mainly in the province of Québec – (Statistics Canada 2000)). Canada has 10 provinces and 3 territories (Figure 1.2), each with its own capital city (in brackets): Alberta (Edmonton); British Columbia (Victoria); Prince Edward Island (Charlottetown); Manitoba (Winnipeg); New Brunswick (Fredericton); Nova Scotia (Halifax); Nunavut (Iqaluit); Ontario (Toronto); Quebec (Quebec City); Saskatchewan (Regina); Newfoundland (St. John's); Northwest Territories (Yellowknife); and Yukon Territory (Whitehorse).

Fig. 1.2. Canada: the political context

---

[1] Elizabeth II, Queen of the United Kingdom, is also Canada's Queen and sovereign of a number of realms. In her capacity as Queen of Canada, she delegates her powers to a Canadian Governor General. Canada is thus a constitutional monarchy: the Queen rules but does not govern.

Management of freshwater resources in Canada depends in important respects on co-operation between these two levels of government (see Chap. XII on Institutional and Legal Framework for Water Management). More recently, many water management responsibilities have been downloaded to municipal governments and a high level of public participation in many aspects of water use decision-making is occurring. Because many of Canada's water resources lie in basins that are shared with the United States, international agreements play an important role in the management of water resources (see Chap. XII).

With 30.6 million people in 1999 (an average of only about 3 people per km^2) Canada is home to just 0.5% of the world's population. However, Canada's population is not distributed evenly throughout the country. Indeed 80% of Canadians live in urban areas, which cover only 0.2% of the country, and nearly 60% of Canada's urban population lives in centres of 500,000 or more (Figure 1.3).

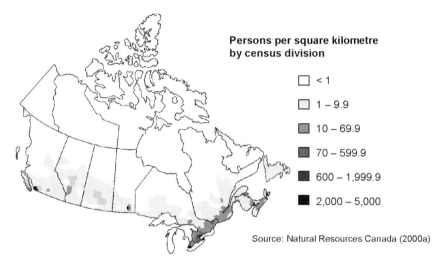

Persons per square kilometre by census division

☐ < 1

☐ 1 – 9.9

▨ 10 – 69.9

▨ 70 – 599.9

■ 600 – 1,999.9

■ 2,000 – 5,000

Source: Natural Resources Canada (2000a)

Fig. 1.3. Population density in Canada

Canadians, have a strong association with water for their recreational and cultural uses (see section on instream uses in Chap. IV: Water uses in Canada). Activities such as swimming, boating, canoeing, fishing, and camping allow Canadians to experience the beauty of Canada's lakes and rivers. Expenditures on water-related recreational activities and tourism also contribute billions of dollars per year to the national economy. For example, according to Department of Fisheries and Oceans and Statistics Canada, in 1990, Canadians spent about $5.9 billion[2] on goods and services directly related to sport fishing.

---

[2] In the report the $ symbol is used in term of Canadian dollars.

But living in relative affluence among seemingly abundant natural resources, the population has adopted a lifestyle that takes the business of *consuming* very seriously. Every day, a Canadian uses the equivalent of nearly 24.64 L of gasoline, the most energy per capita of any country; withdraws 343 litres of water for household use alone, helping to make Canada the world's second largest user of water per capita; and throws away about 1.8 kg of residential waste, assuring Canada's place among the leading *garbage producers* on a global basis.

1.3 Economic Context

More than any other industrialised nation, Canada's economy still depends on the land and natural resources. One in three workers is employed directly or indirectly in agriculture, forestry, mining, energy, fishing or other resource-based activity (see Table 1). Each year, these same activities account for approximately half of the value of exports. This is also very well characterized if we look at the country's land-use situation. Indeed, as shown in the predominant land sectors are forestry, recreation, conservation, agriculture, energy development, transportation, and all of the activities that are carried out in urban settings, including manufacturing and residential uses. In Canada, the GDP in 1999 was $880,254 and was growing at 4.5% per annum (Statistics Canada 2000). Thus the GDP per capita in 1999 was $29,050. If we look at the contribution by sectors to GDP: Agriculture contributes 3%, Industry 33% and Services (commerce, transport, finance, insurance, health care, education, science, technology, public administration, tourism) 64 %.

Table 1.1. Predominant land activities in Canada

Land-use class	Predominant activity in the class	Area[1] [000 km^2]	%[2]
Forestry [3]	Active forest harvesting or potential for future harvesting	2,440	24
Recreation and conservation [4]	Recreation and conservation within national, provincial, and territorial parks, wildlife reserve, sanctuaries, etc.	756	8
Agriculture [5]	Agriculture on improved farmland (cropland, improved pasture, summer fallow) and unimproved farmland	680	7
Urban	Built-up areas	20	< 1
Other activities	Includes hunting and trapping, mining, energy, developments and transportation	6,074	61
Total		9,970	100

[1] Includes the area of all land and fresh water
[2] Rounded to the nearest percent
[3] Canadian Council of Forest Ministers (1995)
[4] Environment Canada
[5] Statistics Canada (1994, 1997) http://www.statcan.ca
Source: Environment Canada (1996a)

1.4 Climate

Even though Canada is generally thought of as a cold country, the climate is so variable that there are rainforest conditions on the west coast and semi-arid conditions in the driest parts of the Prairies and British Columbia. As shown in Figure 1.4 some parts of British Columbia and the Prairies are very dry. Other parts of the country, such as southern Ontario, suffer water shortages only during extended dry periods. The east and west coasts receive a good deal of rain. In winter, large portions of the country have precipitation in the form of snow, much of which remains on the ground until Spring thaw. Snow accumulation over the winter has significant implications for hydrology, wildlife, vegetation, and socio-economic activities.

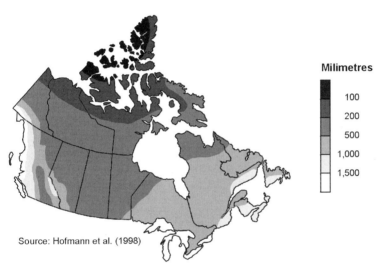

Milimetres

100
200
500
1,000
1,500

Source: Hofmann et al. (1998)

Fig. 1.4. Average annual precipitation in Canada

As mentioned in Hofmann et al. (1998), "the temporal and spatial distribution of precipitation is a major determinant of habitability of land for humans (Rogers 1994) and the development of characteristic biophysical systems". The distribution of precipitation is variable across Canada (see Figure 1.4). These regional differences are due to the size of, and heterogeneity in, the Canadian landscape and the diverse weather systems which influence the nation. For example, the coastal regions of British Columbia receive the highest precipitation primarily in the winter. Meanwhile, the Prairie region has less precipitation due to the *rain-shadow effect* of the Rockies. The Arctic region also receives little precipitation. Eastern Canada receives an even seasonal distribution of precipitation. In an eastward progression from the eastern Prairies in Winnipeg (500 millimetres (mm)) through Ontario and Québec to Halifax, Nova Scotia (1,500 mm), precipitation increases by 40 mm/100 kilometres (km) in a south-east trend. The Atlantic region receives a significant amount of precipitation (Phillips 1990).

1.5 Canadian Ecozones

Canada has a mosaic of 20 major distinctive ecosystems, many of which are unique: 15 terrestrial Ecozones and 5 marine Ecozones. The legend in Figure 1.5 arranges the Ecozones according to general shared properties. This classification established for the State of the Environment (Environment Canada 1996a) is also used in this report to describe the main Canadian ecosystems and later the water resource characteristics of the country. Ecozones are arranged into seven units (Figure 1.6). The intent of organizing them this way was to provide a slightly more familiar presentation for reporting purposes. Some of the Ecozones roughly coincide with particular provinces and territories and so were grouped on that basis. For instance, the five Ecozones mainly associated with British Columbia and the Yukon (Taiga Cordillera, Pacific Maritime, Montane Cordillera, Boreal Cordillera, and Pacific Marine) have been grouped as Pacific and Western Mountains Ecozones. Similarly, the Prairies and the Boreal Plains Ecozones are linked to the Prairie provinces and have been grouped as Central Plains Ecozones. In other cases, such as the Boreal Shield Ecozone, there was no particular coincidence with specific provinces or territories.

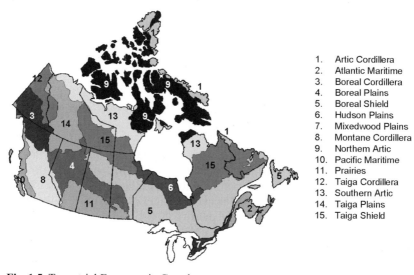

1. Artic Cordillera
2. Atlantic Maritime
3. Boreal Cordillera
4. Boreal Plains
5. Boreal Shield
6. Hudson Plains
7. Mixedwood Plains
8. Montane Cordillera
9. Northern Artic
10. Pacific Maritime
11. Prairies
12. Taiga Cordillera
13. Southern Artic
14. Taiga Plains
15. Taiga Shield

Fig. 1.5. Terrestrial Ecozones in Canada

Canada's three Arctic Ecozones, composing the Arctic Ecozones Unit, occupy diagonal bands across the North, and collectively cover an area more than four times the size of France. The summer land cover ranges from Arctic tundra, with low plants and wetlands, to rocky, barren land, with extremely sparse vegetation or no vegetation at all. Some places are covered with snow and ice year-round (Environment Canada 1996a).

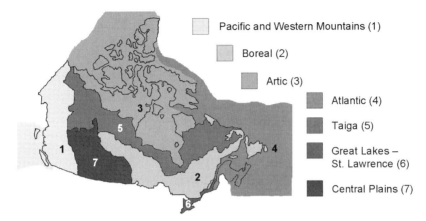

Fig. 1.6. Seven ecozones units

The Taiga Ecozones unit is composed of the Taiga shield, the taiga plains, and the Hudson Plains Ecozones. The Taiga Ecozones cover a vast territory, 2,375 million km^2 in area (almost one-quarter of Canada), which arcs across the nation from the Mackenzie Delta in the extreme Northwest to the Labrador coast in the East. The dominant vegetation cover is an open forest of small, slow-growing conifers, transitional between the closed-canopy boreal forest to the south and the treeless tundra to the north. The characteristic features of the Taiga Eco-zones are: a sub-arctic climate, marked by long, cold winters with persistent snow and ice cover; discontinuous permafrost; numerous lakes and rivers and extensive wetlands; a transition from sub-Arctic to Arctic flora and fauna; a small human population, comprising about 60% Aboriginal people; and an economy based on traditional resource use, with areas of intensive natural resource development, notably hydroelectric developments and mine sites. The Hudson Plains Ecozone, which is the world's largest wetland area, is larger than Italy.

The Boreal Ecozone Unit only contains the Boreal Shield Ecozone which is the largest in Canada. It is an irregularly shaped area stretching across parts of six provinces (from Alberta to Newfoundland). Actually, it covers the entire Atlantic Province of Newfoundland. With 1.8 million km^2 (20% of Canada's land mass) the Boreal Shield is bigger than either the state of Alaska, the country of Mongolia, or the province of Quebec. Studded with evergreens, the granitoid rocks of the Boreal Shield Ecozone stretch 3,800 km from the east coast of Newfoundland to northern Alberta. This rugged, rolling landscape of bedrock and forests is situated at the geographical overlap of two of the biggest physiographic areas on Earth: the Canadian Shield and the boreal forest. The largest of Canada's 15 terrestrial Ecozones, the Boreal Shield Ecozone includes parts of six provinces, covers over 1.8 million km², and encompasses almost 20% of Canada's land mass.

The Pacific and western mountains Ecozones unit is composed of the Pacific Maritime, the Montane Cordillera, The Boreal Cordillera, and The Taiga Cordillera Ecozones. This unit contains very diversified ecosystems. For example, in the Pacific Maritime Ecozone temperate rain forest dominates much of the lower and mid elevations, but the base includes areas such as semi-Mediterranean arbutus and Garry Oak woodlands, whereas alpine meadows and snow packs dominate several layers higher up.

The Central Plains Ecozones unit comprises the Prairies Ecozone and the Boreal Plains Ecozone. The Central Plains Ecozones unit is less diverse than the Pacific and western mountains Ecozones unit and consists of a flat to gently rolling landscape underlain by deep glacial deposits. Prior to European settlement, the Prairies Ecozone consisted largely of dry mixed grasslands in south-eastern Alberta and south-western Saskatchewan, aspen parkland and fescue prairie through the transition to the Boreal Plains Ecozone in the north, and tallgrass prairie in the south-eastern prairies of Manitoba. Since European settlement, the Ecozone has become one of the most extensive agricultural regions in the world. Its total land area is about 47 million hectares. Of that, 70% is classified as cropland and 27% as rangeland and pasture. The remaining 3% is deciduous forest.

The Great Lakes - Saint Lawrence basin has been shaped by the repeated advances and retreats of continental ice sheets, which gouged out deep trenches in the relatively soft sedimentary bedrock on the edge of the Canadian Shield. The last glaciation to cover the region, the Wisconsin, melted about 11,000 years ago. The meltwaters pooled into the precursors of the present lakes and drained south into the Mississippi basin and south-east into the Atlantic Ocean. The Champlain Sea covered the St. Lawrence and lower Ottawa valleys until about 9,800 years ago. The Great Lakes in their present form, draining through the St. Lawrence River, appeared about 2,000-3,000 years ago (Francis and Regier 1995). In the north, the glaciers scoured the ancient granitic bedrock of the Canadian Shield, leaving large expanses exposed. To the south, the glaciers deposited silt, sand, and gravel, forming tills. Moraines, such as the Oak Ridges Moraine near Toronto, form some of the most prominent reliefs in the otherwise level to rolling landscape. The Niagara Escarpment, which extends north-west from Niagara Falls to Georgian Bay, and the Monteregian Hills around Montreal form the only other reliefs. In the east, glacial meltwater lakes and the Champlain Sea left deposits of clay and silt, forming today's outwash plains and kettle lakes. All these deposits provided the basis of the fertile soils of the Mixed-wood Plains Ecozone. Soils in the northern portions of the basin are shallow, often acidic, and generally much less fertile.

The Atlantic Maritime unit also contains only one Ecozone; the Atlantic Maritime Ecozone. It includes Nova Scotia, New Brunswick, Prince Edward Island,

and a portion of Quebec. It represents 2% of Canada's total land and freshwater area and includes 11,200 km of coastline. Forests make up 90% of the total land cover and are the predominant ecosystems in the Atlantic Maritime Ecozone. Until recently, fishing was an economic mainstay of the Atlantic Ecozones. This remark also applies for Newfoundland which is actually covered by the Boreal Shield Ecozone (see above). By the late 1980s, however, cod and other Atlantic ground-fish stocks were in serious decline. A two-year moratorium on commercial fishing, introduced in 1992, has been extended indefinitely. Unlike groundfish, landings of pelagic and invertebrate species have been stable over the past 10 years. Aquaculture production is growing in value. However, municipal wastewaters cause approximately 20% of shellfish closures in the Maritime provinces (Environment Canada 1996a).

1.6 Surface Water

Lakes and rivers cover 7.6% of Canada's surface, and 9% of all fresh water discharged into the world's oceans is from Canada. Despite this relative abundance, the Canadian prairies and some of the valleys in the interior of British Columbia have a limited supply of water and suffer from periodic shortages.

1.6.1 Five Drainage Basins

Rivers in Canada flow into 5 major ocean drainage basins: the Pacific, Arctic and Atlantic oceans, Hudson Bay and the Gulf of Mexico (see Figure 1.7). The river system with the largest drainage area is the Mackenzie River, with 1,805,200 km^2 and the river with the greatest annual discharge is the St. Lawrence River at 9,850 m^3/s.

In Canada few studies have been undertaken at the level of the five national basins. In the late 1980s, as mentioned in Linton (1997), Pearse (1988) offers a study on the state of fish stocks in each of the majors basins in Canada. According to Pearse (1988) generalising about an entire basin was sometimes problematic because of the different conditions and trends affecting fish stocks within each basin. Considering this later remark another exercise to asses the late 90s status of fisheries was established by Jamie Linton of WWF Canada, in his report "Beneath the Surface: The state of Canada's water". The basin geographical unit has not been used for the description below of the status of water resource in Canada. Due to the very limited time available to write this analysis, it was more practical to describe the main water features in each Ecozone unit – the geographical entity used in the State of the Environment report where most of the information concerning water resources comes from. It would have been more appropriate to describe the status of waters at the basin level. This could certainly lead to an appropriate recommendation to the responsible Canadian water authorities.

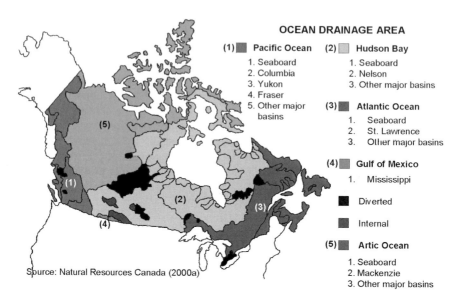

Fig. 1.7. Five drainage basins

1.6.2 Water in the Arctic Ecozones

Hydrological processes vary from place to place. Within the Arctic Ecozones, freshwater ecosystems are characterized by numerous lakes, many concentrated on the Canadian Shield, and an intricate network of surface drainage, culminating mostly in small rivers. Snow, ice, and permafrost determine storage and circulation of water (Woo 1993). The major form of water storage in the Arctic Ecozones is snow, although glaciers also store water in the eastern Arctic. Ice cover on lakes lasts longer than in adjacent Ecozones, owing to reduced solar radiation and low air temperatures in winter. In general, spring comes late, releasing six to nine months of accumulated precipitation within several weeks and producing a high runoff (Environment Canada 1996a).

1.6.3 Water in the Taiga Ecozones

The Taiga Shield Ecozone is where the largest hydroelectric projects have been constructed and where more water than in any other country (Day and Quinn 1992) has been diverted, especially for three large hydroelectric projects: the Churchill Falls project in Labrador; the La Grande (James Bay) complex in Quebec; and the Churchill-Nelson development in Manitoba. Reservoirs associated with the two first projects have flooded a total of approximately 14,150 km² in the Taiga, creating water bodies that cover about 21,300 km². Together, these projects have a capacity of close to 23,500 MW. There is also some small-scale hydroelectric generation in the western part of the Taiga. In the Northwest Territories, there are developments at Taltson River, on the Snare River, and at Blue Fish near Yellowknife. None of these sites includes diversions. There is a small hydroelectric facility north of Lake Athabasca, Saskatchewan, near Tazin Lake where water is diverted from Tazin Lake into Lake Athabasca.

1.6.4 Water in the Boreal Shield Ecozone

The Boreal Shield Ecozone rivers and lakes account for 22% of Canada's freshwater surface area. Rivers that have their headwaters in the Boreal Shield Ecozone include the Saguenay, Albany, and Moose rivers. Huge bodies of fresh water, such as Lakes Winnipeg, Superior, and Huron, lie along its borders. Within it are countless other lakes, some big, such as Lake Nipigon and Lac Saint-Jean, and others too small to be named (Environment Canada 1996a). Acidic precipitation from both local and distant sources may have weakened the general vigour and growth rate of trees in sensitive areas. Certainly acid precipitation has had a profound impact on the fauna of freshwater lakes in this Ecozone, in some cases resulting in the complete loss of all fish life and many invertebrates. Reductions in inputs of acidic materials to the atmosphere have been legislated and are being implemented. Nevertheless, a substantial portion of this Ecozone continues to receive elevated levels of acidic deposition. Thirty-nine percent of Canada's hydroelectric capacity is located on rivers arising in or flowing through the Boreal Shield Ecozone. Of Canada's 662 large dams, 279, or 42%, are located here (Environment Canada 1996a).

1.6.5 Water in the Pacific and Western Mountains Ecozones

The four terrestrial Pacific and western mountains Ecozones have an abundance of clean, fresh surface water and groundwater. Both water supplies and water quality appear to be good (Environment Canada 1996a). There are, however, reasons for localized concerns. Indeed, there is some evidence of long-term changes in water supplies. Data from around the region suggest decreasing average annual stream flows at more southerly sites and increasing flows at some northern BC sites (Leith 1991). Similar trends have been observed in snowpack water storage. However, only 5 of the 17 stream flow stations and 4 of the 11 snowpack monitoring stations showed statistically significant trends. Climate models suggest that global warming may further exacerbate this situation by significantly reducing summer precipitation in southern parts of British Columbia while increasing summer evapotranspiration.

1.6.6 Water in the Central Plains Ecozones

In the Prairies Ecozone annual precipitation ranges from about 500 mm in eastern and northern regions to about 300 mm in the south-west. Most of the major rivers originate in the Rocky Mountains, which constitute the western boundary of the Prairies and Boreal Plains Ecozones. These rivers, fed mostly by snowmelt, glacial runoff, and rainfall, flow in an easterly direction across the Prairies. All of the major rivers are regulated and used for power generation, water supply, and recreation. Depending on the weather conditions, the Prairies Ecozone can contain between 2 and 7 million wetlands. The greatest number of wetlands occurs along the sub-humid northern grasslands and adjacent aspen parkland, where 25-50% of the land surface is wetlands (Morrison and Kraft 1994). Many former wetland areas have been converted to agricultural production.

1.6.7 Water in the Atlantic Maritime Ecozone

The Atlantic Maritime Ecozone generally has an abundance of fresh water, although local supply problems do occur (Environment Canada 1996a). The Maritime provinces have more than 25,000 lakes, ponds, and rivers, with 13,113 in Nova Scotia, 11,335 in New Brunswick, and 1,105 in Prince Edward Island. Many lakes are small, less than 5 ha, especially in New Brunswick and Prince Edward Island (Eaton et al. 1994). The Ecozone has one of the highest runoff and groundwater baseflow ratios (ratio of precipitation entering surface water and groundwater sources to total precipitation) in the country. Indeed, almost 83% of the precipitation appears in rivers and streams. The geology, slope of the terrain, high annual rainfall, and low rates of transpiration and evaporation all contribute to this high ratio. The quality of the Ecozone's water, however, has come under increasing threats in recent years from contaminants emanating from municipal, agricultural, and industrial sources. Groundwater, a primary source of drinking water, has suffered considerable contamination, with corresponding risks to humans.

1.6.8 The Great Lakes - Saint Lawrence Basin

The Great Lakes, with about 20% of the world's fresh surface water, together make up the largest freshwater lake system in the world. The combined drainage area of all the lakes, including the US and Canadian sections, is 766,000 km², 32% of which is water. Average depths of the Great Lakes range from 19 m for Lake Erie to 147 m for Lake Superior. Their combined volume is 23,000 km³. This amount of water would be sufficient to keep the St. Lawrence River flowing for 100 years. Retention times range from 2.6 years for Lake Erie to 191 years for Lake Superior. The lakes are not large enough to have noticeable tides, but winds blowing the length of a lake sometimes cause water levels to differ by more than 2 m from one end of the lake to the other. Such phenomena (seiches) are most common in shallow, open lakes, such as Lake Erie.

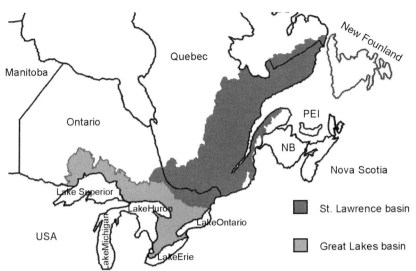

Fig. 1.8. The Great Lakes – Saint Lawrence Basin

Water flows out of Lake Ontario into the St. Lawrence River at a rate of about 6,850 m³/s. By the time the river passes the mouth of the Saguenay River, about two-thirds of the way to the Gulf of St. Lawrence, water from numerous tributaries (e.g., the Ottawa, Richelieu, St. Maurice, Chaudière, and Saguenay rivers) has increased the flow to 10,000 m³/s. At this point, the St. Lawrence is the second largest river in North America, after the Mississippi. Between Lake Ontario and the Gulf of St. Lawrence, the river drops in abrupt steps, creating a series of fluvial lakes (Lake St. Francis, Lake St. Louis, the La Prairie Basin, and Lake St. Pierre) separated by narrow sections of faster flow (see Figure 1.8).

In the estuary, which begins at the outlet of Lake St. Pierre and ends 555 km further downstream in the Gulf at Pointe-des-Monts, three distinct zones can be observed: the freshwater estuary to Île d'Orléans; the brackish middle estuary to the Saguenay River, a biologically productive transition zone where fresh water and seawater mix; and the saline estuary, a marine environment. All three zones are subjected to tides, which cause the water levels at Île d'Orléans to fluctuate by 5 m and which twice daily reverse the direction of flow as far inland as Portneuf. At the boundary between the middle and maritime estuaries, the Saguenay River fjord forms three basins of very cold water, which have maintained enclaves of Arctic ecosystems.

1.7 Groundwater

Canada enjoys relatively high-quality water. Nevertheless, problems remain. Among them is contamination of groundwater, which more than six million Canadians rely on for their water supplies. Groundwater contamination often results from earlier inadequate management of wastes or industrial chemicals. Because groundwater moves so slowly, it may be decades before a problem is detected — possibly long after the source of the contamination has disappeared. Well water, though, is most often contaminated by fecal coliform bacteria and nitrates. Recent surveys suggest that 20–40% of all rural wells may be affected (Environment Canada 1996a).

Groundwater is a critical resource in the Atlantic provinces where 91% of rural residential water users rely on it, compared with the national figure of 82%. In total, nearly 1.2 million people rely on groundwater for domestic needs, and municipal water use in the Atlantic provinces is increasing. In 1981, an estimated 307 million cubic metres of water were drawn from municipal supplies. Groundwater also supplies 90% of all water used for agriculture (Eaton et al. 1994). The major commercial users of groundwater include manufacturing, mining, agriculture, and thermoelectric plants. Thermoelectric plants are the single largest user of water, accounting for more than 60% of water withdrawal; however, water is returned unpolluted, except for added heat.

1.8 Wetlands

Wetlands merit special mention here because altogether, they comprise 14% of Canada's area. With an estimated 1.27 million square kilometres this surface represents nearly 26% of the world's wetlands. Most of Canada's wetlands are located in Manitoba, Ontario, and the Northwest Territories. In Manitoba and Ontario, wetlands represent respectively 41% and 33% of the land area in the province. With 4,000 hectares Prince Edward Island is the *poorest* province in terms of wetlands (see Chap. X. Water and Nature). Lands that have the water table at, near, or above the land surface or that are saturated for extended periods, wetlands bridge many of the different types of landforms and are thus important ecosystems. In Canada a severe loss of wetlands has occurred through the period of agricultural, industrial and population expansion, mostly during the 20th Century. This issue is examined in some detail in Chap. X on Water and Nature.

2. Water Quantity

Canada is blessed with an abundance of water. However the availability of water varies from season to season and year to year. The construction of dams and the development of storage reservoirs have provided the means to manage this variability and generate hydroelectric power. When first dams were built often the primary objective was to maximise power production or flood control. Fish passage and migration were blocked; mercury was released from flooded soils; water temperature, quantity, and sediment transport changed, all having significant impacts on species, wetlands, and users. Water managers in Canada now recognise these problems, but it is more than ever needed to put increasing emphasis on conservation and demand-management with respect to all uses.

2.1 Water Quantity Monitoring

The Water Survey of Canada, established in 1908, operates the national hydrometric network (2,900 active stations and some 5,100 discontinued sites across Canada) under federal–provincial agreements, and supplies daily, monthly, and/or instantaneous information for streamflow, water level, suspended sediment concentration, sediment particle size, and sediment load. Modernisation of the network permits users access to real-time data via satellites or telephone lines. Since 1991, most of this data has been available on CD-ROM which has replaced data publications. Water quantity data can be integrated with other environmental data in geographic information systems to assist water resource managers. Models are applied to watercourses to manage and apportion the flow, to forecast floods and supplies, and to predict the impacts of changes on flow regimes to human and aquatic health and economic activity (Environment Canada 1996b; Environment Canada 2000).

2.2 Inter Basin Water Transfer

Canada is blessed with an abundance of rivers with waterfalls and rapids suitable for the generation of hydroelectric power. In 1994, about 61% of the electricity generated in Canada was hydroelectric (Environment Canada 1996a). Since the mid 20th century, Canada has diverted more water than any other country (Day and Quinn 1992), most of it for three large hydroelectric projects: (1) the Churchill

Falls project in Labrador; (2) the La Grande (James Bay) complex in Quebec and (3) the Churchill–Nelson development in Manitoba. Reservoirs associated with the two first projects have flooded a total of approximately 14,150 km², creating water bodies that cover about 21,300 km². Table 2.1 presents the main cause of water inter-basin transfer in Canada.

Table 2.1. Inter basin water transfer in Canada

Province	No. of diversions [a]	Average annual flow [m³/s]	Major use	Flow diverted for major use as a % of total
Newfoundland	5	725	Hydro	100
New Brunswick	2	2	Municipal	70
Nova Scotia	4	18	Hydro	100
Québec	6	1,854 [b]	Hydro	100
Notario	9	564	Hydro	89
Manitoba	5	779 [c]	Hydro	97
Saskatchewan	5	30	Hydro	85
Alberta	9	117	Irrigation	53
British Columbia	9	361	Hydro	89
Total	54	4,450		96

[a] Includes only diversions that meet the following criteria: the diverted flow does not return to the stream of origin within 25 km of the point of withdrawal, and the mean annual diverted flow is not less than the rate of 1 m³/s. There were no inter-basin transfers in Prince Edward Island, Yukon and Northwest Territories.
[b] Excludes Beauharnois Canal flows from the St. Lawrence River.
[c] Excludes floodway flows (Portage Diversion, Winnipeg Floodway, Seine Diversion) of short duration.
Adapted from Environment Canada, 1996a; original source: Day and Quinn, 1992

Newfoundland's Churchill Falls generating station in Labrador is the third largest hydro development in Canada. It was constructed between 1966 and 1974 and at that time was the largest civil engineering project ever undertaken in North America. Eighty dikes were constructed to create the Smallwood and Ossokmanuan reservoirs. A huge underground powerhouse was built, and power began flowing from the site in 1971.

The La Grande hydroelectric complex has been developed in two phases and has entailed the diversion of flows from the Caniapiscau River to the east and the Eastmain and Opinaca rivers to the south. These diversions have almost doubled the La Grande River basin area to nearly 177,000 km². A land surface area of 11,505 km² was flooded in the process of creating seven reservoirs. Construction is completed on La Grande Phase 1 and continues on Phase 2. The project incorporates the largest underground powerhouse (LG-2) in the world. In total, there are nine generating stations.

The Churchill-Nelson hydro development in northern Manitoba has, to date, involved the construction of five generating stations on the Nelson River. The lowest two facilities, Long Spruce and Limestone, are within the Hudson Plains Ecozone. In addition, most of the Churchill River flow has been diverted to the Nelson River. This was accomplished by building a control structure at the outlet of Southern Indian Lake on the Churchill River, thereby elevating the lake surface 3 m. This allowed water to flow into a diversion channel and from there via the Rat and Burntwood Rivers to the Nelson River. Diversion of flow from the Churchill to the Nelson began in 1976 and averages 765 m^3/s.

In northern Ontario, there are hydroelectric developments on the Abitibi and Mattagami Rivers of the Moose River basin. These are comparatively small facilities that operate as run-of-the-river systems with no major reservoirs There is a small hydroelectric facility north of Lake Athabasca, Saskatchewan, near Tazin Lake where water is diverted from Tazin Lake into Lake Athabasca.

As shown in Table 2.1, hydroelectricity is one of the reasons for inter-basin water transfer in two provinces. Indeed in Alberta inter-basin transfers are occurring for irrigation and in new Brunswick water transfers are caused by municipal activities. As of 1994, there was no inter-basin transfer in Prince Edward Island, Yukon, the Northwest Territories or Nunavut. In the Northwest Territories, there are hydroelectric developments at Taltson River (Twin Gorges and four small units), on the Snare River (Snare Rapids, Snare Falls, and Snare Forks are owned by the Northwest Territories Power Corporation and Snare Cascades by the Dogrib Nation), and at Blue Fish near Yellowknife. None of these sites includes diversions.

Inter-basin transfers are the subject of many environmental concerns. Aquatic habitat and productivity losses in the watersheds from which the water is extracted are frequently matched or exceeded by the impacts due to increased flooding, erosion and instream flows in the receiving watershed. In addition, these transfers of water are accompanied by a transfer of pollutants, alien species and occasionally disease organisms into recipient system causing great impact on the ecosystem. Though environmental impact assessments that have been done in Canada to address these issues shed some light on this concerns there is still a great deal more to learn about the individual and cumulative effects of these changes over time on the aquatic environment and resources.

2.3 Flood Management in Canada [3]

Canadians may be a little perplexed about the state of their flood management programs. On the one hand, *Canada established a world leadership role* in 1975 with the development of the Flood Damage Reduction Program (FDRP) (Bruce

---

[3] This section was provided by Dan Shrubsole, Associate Professor, Department of Geography, University of Western Ontario.

and Mitchell 1995). Handmer and Parker (1992) identified Emergency Prepared-ness Canada (EPC) as a model that Britain could follow in enhancing its institu-tional arrangements for emergency management. Despite these accolades, the ef-fectiveness of past approaches must be questioned. Between 1975 and 1999, the federal government provided almost $720 million (1999 figures) in disaster relief (EPC 2000). Between 1984 and 1998, an additional $750 million (1999 figures) was paid by the insurance industry (Insurance Council of Canada 1998). Major flood events in the 1990s have left some people to question the efficacy of current programs. In Quebec, the Saguenay River flood in 1996 resulted in 10 deaths and $800 million in damages. In Manitoba, approximately $500 million in damages were inflicted during the 1997 Red River flood. Thus, questions regarding the ef-fectiveness of existing programs are being raised.

The present practice of flood management in Canada is characterised by at least three realities. First, it is impossible to provide absolute protection to people and communities. Second, a mix of structural and non-structural adjustments that cover the entire range of protection, warning, response and recovery are needed to effectively protect lives and property. Third, implementation of flood adjustments requires the effective participation of all levels of government and the public. A central question related to the recent trend of increasing losses pertains to whether they are: (1) the result of isolated and improbable large weather events; (2) reflect-ing an increased level of wealth (and damages) by Canadians; or (3) indicating fundamental problems with present practice of flood management. This section addresses the last point.

2.3.1 Flood Damage Reduction Programme

When the FDRP was initiated in 1975, its primary and innovative intent was to promote floodplain-mapping studies. It also prohibited federal and provincial gov-ernments from engaging in, or providing assistance for undertakings that were in designated high-risk areas. In addition, it restricted federal and provincial disaster assistance only to those structures built before designation and encouraged mu-nicipalities to enact zoning regulations based on engineering studies and mapping. Flood hazard information was also to be communicated to the public, municipali-ties and industry (Bruce 1976). Expertise and available funds within Environment Canada made it the leader of flood management in Canada.

By 1999, 982 communities were mapped and designated under FDRP (Envi-ronment Canada 1999a). Where floodplain regulations were implemented, there is a diversity of opinion concerning their effectiveness. Several studies concluded that Ontario's approach to floodplain regulation has been generally effective (Boyd 1997; Brown et al. 1997; Shrubsole et al. 1995, 1996, 1997a). Other authors have identified difficulties in managing land uses on designated floodplains. Roy et al. (1997) commented that the initial floodplain protection provisions within the Canada-Quebec FDRP agreement were relaxed through policy exemptions. Forget et al. (1999) found that designation failed to prevent inappropriate development in Montreal. They also concluded that dykes, mainly constructed in developed areas, may have promoted a false sense of security and noted a highly variable level of structural integrity, design and maintenance. Cardy (1976) commented on the false sense of security that was associated with structural adjustments in St. John, New Brunswick. Day (1999) maintained that comparatively little mapping occurred in the lower Fraser River basin of British Columbia. Instead, a dyking program there consumed a very large proportion of all funding under the Canada Water Act. In 1979, Manitoba established a Red River Designated Flood Area that saw many communities implement floodplain regulations that had been supported through FDRP mapping. Unfortunately, due to a lack of enforcement, only 63% of new homes in the designated flood areas complied with that regulation (IJC 1997). Collectively, these studies suggest that local governments have not always effectively managed floodplain development. This is partly explained by a lack of political will, inadequate capacity, or the absence of mechanisms that promote watershed-based responses. Local governments usually promote economic development and must provide other important services (e.g. fire, police, parks, water) to its citizens. Flood and other hazard management initiatives must compete with these needs for the attention of local government. Typically, the need for flood management is only appreciated after an extreme event flood with high losses. At a practical level, many flood mitigative designs cannot be feasibly incorporated into building additions or renovations (Shrubsole et al. 1995, 1996, 1997a).

2.3.2 Public Education

Public information was another important aspect of the FDRP. Handmer (1980) assessed the efficacy of FDRP maps in changing people's attitudes to floods. He concluded that although there was an increase in flood awareness, this change could not be attributed solely to the maps. Kreutzwiser et al. (1994), Shrubsole and Scherer (1996), and Shrubsole et al. (1997b) surveyed the perceptions of floodplain residents in three southwestern Ontario watersheds to the flood hazard and flood adjustments. Generally, residents did not perceive a significant risk of future flooding. There was a poor understanding of floodplain regulations, and structural adjustments were viewed as the most effective approach. Thus, although floodplain regulations are supposed to be the most effective mechanism in reducing future flood damages, residents preferred other measures.

Providing information about the likely hazards associated with a particular property during real estate transactions could better inform residents. The mandatory disclosure of this information is part of the approach to flood management in the United States (Platt 1999). Shrubsole and Scherer (1996) obtained the views of home mortgage lenders, real estate agents and land appraisers in portions of the Grand River watershed (Ontario). They concluded that although formal training pertaining to floods and regulations was limited, the real estate sector was aware of the need to disclose this information to potential buyers. However, this was pursued in neither an effective nor consistent manner. Real estate agents most often provided this information to potential purchasers late in the purchase process but prior to an offer to purchase. In the United States, this information is to be provided early in the purchase process (Platt 1999).

2.3.3.Limitations of the Federal Flood Damage Reduction Programme

One shortcoming with the FDRP was the varied and limited mapping of aboriginal lands (Watt 1995). By 1999, 40 Aboriginal Reserves and communities were mapped (Environment Canada 1999a). However, the procedure for designation was not part of the agreement. The implementation of any adjustments was also impeded because the institutional arrangements failed to reflect important socio-political differences between native and non-native communities. For instance, traditional benefit-cost studies that are frequently used to prioritise projects were inappropriate for use in aboriginal areas where there is communal ownership. There were also insufficient funds targeted for flood management on native lands. The Federal Departments of Indian and Northern Affairs, and Public Works and Government Services Canada, native communities and provincial governments through initiatives such as Flood Damage and Erosion Mitigation Plan are now addressing this problem.

By the late 1980s, the engineering expertise within and the spending powers of Environment Canada had been eroded significantly (Bruce and Mitchell 1995). Environment Canada is not renewing the 10-year General and Mapping Agreements, and other levels of government have provided neither effective leadership nor action. At this time, the federal government was no longer a leader in flood management, and the provinces and municipalities were not effectively filling the void.

2.3.4 The Red River Flood: The Event, Current Cultures and their Implications

The 1997 Red River has a very long history with the significant floods occurring in 1826, 1852, 1861, 1950, 1966, 1979 and 1997. The 1997 flood was one of the worst on record. The primary response to previous floods had been to build structures such as the Winnipeg Floodway (Red River Floodway) in 1950 and dykes in 1966 and 1979 (Haque 2000). Floodplain regulations were introduced following the 1979 flood but were ineffectively implemented. During the 1997 flood, an area in excess of 1,945 km^2 was inundated, sometimes extending over 40 km in width (Rahman 1998). Over 2,500 homes were flooded, and in excess of $500 million in damages resulted. In 1997, floodwaters remained in homes for one week.

The International Joint Commission (see Chap. XII on Institutional and legal framework for water management) released its final report on the Red River flood in December 2000. Its key conclusions were:

- While the 1997 event was a natural and rare event, floods of the same or greater magnitude could be expected in the future;
- The people and property of the Red River basin will remain at undue risk until comprehensive, integrated and binational solutions are developed and implemented.
- Since there is no single solution to the problem and reduce vulnerability, a mix of structural and non-structural adjustments are required.
- Specific flood management activities were required in several U.S. and Canadian communities.
- Ecosystem considerations must be a more important consideration in managing future flood damage reduction initiatives. In this regard, hazardous materials must be carefully controlled and banned substances removed from flood-prone locations.
- A culture of preparedness and flood resiliency must be promoted throughout the basin.

Based on the previously described practice of flood management, a number of cultures that inhibit effective flood management are identified below. In short, past and current flood management efforts have fostered a set of cultures that are at the root of the current predicament. Collectively, they have increased vulnerability – "those circumstances that place people at risk while reducing their means of response or denying them available protection" (Comfort et al. 1999). Comments provided by the IJC (2000a) that pertain directly to promoting a culture of preparedness and flood resiliency in Canada will be noted.

A culture of conflict has, at times, been generated between upstream and downstream communities over the construction and operation of structural adjustments. While the operational procedures for the Winnipeg Floodway provided an increased level of protection to its residents, this came at the cost of increased flood levels for upstream residents. According to the IJC (2000a), many upstream residents "were troubled by what they perceived as a lack of concern" by the City and residents of Winnipeg about what happens outside its boundaries. The IJC (2000a) echoed this view when it maintained that "many residents upstream of the Winnipeg Floodway who were harmed by increased water levels caused by the way in which the Winnipeg Floodway was operated to save Winnipeg feel that [compensation] matter still has not been satisfactorily addressed by the government of Manitoba" (International Joint Commission 2000a). Addressing the culture of conflict will require that future initiatives are not only assessed on the traditional mix of economic, social and environmental criteria that are supported through existing impact assessment and planning legislation, but that proposals are grounded in the principles of trust, equity and participation among all participants. Given the long history of upstream-downstream conflicts in the Red River, engendering these attributes is a very significant challenge.

The emphasis on structural adjustments has nurtured a culture of land development that is based on a false sense of security. Short-term economic gains are more valued than the long-term costs associated with flood damages. The IJC (2000a) recognised this problem and suggested that flood preparedness must be improved through the development of contingency plans that deal with the over-topping and breaching of levees. The IJC (2000a) also maintained that resiliency to floods could be enhanced through measures that reduce future impacts. Timely warnings, flood control measures, the designation of flood-prone areas as open space, flood-resistant construction and storm water management were among some of the preferred adjustments. The need to monitor indicators of resistance, such as the extent of floodplain development, was also noted.

The implementation of any adjustment on aboriginal lands was relatively slow because planning failed to reflect important socio-political differences between aboriginal and non-native communities. This bias increased the aboriginal communities vulnerability because they have not had the same level of access to flood relief and FDRP funds. Although the IJC (2000a) did not specify the difficulties experienced by native communities, its recommendations to improve the nature of existing institutional arrangements should have positive implications for them.

A culture of institutional fragmentation works against effective flood management as illustrated by the lack of integration between and within structural adjustments and floodplain regulations. The Natural Hazards Centre and the Disaster Research Institute (1999) supported this view when it concluded that the removal and repair of structures that would reduce future flood damage did not occur systematically after the Red River flood. They also maintained that recovery assistance did not foster strong individual responsibility or self-sufficiency. In their view, communities will not be adequately prepared for a future flood that might exceed design standards. Their report suggested that some measures might work against the long-term resilience of the residents. The IJC (2000a) supported integrated approaches to flooding and recommended "the Canadian federal government should establish a national flood mitigation strategy, or a broader disaster mitigation strategy, and support it with comprehensive mitigation programs." The IBC (1999) and EPC (1998) have been promoting this type of multi-hazard program to the federal government. In this manner, public and government support could be sustained between flood events rather than be the short focus of attention immediately after an event. A comprehensive policy would also embrace the range of structural and non-structural adjustments in order to promote the integration of activities within communities.

A related concern pertains to the disparity in the level of commitment between non-structural and structural approaches, and the lack of capacity within governments to implement floodplain land-use regulations. The rhetoric surrounding

FDRP emphasised the mapping of flood-prone areas. Implicitly, the FDRP assumed that once municipalities were made aware of the flood hazard, they would establish floodplain regulations. However, the traditional focus on structural adjustments consumed over 50% of the all expenditures made under the Canada Water Act and were concentrated on relatively few, expensive projects in British Columbia, Manitoba, Ontario and Quebec. The FDRP did little to build institutional capacity at the local level so that regulations could be developed and implemented (Handmer 1996). An exception appears to be in Ontario where conservation authorities had adequate technical and institutional capacities (Shrubsole et al. 1995, 1997). In Manitoba, "a lack of enforcement of floodplain regulations, zoning by-laws and ordinances resulted in significantly higher damages that would have occurred with more effective enforcement" (International Joint Commission 2000a). The IJC (2000a) commented that fully integrating building codes with floodplain regulations was required to achieve successful flood damage reduction programs.

Rising flood damages could occur in the future without a renewed effort to map flood-prone areas. The IJC concluded that since the demise of FDRP in 1993, there is a "lack of current floodplain information, and there are currently no incentives to obtain such information needed to assess the overall impact of flood risk" (IJC, 2000a). In this regard, the IJC recommended that the feasibility and desirability of adopting a flood insurance program in Canada be considered. This could, among other measures, promote the early disclosure of any flood-hazard information to potential purchasers.

The IJC (2000a) indirectly addressed issues of capacity. Emergency plans should be reviewed and modified. This might result in governments increasing their capacity to prepare, warn, respond and recover. The lack of capacity within local government is a very significant problem because it promotes a culture of dependency. Local governments appear to rely excessively on senior governments for technical, financial and policy assistance to conduct what should be routine activities. This establishes an inevitable cycle of escalating flood losses and 'passing the buck'. The cycle begins with significant flood damages being inflicted on a community that is located on a well-known flood-prone area. Past flood events have prompted the construction of structural works and the establishment of a flood warning system and information campaigns. If floodplain regulations exist, they have likely been implemented poorly. The news media report on the flood, its damages and the emergency response efforts to the nation. Relief programs, largely funded by senior governments and NGOs immediately, respond to this event. The public places much of the blame for the flood on inadequate government effort. In response, bigger and more structures are built with most of the funding coming from senior governments. Commercial properties and residences are refurbished, in part through the DFAA, to pre-flood conditions. Flood warning systems and information programs are improved. Since senior governments provide neither consistent nor strong signals on the need to truly integrate structural and non-structural adjustments, intensive development continues on flood-prone areas. When these developments are flooded, primary blame is often placed upon municipalities. However, it is the previous steps that implicitly support this cycle of escalating economic losses. This cycle reinforces a culture of dependency for future structural adjustments and disaster relief payments

Communities are not held effectively accountable for poorly implementing floodplain regulations because senior government funding will *bail them out*. Hunt (1999) observed that the FDRP encouraged local governments and individuals to rely on senior governments for leadership and financial support. In this context, the generally limited support afforded floodplain regulations by many municipalities and property owners is understandable.

The rising trend in flood and other hazard damages has made the reduction in disaster losses more important than ever. It will likely require more than a public education campaign about the nature of flood risks to ensure that "appropriate decisions will be taken to make developing communities resilient to flooding" (IJC, 2000a). While past efforts have realised real cost savings, future efforts must overcome the current set of cultures. Failure to meet this challenge might see governments becoming a more frequent provider of disaster assistance than is desirable and feasible.

An important aspect in achieving a culture of preparedness or safety is ensuring that our daily activities and programs are intentionally designed to reduce risk and/or raise resistance or resiliency to hazards. The IJC (2000a) report makes some progress in these areas. The relatively easier component of changing cultures involves improving the implementation of existing technologies and programs, such as flood prediction and floodplain regulation. The more difficult aspect pertains to changing the existing paradigm within public and private flood and emergency management agencies which views themselves as responding to crises rather than to being concerned with the broader issues of vulnerability and its management.

However, stating policies is one thing, implementing them will require a level of leadership that has not been seen in the Canadian hazard policy arena for over 10 years. It is incumbent upon all levels of government, particularly senior governments, to commit to making real and measurable impacts on Canada's risk profile. In this regard, the call for the adoption of a National Mitigation Strategy by the Insurance Bureau of Canada (1999) and Emergency Preparedness Canada (1998a) might be provide an opportunity to debate and discuss the principles, mechanisms, processes and partnerships that might be required to address Canada's less-publicised cultural crisis.

3. Water Quality

3.1 Surface Water Quality

3.1.1 Current Status of Surface Water Quality in Canada

In Canada, surface water is generally plentiful and clean. However, it is sometimes locally or regionally polluted. Pollution enters water bodies in a number of ways, including industrial and municipal discharge, runoff, spills, and deposition of airborne pollutants. In the last half-century, increased amounts of industrial, agricultural, and municipal wastes entering Canadian rivers, lakes, and marine areas have had a serious impact on water quality. Wetlands, which act as natural storm buffers, sinks for pollutants and heavy metals, and regulators of flood water, are being lost across southern Canada (Environment Canada 1998a).

The Great Lakes and the St. Lawrence River basins continue to suffer from industrial and municipal pollution, urban and agricultural runoff, and deposition of airborne pollutants. The Red River and other prairie rivers are being degraded by agricultural runoff and inadequately treated sewage. The Fraser River is under pressure from industrial effluents, landfill pollutants, wood treatment chemicals, and forestry and agricultural runoff (Environment Canada 1998a).

Impacts of pollution include threats to drinking water in certain areas (e.g., causing recently in Walkerton, Ontario, the death of seven people due to illness caused by contamination of water with *Esherichia coli* bacteria), closures of shellfish harvesting areas on the Atlantic and Pacific coasts, the loss of part of the Great Lakes fishery, reduced ecosystem diversity, and fewer recreational opportunities. In Kejimkujik National Park, Nova Scotia, long-range pollutants were identified as a contributing cause of blood mercury levels in loons at least twice as high as anywhere else in North America (Environment Canada 1998a). Impacts of certain chemicals that interact with endocrine systems, potentially impacting growth, development, and reproduction, is an increasing concern within the Canadian Scientific community and is now attracting great public concern as well (Environment Canada 1998a). The agriculture sector has been identified as a potential source of environmental endocrine-disrupting chemicals (EDCs) through the use of pesticides, land-applied sewage sludge, and the production of natural EDCs in livestock wastes (Topp 2000). Such as the urban sector since, as noted by Chambers et al. (1997), municipal effluents are complex mixtures which contain a variety of contaminants which may be capable of endocrine disruption.

On a daily basis, human activities cause natural and synthetic chemicals to be emitted into the atmosphere. Once released, these substances are dispersed throughout the globe by air currents. Canada's waterways continue to be affected by the long-range transport of airborne pollutants (Environment Canada 1999b; Koshida and Avis (eds.) 1998). Acid rain continues to be a problem in Eastern Canada (Environment Canada 1999b). In many places, the deposition continues to exceed critical levels, with potentially serious implications for the health and productivity of aquatic ecosystems and forests in Ontario, Quebec, and New Brunswick. As presented in the Environment Canada's State of the Environment (SOE) Bulletin on Acid Rain (Environment Canada 1999b), of 152 lakes monitored for acid rain effects in Ontario, Quebec, and the Atlantic Region between 1981 and 1997, 41% showed some improvement in acidity. Acidity levels were stable in 50% of the lakes and became worse in 9% (Figure 3.1). Approximately half of the emissions causing this acid rain originate in the United States; co-operation is essential to resolve this problem (Environment Canada 1998a). Acid rain is a less serious problem in western Canada because of lower overall exposure to acidic pollutants and a generally less acid-sensitive environment (Environment Canada 1999b)

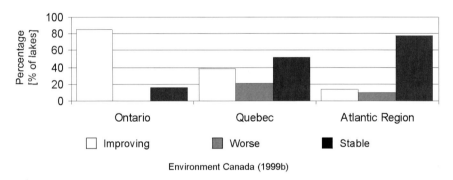

Environment Canada (1999b)

Fig. 3.1. Trends in lake acidity by region from 1981 - 1997

Canada has made progress in reducing some major water pollution problems. Increasingly, Canadians are focusing on preventing rather than remediating pollution. Changing agricultural practices, including the development and use of more environmentally friendly pesticides and fertilisers, and increasing conservation tillage have contributed to improvements in water quality. Sewage treatment has improved. There is a significant decrease in the amount of toxic pollutants coming from industries such as petroleum refining, mining and smelting, and pulp and paper.

3.1.2 Water Quality in British Columbia

British Columbia (BC) is blessed with an abundance of freshwater from streams, lakes, and groundwater. A provincial-federal co-operative water quality monitoring program began in 1985 to collect long-term data suitable for detecting regional trends (Environment Canada 1996a). In 1996 the first Water Quality Status Report for the Province of BC was released by the Ministry of Environment, Lands and Parks of British Columbia (Ministry of Environment, Lands and Parks of British Columbia and Environment Canada 1996). In the first water quality status report 124 water-bodies have been ranked (81 river sections or creeks, 26 lakes, 12 marine bays or inlets, and 5 groundwater aquifers). These are areas where information was collected because problems with the quality of the water were expected. Of the 124 water-bodies considered, 48% are ranked as having fair water quality and 35% as good. Fair means most uses of the water are protected with conditions only sometimes different from natural levels while good indicates all uses are protected with conditions close to natural levels. The remainder are either excellent (7%), borderline (5%), or poor (5%).

More recently, the data from 133 monitoring stations in 68 water-bodies (49 river sections or creeks, 14 lakes or reservoirs and 5 groundwater aquifers) were examined for trends. The results are presented in a report which is the compilation of the trend reports for each water-body (Ministry of Environment, Lands and Parks of British Columbia and Environment Canada 2000). The trend report is based on regular and consistent long-term monitoring (5 to 10 years or more), and shows whether the water quality is changing over the longer term, contrary to the first status report which was based on the degree of attainment of water quality objectives during a critical month for at least three consecutive years. The water-bodies assessed range from those that are relatively pristine to those that are heavily impacted by human activity. According to the authors, because more monitoring is done in areas where people are active, this analysis gives a view of water quality in developed areas, rather than of undeveloped watersheds where water is still in a largely natural state. For surface water (lakes and streams) in BC, 59% of the stations had no observed changes in water quality, 31% had improving trends, and 10% had deteriorating trends. For groundwater, 53% of the stations had no observed changes, 27% had improving trends, and 20% had deteriorating trends. According to the Ministry of Environment, Lands and Parks of British Columbia and Environment Canada (2000), these trends are not considered to be representative of the water quality trends in the province as a whole. They are reporting the situation in developed areas and watersheds.

3.1.3 Water Quality in the Prairies

In Manitoba and Saskatchewan Water Quality Status is described by majors Eco-zones (Government of Saskatchewan 1999; Government of Manitoba 1997). For example, as reported in the State of the Environment Report for Manitoba (Government of Manitoba, 1997), the Boreal Shield Ecozone has overall "good" water quality, the Boreal Plains Ecozone has overall "fair" water quality. Overall, the water quality for the Taiga Shield Ecozone is "fair" and remained relatively constant over the five years (from 1991 to 1995) data where collected. For the Hudson Plains Ecozone where small amount of data is available it is difficult to determine a clear trend for water quality. In the Prairie Ecozone overall water quality is fair and has shown little change across the from 1991 to 1995. Water quality reflects agricultural activities, natural sediment load carried by prairie rivers and streams, and seasonal variation of prairie rivers. Several herbicides (dicamba, MCPA, bromoxynil, simazine and trifluralin) evaluated were found to exceed the water quality guidelines in the Ecozone. Dicamba exceeded the guidelines at every location sampled, while MCPA exceeded guidelines at approximately half the locations sampled. The presence of a variety of herbicides in the watercourses reflects the high degree of agricultural activity in the Prairie Ecozone. Exceedances are most frequent at the Red River north of Winnipeg and in the La Salle River. Guidelines for the Red River downstream of Winnipeg were exceeded on occasion, for almost all substances analysed over the period 1991 to 1995. Water Quality for the Assiniboine River at Headingley has been ranked as *fair*. Periodically, fecal coliform levels higher than the water quality objective appear at each monitoring location in the Ecozone. They occur more frequently in the Red River downstream of Winnipeg where 67% of all measurements exceed the guidelines. Fecal coliform densities reflect the impact of population centres and agricultural activities near watercourses.

The Alberta Ministry of Environment (1999) reports that as water demand increases, water quality becomes more prominent in defining water management issues. River flows determine in-stream temperature and oxygen regimes important to fish and other aquatic life. Particularly for southern rivers in summer and in all rivers during winter, in-stream water quality by augmenting river flows during natural low flow periods. From the Red Deer River south, licensed water withdrawal adds a second dimension to the management of water quality. These river systems are less turbid, and have a heavier nutrient loading. Recently steps have been taken to reduce or remove industrial discharges into rivers, but below each major city, municipal discharges still have an impact. During summer in early morning hours, these discharges cause a significant dissolved oxygen sag which may affect fish and invertebrates. By comparison, the northern systems (Peace and Athabasca) are much larger and are not subjected to significant water withdrawals. These rivers tend to be deeper, cooler, more turbid and biologically less productive. The major water quality issues are associated with the discharges of industrial and municipal effluent, which are regulated by a system of licences and permits. Seasonal patterns of river discharge, in particular low winter flows coupled with ice-cover, dictate specific management approaches which have focused recently on the maintenance of minimum winter dissolved oxygen concentrations. The Peace River is regulated by the WAC Bennett dam in British Columbia and Alberta is negotiating guaranteed in-stream flows with BC to protect water quality and other uses. River water quality management is progressing from a system based on available treatment technology to one based on receiving water conditions.

3.1.4 The Health of the Great Lakes

According to the State of the Great Lakes 1995[4] (Governments of Canada and the United States 1995), the health of the Great Lakes basin ecosystem is variable. By some measures, the health is good/restored; such as the aquatic community in Lake Superior. At the other end of the spectrum, a number of indicators show that some aspects of ecosystem health are poor such as habitat loss, encroachment and development in wetlands, and the imbalance of aquatic communities in Lakes Michigan, Ontario and the eastern basin of Lake Erie. Other indicators show conditions between these extremes.

The state of aquatic communities, including native species loss and ecosystem imbalance and reproductive impairment of native species shows generally mixed/improving conditions throughout the Great Lakes. Lakes Michigan, Ontario and the eastern basin of Lake Erie rated somewhat lower, but Lake Superior rated higher because of fewer numbers of native species lost to extinction and to healthier lake trout populations. Aquatic habitat and wetlands have been given an overall rating of poor. This is because of the huge losses of wetlands and other habitats, in quality as well as quantity. There are a few bright spots however, for example brook trout stream habitat in the upper lakes is in relatively good condition, and there are programs which exist to protect remaining habitat. Nutrients such as phosphorus are no longer the widespread problem they once were in the early 1970s thanks, in large part, to the efforts made under the United States/Canada Great Lakes Water Quality Agreement (GLWQA). Nutrient stresses have been given an overall good/restored rating in the Great Lakes basin. Indicators such as phosphorus concentrations and loadings must still be monitored however, to ensure that elevated nutrient levels do not throw the ecosystem out of balance again, and because they can still create local problems. Loadings of persistent toxic contaminants, levels of chemical contaminants in fish and herring gulls, and concentrations in water were rated as mixed/improving. This rating is based on the positive response of reductions from peak levels, but many contaminants still need further reductions to reach acceptable levels of risk (Governments of Canada and the United States 1995).

---

[4] The restoration and maintenance of the chemical, physical and biological integrity of the waters of the Great Lakes basin ecosystem is the purpose of the United States/Canada Great Lakes Water Quality Agreement (GLWQA) (Governments of Canada and the United States, 1995). In support of that purpose the governments of the United States and Canada sponsored a State of the Lakes Ecosystem Conference (SOLEC) in October of 1994 which leads to the establishment of the State of the Great Lakes 1995 document.

3.1.5 Water Quality in Quebec[5]

In Québec, water is present in abundance. Renewable freshwater resources make up one-third of the water resources available in Canada as a whole, and roughly 3% of the resources of the planet. Despite its abundance, the quality of Québec's water has been affected by urbanisation, industrialisation and the intensification of agriculture. From the early 19th century to the middle of the 20th century, the contamination of the St. Lawrence by wastewater led to serious sanitary problems, and in particular to epidemics of cholera and typhus, and the period of strong economic, demographic and agricultural growth that followed the Second World War resulted in an accelerated deterioration in water quality. A study of the state of the St. Lawrence in the 1970s first raised awareness of the low quality of Québec's surface water, and led in 1978 to the launching of the *Programme d'assainissement des eaux du Québec* (Québec Wastewater Treatment Program (known by its French acronym PAEQ)).

Since then, a large number of municipal and industrial clean up programs and action plans have been implemented, and the regulatory framework has been made considerably stricter. In twenty years, an appreciable improvement in water quality has been observed, and the overall particle and organic load from municipal sewage systems has been reduced significantly. In addition, the effluents from a growing number of industrial facilities is now processed in water treatment plants. Another important aspect is the work carried out to treat the industrial effluents that do not pass through the water treatment system, and that are pumped directly into watercourses. For example, the pulp and paper industry has invested millions of dollars in treating its wastewaters.

Overall, water quality in Québec can now be considered to be relatively high. However, over time, other environmental problems have become more acute, especially non-point source pollution and toxic pollution. Certain toxic substances and pesticides are still found in some rivers, where high levels of nutrient elements may be recorded. A water quality index, developed to reflect the range of values observed in Québec, has shown that in river basin headwaters and in outlying regions, water quality is generally high.

Water quality decreases in the south-west of Québec, especially in the agricultural areas of the St. Lawrence Lowlands, mainly because of the non-point source pollution resulting from the use of fertilisers and pesticides, and animal waste. Studies of biological communities and the quality of riverbank vegetation also reveals the impact of agricultural activities. However, much recent work has been carried out in this area, in particular since the 1997 implementation of the new Regulation respecting the reduction of pollution from agricultural sources and the Agro-environmental Investment Assistance Program, which has a budget of

---

[5] This section is taken from a contribution made by Environment Quebec that is presented in the appendices.

$400 million. The contamination of the environment by toxic substances is another point that cannot be ignored. The data shows that this type of contamination has decreased considerably since the 1970s, but is still significant.

With regard to lake water acidification, data collected for over 1,500 lakes by Québec's spatial lake acidity monitoring network shows that almost 20% of lakes are acid, and that half of all lakes are likely to suffer biological damage relating to the acidification of surface water. The acid-rain problem is far from solved, but recent reductions in SO_2 emissions will probably result in a partial improvement in the quality of the ecosystems affected.

In the St. Lawrence, water quality has also improved in response to the water treatment processes implemented in recent years. For example, the quantity of fecal coliforms measured off Contrecoeur dropped significantly from 1992 when the wastewater treatment plant for the south shore came into service. Levels of several other variables (phosphorous, coliforms, suspended solids, ammonia nitrogen) also decreased between 1990 and 1997.

A substantial difference in water quality can be observed between Lake Saint-Louis, where water quality is high, and the river downstream from Montréal. The microbiological traces left by the effluent from the *Communauté urbaine de Montréal* wastewater treatment plant can be observed in the centre of the river and along the northern bank as far downstream as Bécancour. Although its overall water quality exceeds that of most of the world's great rivers, the St. Lawrence still receives contaminant discharges that limit use of the river in certain sectors.

3.2 Groundwater Quality

Six million inhabitants (100% of the population of Prince Edward Island and over 60% of the population of New Brunswick and the Yukon) rely on groundwater for their water supplies. Most of the people who rely on groundwater in Canada live in rural areas and draw their water from private wells Although the quality of the groundwater is generally very high, locally problems, due to contamination, remain. Groundwater contamination often results from earlier inadequate management of wastes or industrial chemicals.

The kinds of chemical constituents found in groundwater depend, in part, on the chemistry of the precipitation and recharge water. Near coastlines, precipitation contains higher concentrations of sodium chloride, and downwind of industrial areas, airborne sulphur and nitrogen compounds make precipitation acidic.

One of the most important natural changes in groundwater chemistry occurs in the soil. Soils contain high concentrations of carbon dioxide which dissolves in the groundwater, creating a weak acid capable of dissolving many silicate minerals. In its passage from recharge to discharge area, groundwater may dissolve substances it encounters or it may deposit some of its constituents along the way. The eventual quality of the groundwater depends on temperature and pressure conditions, on the kinds of rock and soil formations through which the groundwater flows, and possibly on the residence time. In general, faster flowing water dissolves less material. Groundwater, of course, carries with it any soluble contaminants which it encounters.

The suitability of water for a given use depends on many factors such as hardness, salinity and pH. Acceptable values for each of these parameters for any given use depend on the use, not on the source of the water, so that the considerations important for surface water are equally applicable to groundwater.

The natural quality of groundwater differs from surface water in that:

- for any given source, its quality, temperature and other parameters are less variable over the course of time; and,
- in nature, the range of groundwater parameters encountered is much larger than for surface water, e.g., total dissolved solids can range from 25 mg/L in some places in the Canadian Shield to 300,000 mg/L in some deep saline waters in the Interior Plains.

At any given location, groundwater tends to be harder and more saline than surface water, but this is by no means a universal rule. It is also generally the case that groundwater becomes more saline with increasing depth, but again, there are many exceptions.

As groundwater flows through an aquifer it is naturally filtered. This filtering, combined with the long residence time underground, means that groundwater is usually free from disease-causing micro-organisms. A source of contamination close to a well, however, can defeat these natural safeguards. Natural filtering also means that groundwater usually contains less suspended material and undissolved solids than surface water.

However, when groundwater becomes polluted, its inaccessibility makes it difficult to clean up. In Canada, there are several locations where problems with groundwater have been observed as a result of the presence of naturally occurring contaminants, such as salt, arsenic and fluoride. Pollutants resulting from human activities may also come from a range of sources, including septic systems, leaky storage tanks, municipal landfills, industrial discharges and land sprayed with pesticides and other agricultural chemicals like fertilizer (causing, for example, high level of nitrate concentration).

From 1960 to 1985, the annual consumption of fertilizers in Canada increased from one million tonnes to four million tonnes, although fertilizer use remained steady for the rest of the 1980s. In the same period, the nitrogen content of fertilizers increased from about 10% to 30%. As a result of such trends, the amount of nitrates present in Canadian water bodies has risen, particularly in rural areas that rely on groundwater. In a 1993 survey of Ontario groundwater, 15% of 1,300 rural wells contained nitrates at potentially harmful levels.

In the 1980s, atrazine was widely used in Canada due to the limited availability of other herbicide (weed-destroying) products. As well, the use rates at that time were considerably higher than is the case today. Groundwater testing conducted during that time period frequently detected the presence of atrazine, thereby raising public awareness of the issue of pesticides in groundwater. Atrazine is used in Canada as a weed control agent in corn and blueberries, with the latter representing only marginal use of the product. Atrazine has been detected in surface and well water supplies in several provinces owing to its rapid movement with surface runoff and through soil into groundwater. In response to these concerns, a reevaluation of the regulatory status of this compound was initiated in 1988 by the Pesticides Directorate, Agriculture and Agri-Food Canada, now the Pest Management Regulatory Agency (PMRA). In conjunction with the reevaluation and in an effort to address the groundwater contamination issue, all registered atrazine products have been the focus of a label improvement program, initiated by the basic manufacturers, in which both the use pattern (the types of sites on which it can be used) and registered rates (the amount that can be used per hectare) for this compound have been reduced. In addition, buffer zones (for both mixing and spraying) were established for wells and water sources. As a result, Atrazine use in the province of Ontario (a major corn production area) declined by 66% over the 10-year period from 1983 to 1993. Accordingly, more than 99% of well water samples tested contain atrazine at levels below the established drinking water quality guideline. The situation is continuing to be monitored to ensure the protection of our water resources.

3.3 Water Quality Monitoring

3.3.1 National Monitoring

The federal government first became involved in water quality monitoring in 1934, emphasising water used by municipalities and industry. In the 1960s, a national water quality network has being established to assess the impact of human activities on rivers and lakes throughout the country. Data collected across Canada were compiled by Environment Canada in the National Water Quality Database (NAQUADAT). This database provided a comprehensive overview of the quality of surface waters in the country and was used with national water quality guidelines to establish water quality objectives for lakes and rivers (Harker et al. 2000).

The federal government monitors the quality of surface water under the Canada Water Act and the Canadian Environmental Protection Act through Environment Canada. Since the late 1970s, Environment Canada's role in monitoring water quality has diminished markedly, with remaining programs focused on (1) boundary and transboundary waters, including the Great Lakes and its connecting channels (e.g., Niagara, Detroit, and St. Clair rivers); (2) major rivers crossing the boundaries of the Prairie provinces (e.g., North and South Saskatchewan rivers) (3) cases that fit closely with the requirements of pertinent federal legislation and with Environment Canada's mandate, science priorities, and ecosystem initiatives.

In some regions, provincial governments have taken over monitoring the water quality of inland rivers and lakes in areas in which they have an interest (Harker et al. 2000). National coverage is therefore incomplete and inconsistent. At present, in Canada no single agency operates a general, nationally integrated monitoring program for water quality.

Because groundwater is legislated under provincial jurisdiction, the federal government generally does not undertake programs to routinely assess the quality of groundwater. However, it does become involved in groundwater concerns when they relate to federal responsibility or federal interest, such as when groundwater contamination threatens to cross an international border. In such cases, groundwater monitoring is undertaken at a specific location by the regional offices of Environment Canada in collaboration with provincial government agencies. Natural Resources Canada's Geological Survey of Canada is involved in groundwater quantity research and monitoring in several regions of Canada.

3.3.2 Provincial and Local Monitoring

Most provinces and some regional administrations and municipalities are involved in monitoring water quality (or flow rates). Liebsher et al. (2000) provide few examples of the water quality monitoring activities that are being undertaken by provincial and local governments, river basin authorities, and universities. These examples have been selected to provide the reader with an idea of the scope of different monitoring programs.

British Columbia

Provincial agencies are co-operating with Environment Canada and U.S. agencies to conduct a groundwater monitoring program of the transboundary Abbotsford–Sumas aquifer in southern British Columbia. A network of 40 wells has been monitored monthly since the late 1970s. Analysis has focused on nitrate levels in the groundwater, but samples are occasionally analysed for pesticides. Reduced funding in recent years has resulted in a reduction in the number of wells being sampled and the frequency of sampling. Concerning surface water monitoring, as mention earlier a provincial-federal co-operative water quality monitoring program began in 1985 to collect long-term data suitable for detecting regional trends.

Prairie Provinces

Between 1994 and 1997, the Canada-Alberta Environmentally Sustainable Agriculture Agreement (CAESA) provided federal and provincial support for monitoring to determine the impact of agriculture on water quality in Alberta. Wells, dugouts, streams, and lakes were checked for nutrients, pesticides, and bacteria. Federal, provincial, private sector, and university researchers were involved. Monitoring continues under other programs.

Central Canada

Ontario has 200 stations to monitor the water quality of streams and rivers, down from a peak of about 2,700 stations in the 1970s. Sampling is currently carried out eight times a year, and the water is analysed for major ions, including nitrate. Pesticide analysis is not done routinely. A subset of this monitoring network focuses on five watersheds: those of the Grand, Thames, Saugeen, Humber, and Don rivers. These watersheds are monitored in more detail, including pesticide analysis of surface water samples. Provincial agencies do not publish long-term trend data or assessments.

Quebec is one of the few provinces that has maintained a year-round, long-term monitoring program since the late 1970s, at a network of water quality stations. Detailed reports on water quality and land use are published.

Atlantic Provinces

Prince Edward Island's Department of the Environment carried out a 3-year (1996–1998) monitoring program consisting of three components: (1) regular sampling of 30 wells located in areas of intense agriculture, targeting pesticides; (2) regular sampling of 30 wells located across the province in different land use areas (agricultural, residential, industrial, urban) to access the state of groundwater quality; (3) sampling of wells installed by the department to identify and track the movement and persistence of target pesticides; these wells were installed in fields where a known pesticide was applied, to assess the potential for pesticides to contaminate groundwater under field conditions. The program is expected to be extended, focusing on different pesticides.

3.4 Water Quality Guidelines

Development of the Canadian Water Quality Guidelines began in 1984. In 1987, the Canadian Council of Resource and Environment Ministers (CCREM), the forerunner of the Canadian Council of Ministers of the Environment (CCME) published the Canadian Water Quality Guidelines (CCREM 1987). The Canadian Water Quality Guidelines are used by provincial, territorial, and federal agencies to assess water quality problems and to manage competing uses of water resources. They are based on the best scientific information available at the time and are subject to periodic re-evaluation as new information becomes available (Environment Canada 1999c). They have received international recognition by the United Nations and the World Health Organization as models for harmonised ambient environmental quality standards and are currently being used in more than 45 countries around the world.

Recently, in 1999, a new release of the Guidelines which includes all media (i.e., water, soil, air, sediment and tissue) has been completed and includes about 550 guidelines for more than 200 priority substances and parameters. Canadian Water, Sediment, Soil, and Tissue Quality Guidelines are being developed to protect and sustain specific uses of land, water, and biota (CCME 1999). They recommend levels in the environment that should not be exceeded in order to prevent negative environmental effects. They include the Canadian Water Quality Guidelines, which have been developed for various substances in raw untreated drinking water, recreational water, water used for agricultural purposes, and water to support aquatic life (CCME 1999; Environment Canada 1999c). Also included are guidelines developed by Health Canada for finished treated drinking water, entitled Guidelines for Canadian Drinking Water Quality (Health Canada 1996).

4. Water uses in Canada

4.1 Withdrawal Uses

The term *withdrawal uses* defines the water that is withdrawn from its source (a river, lake or groundwater supply), piped or channelled to many different locations and users (e.g., household and industrial uses, thermal and nuclear power generation, irrigation and livestock watering), and then is collected again for return to a lake, river or into the ground. Withdrawal uses are measurable as quantities of intake, discharge, and consumption. Water intake is the amount withdrawn from the source for a particular activity over a specific period of time. It represents the demand imposed by that particular use on the water source at a given location. Table 4.1 shows the regional water intake in Canada for each major activity. Usually, however, most of the water taken out is returned at or near the source. This is called water discharge. Water consumption is the difference between water intake and water discharge. Consumption removes water from a river system and makes it unavailable for further use downstream (Environment Canada, 1993a).

Table 4.1. Water intake in Canada, 1991

Region	Thermal power	Manufac- turing	Munici- pal [a b]	Agricul- ture	Mining	Regional Total
Atlantic	2,126	601	356	15	77	3,175
Quebec	1,005	1,616	1,703	100	74	4,498
Ontario	23,095	3,457	1,660	186	87	28,485
Prairies	2,025	447	685	3,014	50	6,221
British Columbia	106	1,161	698	676	75	2,716
National total	28,357	7,282	5,102	3,991	364	45,096
Percent of total	63.0	16.0	11.0	9.0	1.0	100.0

Source: Environment Canada, 1993b
[a] Municipal data exclude water supplied to industry.
[b] Municipal data include estimates for rural residential water use.
[c] Sectorial data for the Yukon and Northwest Territories are included with British Columbia.

According to Environment Canada (1993a), how efficiently water is used in a particular process or economic sector can be determined with the help of two additional measurements: (1) gross water use and (2) the amount of water that is recirculated. Gross water use represents the total amount of water used during a process. This would normally be equal to the water intake, except that more and more users (especially industries) reuse the same water one or more times. In such cases, the gross water use could be equal to several times the water intake. The difference between gross water use and water intake is the amount recirculated, which can be expressed as a use rate. This is the number of times that the water is recirculated and indicates how efficient a particular water use is.

In 1991, five main withdrawal uses (Thermal power generation; Manufacturing; Municipal Use; Agriculture and Mining) accounted for a gross water use in Canada of 57,935 million m^3, made up of intake (45,096 million m^3) and recirculation (12,839 million m^3). About 10% of the intake was consumed and the rest was discharged back to receiving waters. Figure 4.1 taken from Environment Canada (1993b) shows the importance of the main water uses in Canada in 1991. Details are presented in the five following sections.

4.1.1 Thermal Power Generation

This industrial sector includes both conventional and nuclear power generating plants. It withdrew slightly more than 63% of the total water intake in 1991. Next to fuels, water is the most important resource used in large-scale thermal power production. Production of one kilowatt-hour of electricity requires 140 litres of water for fossil fuel plants and 205 litres for nuclear power plants. Some of the water is converted to the steam which drives the generator producing the electricity. Most of the water, however, is used for condenser cooling.

4.1.2 Manufacturing

Water is used in industry as a raw material, a coolant, a solvent, a transport agent, and as a source of energy. Manufacturing accounted for 16% of water withdrawals in 1991. Paper and allied products, chemicals, and primary metals were the three main industrial users.

4.1.3 Municipal Use

Municipal use includes domestic use of water: drinking, cooking, and other household needs. Water is also needed to clean streets, fight fires, fill public swimming pools, and water lawns and gardens. These residential, commercial, and public uses, and the water lost from reservoirs and pipes amounted to about 11% of all withdrawals in Canada in 1991. This figure does not include rural areas where water use is not measured. If rural domestic uses were included, this figure would rise to about 13%.

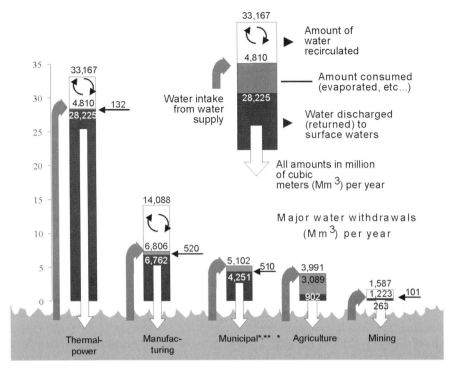

* Municipal data exclude water supplied industry.
** Municipal data include estimates for rural residential water use.
1) Intake + Recirculation = Gross Use
2) Intake – Discharge = Consumption (except in the municipal sector, where consumption has been estimated at 510 or 10% of intake).
3) Municipal consumption is an uncertain figure, but has been estimated.
   However, the difference between intake and discharge is not consumption, but non-metered sewage. If this non-metered sewage (including rainfall) was know, then the municipalities would be *net producers* of water, not consumers.
4) Data for some sectors have been extrapolated and roundede.

Source: Environment Canada (1993b)

Fig. 4.1. Water use in Canada, 1991

4.1.4 Agriculture

Farmers depend on water for livestock and crop production. In 1991 agriculture was still the fourth largest user, accounting for 8.9% of total withdrawals. Water is withdrawn mainly for irrigation (85%) and livestock watering (15%). As mentioned by Connor (1999), the agricultural sector in Canada relies heavily on groundwater for livestock watering, particularly in Alberta and Saskatchewan. Moreover the large number of private wells in rural areas across the country accounts for the relatively high percentage of domestic groundwater use. The relatively small requirement for irrigation water in Canada compared to other countries with a similar scale of agriculture is attributable to the more consistent and reliable seasonal rainfall in this country.

4.1.5 Mining

This category includes metal mining, non-metal mining, and the extraction of coal. Water is used by the mining industry to separate ore from the rock, to cool drills, to wash the ore during production, and to carry away unwanted material. Although the mining industry had a gross use almost as great as agriculture, mining accounted for only about 1% of all water withdrawals in 1991. This was the smallest withdrawal use, but mining recirculates its water intake to a greater extent than any other sector.

4.2 Instream Uses

Unlike withdrawal uses, instream uses cannot be measured quantitatively because the water is not removed from its natural environment. Instead, instream uses are described by certain characteristics of the water or by the benefits they provide to the human population and the ecosystem (Environment Canada 1993a).

Flow rates and water levels are very important factors for instream uses. When these conditions are changed by a dam, for example, it is easy for conflicts to arise. The most common conflict is between hydroelectric development and other uses with respect of aquatic life, wildlife, water supply and water transportation. Storage of the spring freshet (a high river flow caused by rapidly melting snow) removes the natural variability of streamflows on which many life processes depend, in particular, the highly productive ecosystems of deltas, estuaries and wetlands. To make the best use of water, all needs must be carefully assessed and taken into account.

4.2.1 Hydroelectric Power Generation

This water use is the principal source of electricity in Canada today. Billions of dollars have been invested in its development. With large undeveloped hydroelectric sites still available in Quebec, Newfoundland, Manitoba, British Columbia, and the three northern territories, this form of energy development could retain its prominent position for years to come unless public reaction to the negative impacts of hydropower development cause a change in public policy. In any event, the environmental and human effects to be avoided or mitigated in such large projects make them increasingly difficult and costly to plan and build.

4.2.2 Water Transport

Inland waterways in Canada have historically played a major role in getting Canadian goods and raw materials to market. Some traditional uses, such as log driving, have now disappeared. However, water transport is still the most economical
means of moving the bulky raw materials which are Canada's main exports:
wheat, pulp, lumber, and minerals. The main transportation waterways are the St.
Lawrence River, which allows passage of ocean-going ships from the Atlantic
Ocean deep into the heart of North America, nearly as far as the prairie wheat
fields; the Mackenzie River, which is a vital northern transportation link; and the
lower Fraser River on the Pacific Coast. Cargo in the hundreds of millions of tonnes is transported along these routes each year. Reliable and predictable lake and
river levels are very important for this use.

4.2.3 Freshwater Fisheries

Blessed with hundreds of thousands of freshwater lakes and rivers, Canada provides some of the most spectacular sport fishing in the world. According to a 1990
survey, 6.5 million people make use of this fisherman's paradise every year. They
spent about $5.9 billion dollars that year on goods and services directly related to
sport fishing. In addition, inland commercial fisheries employ some 10,000 Canadians, mostly in Ontario and the Prairie provinces. The fish they catch has a market value of about $140 million. Moreover, coastal rivers provide spawning
grounds for salmon and other fish populations which support major saltwater fisheries.

4.2.4 Wildlife

Many wildlife species live in, on, or near the water and require access to it
throughout their lives. Other species may not use water as their primary habitat,
but it is nonetheless essential to their well-being. Watching, photographing and
studying wildlife are all popular forms of recreation for Canadians. About 70% of
Canadians participated in these activities, according to a 1991 survey, and spent
about $2.4 billion that year on them. Hunting attracts nearly one in ten Canadians
and accounts for $1.2 billion of wildlife-related spending each year. The majority
of Canadians believe that it is important to maintain abundant wildlife and to protect declining or endangered wildlife.

4.2.5 Recreation

Canadians have traditionally valued opportunities for outdoor recreation and in recent years have sought the outdoors as never before. Activities such as swimming,
boating, canoeing, fishing, and camping allow Canadians to experience the beauty

of Canada's lakes and rivers. While not all outdoor recreation requires water, the presence of water tends to enhance the experience. Expenditures on water-related recreational activities and tourism also contribute billions of dollars per year to the national economy.

4.2.6 Waste Disposal

It has long been convenient to use lakes, rivers, and oceans as receiving bodies for human and industrial wastes. While water is capable of diluting and *digesting* society's wastes to some degree, there are limits to what even the largest body of water can absorb. The extent to which instream processes can absorb contaminants depends on factors such as the nature of the contaminant, how much of it there is compared to the volume of water, how long the contaminant stays in the water, the temperature of the water, the rate of flow. Many of our waterways are now overloaded with wastes. This problem can best be resolved by increased regulation and/or monitoring.

5. Water and Urban Areas

Although Canada is perceived as a country with abundant water resources, about 60% of this water flows north and is not readily available or easily accessed where it is needed most in the south, home to 90% of the population (Environment Canada 1998b). Urban areas cover only 0.2% of the country but 80% of Canadians are living in them. Moreover, nearly 60% of Canada's urban population lives in centres of 500,000 or more. The environmental pressure caused by urban areas is not negligible. Discharges from wastewater treatment plants, stormwater sewers and combined sewers are still causing adverse impacts on Canadian waterbodies. The impervious surfaces of cities, towns and suburbs increase runoff, which can contain nutrients, pathogens, sediment, industrial chemicals and pesticides, into waterways. The increased runoff can exacerbate erosion and flooding and also threaten groundwater. The International Joint Commission cautions that, although measures have been taken in specific locations, governments at all levels must give adequate attention to the issue of urban sprawl (International Joint Commission 2000b).

5.1 Water Supply Status in Canada

5.1.1 Water Use

About 11% of all water withdrawn from natural sources in Canada enters municipal systems to supply residents, businesses, and some industries. Over half the water in municipal systems supplies residential consumption, which increased by nearly 23% between 1983 and 1994 (Figure 5.1), despite an increase of less than 16% in the municipal population served. Within the home, most water is used in the bathroom (Table 5.1), but residential water use can rise by as much as 50% on peak days in summer owing to lawn and garden watering and car washing. Thus, the extended form of the modern city, with a large portion of the population living outside the urban core, contributes to excessive water use. Table 5.1, describes water use in Canadian homes and is for households that paid for water based on volume used, with the higher rate being households that paid a flat rate, regardless of the volume used.

Urban Canadians on average use almost twice as much water per capita as urban residents in most other industrialised countries except the United States. The average household daily use was 331 L per capita in 1994 (Environment Canada, 1998). This can largely be attributed to low prices and flat-rate pricing (as distinct from charging according to the volume of water used). Water use responds closely to price: of several Western industrialised countries, Canada has the lowest average water price, and Canadian per capita consumption is on average 40% lower where consumers are charged by the amount they use rather than on a flat-rate basis (Environment Canada, 1994). For example, per capita consumption of water is 60% higher in Calgary, which charges a flat rate, than in Edmonton, which charges according to volumes used.

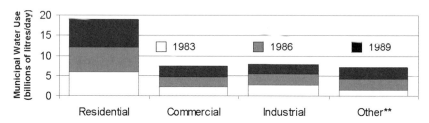

* Water use values are based on (1) municipalities that responded in a given year, and (2) a regional level.
** The *Other* category includes: water lost through leakage; unaccounted water uses, such as water used in firefighting or to flush out pipes; and water that a municipality was unable to assign to one of the other three sectoral categories.

Source: Environment Canada (1998b)

Fig. 5.1. Daily municipal water use by Sector, 1983-1994

Table 5.1. Averge water use in Canadian homes

Showers & baths	35%
Laundry & dishes	20%
Toilets	40%
Drinking & cooking	5%

Taken from Coote and Gregorich (eds.), 2000; original source: Ontario Ministry of Agriculture, Food and Rural Affairs (OMAFRA) ,1994

5.1.2 Water Quality

Water quality in Canada, compared with that in most countries, remains relatively high. However water pollution problems do exist in many parts of the country, usually in association with industrial, agricultural, or mining activities or human sewage. Small lakes and rivers in every province and territory, as well as large water bodies such as the lower Great Lakes – St. Lawrence system have been affected by human activities. Groundwater contamination is also recognised as a widespread and growing concern. Moreover, recently in Walkerton, a small community in Ontario, the Drinking Water Quality issue has been highlighted by a dramatic case, where the failure of the municipality and the province to adequately safeguard the quality of the drinking water led to a number of deaths and a high-profile public inquiry.

Despite many regional and local studies of water quality in Canada, no comprehensive national study has been undertaken. In 1990, Health Canada conducted a limited survey of water treatment facilities. Municipal water supplies of 39% of the population were tested for a limited number of substances. Table 5.2 illustrates some of the survey results.

Table 5.2. Average concentrations of selected sustances in municipal water supplies, 1990

	Concentration [mg/l]					
	Fluoride [a]			Chloride [b]		
Province/territory	Raw	Treated	Dist. [e]	0Raw	Treated	Dist. [e]
Newfoundland	0.10	0.77	0.71	2.98	6.15	6.16
New Brunswick	0.10	0.19	0.28	10.40	12.50	12.40
Nova Scotia	0.10	0.79	0.82	7.63	10.40	11.00
Québec	0.10	0.19	0.17	15.10	17.60	20.50
Notario	0.11	0.50	0.55	17.60	21.00	21.10
Manitoba	0.18	0.84	0.81	33.00	33.50	34.10
Saskatchewan	0.30	1.01	0.78	7.94	10.60	10.70
Alberta	0.18	0.85	0.85	11.80	14.70	14.30
British Columbia	0.24	0.60	0.50	1.96	2.81	2.81
Yukon	0.10	0.52	0.60	1.12	1.30	1.49
NW Territories	0.10	0.65	0.98	1.97	3.48	3.31

Table 5.2. (cont.)

	Concentration [mg/l]					
	Nitrate [c]			Sulphate [d]		
Province/territory	Raw	Treated	Dist [e]	Raw	Treated	Dist. [e]
Newfoundland	0.27	0.30	0.26	1.89	1.88	1.89
New Brunswick	2.32	0.53	0.49	6.52	12.60	12.50
Nova Scotia	0.25	0.27	0.33	7.08	11.70	11.40
Quebec	2.10	2.01	2.22	14.60	25.40	25.60
Ontario	1.94	1.71	1.65	22.90	28.80	29.10
Manitoba	1.29	0.71	0.65	67.70	73.10	72.50
Saskatchewan	3.27	1.59	1.42	154.0	171.00	172.0
Alberta	2.89	1.41	1.87	52.70	68.90	69.30
British Columbia	0.78	0.84	0.80	8.78	8.79	8.71
Yukon	0.10	0.17	0.10	27.70	27.30	27.70
NW Territories	0.10	0.18	0.10	3.08	3.08	3.11

[a] The Canadian guideline value, or MAC (maximum acceptable concentration), for fluoride is 1.5 mg/l;
[b] MAC for chloride is 250 mg/l;
[c] MAC for nitrate is 45 mg/l (equivalent to 10 mg/l nitrate as nitrogen);
[d] MAC for sulphate is 500mg/l;
[e] Water distributed by water supply system.
Taken from: Environment Canada, 1996a. No data are available for Prince Edward Island.

5.2 Municipal Wastewater Status in Canada

5.2.1 Municipal Wastewater Treatment Plants

Nearly 81% of Canadians were serviced by municipal sewer systems as of 1994 (Chamber et al. 1997). Of those serviced, 93% had at least primary wastewater treatment provided in 1994. The remaining 1.6 million Canadians that were serviced by sewage collection systems were not connected to wastewater treatment facilities and discharged their untreated sewage directly into receiving water bodies (Environment Canada 1998b).

In general, the level of municipal wastewater treatment in Canada has been rising steadily (Figure 5.2). In 1983, 18.2 million people, three-quarters of the total population (including all city residents), were served by sewage systems, and 13 million of these (72%) were also served by some form of sewage treatment. The wastewater of the remaining 5.2 million was discharged untreated. By 1994, the number served by sewers had risen to 21.2 million. Of these, 19.6 million (93%) were also served by some form of treatment, while the wastewater of 1.6 million was discharged without treatment. During this period, the numbers served by tertiary treatment rose from 5 million to 8.2 million, 39% of the population served by sewers (Environment Canada 1998b).

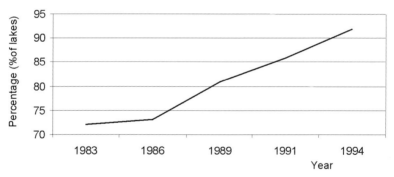

Refers only to municipal population served by a municipal sewer system.

Source: Environment Canada (1998b)

Fig. 5.2. Municipal wastewater treatment 1983 – 1994

Levels of wastewater treatment are mainly primary in British Columbia, secondary in the Prairie provinces, and tertiary in Ontario. Quebec has a mixture of primary and secondary, with some tertiary treatment. In the Atlantic provinces, more than half of the population served by sewer systems released untreated wastewater directly into estuarine and coastal waters. This great variation of level of treatment across Canada has been determined because of the nature and scale of the receiving waters (Figure 5.3). Whereas many coastal cities discharge their wastewater directly into the ocean, and cities on the lower St. Lawrence into that river, most Ontario cities must discharge into smaller rivers or the Great Lakes, and Prairie cities discharge exclusively into rivers. That is why, in 1994, no sewage was left totally untreated in the Ontario or Prairie communities with systems. However, more than 50% of the population of the Atlantic provinces with sewage systems were not served by any form of wastewater treatment. The situation in Quebec has changed sharply in recent years. As of 1994, 84% of the population with sewage systems also had treatment plants; compared to 56% in 1991. It is planned that at the end of the *Programme d'Assainissement des Eaux Municipales* (PADEM) started in 1995, 98% of the population with sewage system will have their waste water treated (Gouvernement du Québec 1999).

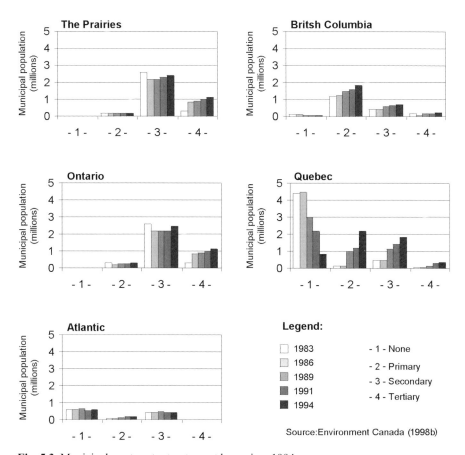

Fig. 5.3. Municipal wastewater treatement by region, 1994

5.2.2 Stormwater and Combined Sewer Overflows

In Canada detailed data on the proportion of population served by storm sewer versus combined sewer system are not available. However, most urban areas developed before 1940 are served by a combined sewer system in which stormwater and sanitary sewage are not separated (Chambers et al. 1997).

Urban stormwater management in Canada has been practised for more than 30 years (Marsalek and Kok 2000). Recently, in Ontario, for example, the Urban Drainage Program of Canada's Great Lakes 2000 Cleanup Fund (see Chap. XIII on Partnership and Community involvement issues) has been supporting the development and implementation of cleanup technologies to control municipal pollution sources, to clean up contaminated sediment, and to rehabilitate fish and wildlife habitat (Sandra et al. 2000). Real time control strategies taking advantage of excess capacity in the sewer network and reducing overflows are in operation for example in the Quebec Urban Community (*Communauté urbaine de Québec* 1999) and under development in several other Canadian Cities e.g. the Region of Hamilton-Wenthort (Stirrup et al. 1997)

However, the situation is not under total control and in most large Canadian cities combined sewer overflows (CSOs) still occur and contribute to broadscale impacts on individual receiving waterbodies. In the future, enhanced effort will be needed in Canadian cities to reduce stormwater runoff and CSO.

5.2.3 Wastewater Treatment in Northern Region

The water and wastewater services provided to Aboriginal people in Canada, particularly those living in the North, deserve special mention. In 1992-1993, about 91% of dwellings on Indian reserves had some form of potable water supply (piped, well, trucked, or other), and almost 83% of dwellings had sewage disposal facilities (Department of Indian Affairs and Northern Development 1993). Although much progress had been made since the 1970s, this still left about 9% of reserve dwellings without a satisfactory supply of potable water and 17% without sewage disposal services, with consequent discharge of raw sewage into receiving waters including vulnerable coastal shellfish growing areas.

5.3 Impacts of Municipal Wastewater Effluents on Canadian Waters

As mentioned by Chambers et al. (1997), in Canada discharges from wastewater treatment plants, stormwater sewers and combined sewers have caused adverse impacts on some lakes, rivers and coastal waters. The most publicly recognised impacts are shellfish harvesting restrictions and beach closures resulting from microbial contamination. Habitat degradation and chemical contamination also occur and these, in turn, have altered the abundance and diversity of aquatic organisms. More recently, the presence of endocrine disrupters in discharged waste-water has become a major concern.

In Canada there is a need to review sewage treatment requirements (Chambers et al. 1997). Further research is also required on the interactive and cumulative responses to habitat degradation and to long-term exposure to persistent and bioaccumulative pollutants. Some studies presenting the use of benthic assessment techniques to determine CSO and Stormwater impacts in the aquatic ecosystem have been recently published (Rochefort et al. 2000). Finally, an integrated approach to wastewater management is needed that addresses loadings from treatment plants, stormwater sewers, combined sewer overflows and other wastewater sources.

5.4 Ageing Infrastructure

In Canada like in many other industrialised countries, the state of urban infrastructure has become of increasing concern in recent years to municipalities, and the provincial and federal governments. The Federal - Provincial Infrastructure Program is a concrete response to renovate and enlarge ageing urban infrastructure and construct new services in urbanised areas where they are lacking.

In this context, many studies have been undertaken to asses the state of urban infrastructure. In 1996, The Federation of Canadian Municipalities released a study that concluded that more than the half of the existing equipment was in an unacceptable state (Federation of Canadian Municipalities 1996). In some provinces such as Quebec, the judgement was not quite as severe. A survey completed by *l'Union des municipalités du Québec* concluded that the general state of the infrastructure was relatively good (UMQ). More recently, in 1998 a study completed by INRS-Eau and INRS-Urbanisation (two research groups associated with the University of Québec) and presented by Trepanier (1998) at the *Symposium on Water Management in Quebec*, showed that the major problem associated with ageing infrastructure in Quebec, excluding Montreal which was not integrated in the study, was in the underground infrastructure. The study showed as well the lack of information available to asses the state of urban infrastructure and the need for a better knowledge of urban infrastructures.

Financially speaking, a projection made by the *Coalition pour le renouvellement des infrastructures du Québec* (2000) indicates that $9 billion in Quebec Province and $36 billions investments in all Canada are needed for the next 15 years to bring municipal infrastructures to an acceptable state according to North-American standards.

6. Water and Agriculture

Although 99% of the farms in Canada depend on natural precipitation, agriculture was still the fourth largest consumptive use of water in 1991, accounting for 8.9% of total withdrawals. Water is withdrawn mainly for irrigation (85%) and livestock watering (15%). Irrigation is needed mainly in the drier parts of Canada, such as the southern regions of Alberta, British Columbia, Saskatchewan, and Manitoba. Irrigation is also used in Ontario and the Maritimes for frost control. Since so much of the water intake evaporates, only a small fraction is returned directly to its source. This is a highly consumptive use. This chapter focuses on the major withdrawal uses of water for agriculture; irrigation, livestock production, and on-farm domestic use. It also briefly summarises the main impacts caused by agricultural activities on Canadian water quality. It is largely inspired by the following document: The Health of our Water; Toward sustainable agriculture in Canada (Coote and Gregorich 2000).

6.1 Irrigation

The benefits of irrigation include: (1) increased stability of production; (2) the potential production of a greater diversity of crops than would be possible otherwise; and, (3) the intensification of production.

As reported in Kienholz et al. (2000), dry regions in the interior of British Columbia and the southern Prairies have severe soil moisture deficits at some time during most summers and can suffer from long-term drought conditions. These areas represent most of the 1 million hectares of irrigated cropland in Canada (Figure 6.1), with Alberta alone accounting for 60%. As shown in the 1996 State of the Environment report (Environment Canada 1996a), and as already discussed earlier in Chapter II on Water Quantity, Irrigation in Alberta is the main purpose for inter basin water transfers.

The relatively moister conditions found in central and eastern Canada reduce the need for supplemental water, but limited irrigation of high-value crops (e.g., fruits and some vegetables) is practised.

The peak design flow rate of an irrigation system varies according to climate, crops, and soil conditions. An estimate can be made from the peak evapotranspiration rate. The amount of water withdrawn for irrigation varies annually and depends mainly on: (1) winter precipitation; (2) weather; and, (3) soil moisture conditions during the growing season.

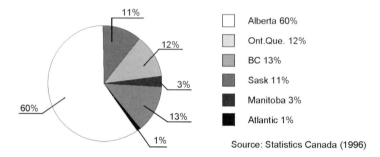

Alberta 60%
Ont.Que. 12%
BC 13%
Sask 11%
Manitoba 3%
Atlantic 1%

Source: Statistics Canada (1996)

Fig. 6.1. Distribution of irrigated land in Canada

In areas such as south-west Saskatchewan, spring runoff determines the amount of water available for irrigation during the following summer. Temperature, the amount and timing of rainfall, wind, and evaporation all influence the need for supplemental water for optimum plant growth.

Water use for larger irrigation projects is often licensed by the province in which they are located, as a means of controlling total withdrawals from a water source and minimising the potential for conflicts among users. The licence stipulates the maximum volume of water that can be withdrawn in a year. The licensed amount is often considerably greater than that withdrawn in an average year (Kienholz et al. 2000).

The expansion of irrigated area depends on both soil characteristics and a secure supply of water of suitable quality. Some provinces require irrigators to undertake a soil water compatibility study before approving irrigation plans. To limit competition with other water users, irrigators, private industry, governments, and researchers have co-operated to introduce greater efficiencies in the way irrigation water is stored, conveyed, and applied in the field (Kienholz et al. 2000). For example:

- Irrigation headworks, main canals, and whole distribution systems have been renovated to minimise water loss.
- Irrigators are encouraged to switch from less-efficient gravity systems to more-efficient sprinkler systems or to highly efficient drip or trickle systems.
- Some irrigators are converting saline land back to dry land.
- Governments and industry are conducting research and demonstration projects to determine the applicability of new irrigation technologies and to identify the actual water requirements of irrigated crops.

Water meters are being used at the district and farm levels to measure water use and charge for water based on consumption.

6.2 Livestock Production

6.2.1 Water use

Livestock production depends on ready access to water of suitable quality. The main use of water is for drinking, but it is also used to clean facilities, sanitise equipment, and dilute manure (Coote and Gregorich 2000).

Water is important to animal growth and maintenance of body tissues, reproduction, and lactation. Animals lose body water in expired air, milk, urine, and faeces, and by evaporation from the skin. Animals whose water intake is restricted, either because of limited supplies or poor quality, will likely eat and grow less and be less productive. In some cases, they may become sick and even die.

How much an animal drinks depends on: (1) the species, (2) physiological conditions, such as age and whether the animal is lactating; and (3) environmental factors, such as temperature, humidity, activity level, and water content of the feed. For example, a lactating dairy cow may drink 70 to 160 litres daily (Table 6.1), whereas a dry cow requires only 35 to 60 litres daily. Table 6.1 taken from Kienholz et al. (2000), shows average daily water requirements for different types of livestock.

Table 6.1. Daily water needs of farm animals

Animal type	Water (L/day)
Beef feeder	35
Beef cow	55
Dairy cow	160
Lactating sow	20
Ewe	7
Chicken layer	0.25–0.30
Chicken broiler	0.15–0.20
Feeder pig	10

For example, 121,480 millions litres/day is required just to provide sufficient drinking water for the number of pigs being reared (total number given by statistics Canada, (2000)) in Canada.
Adapted from Kienholz et al. 2000; original source: University of Saskatchewan, 1984

6.2.2 Water Use Reduction Strategies

As Kienholz et al. (2000) point out, studies on the use of water in hog barns have shown that considerable water savings can be achieved by using bowls and hopper waterers instead of nipple waterers, the most common watering system in hog barns. Growing and finishing pigs may waste up to 60% of the water from a nipple drinker. This wastage adds greatly to the volume of wastewater and the cost of storing and disposing of effluent.

Table 6.2 shows how water is used in a dairy operation. Opportunities exist to reduce the amount of water used, through

- scraping or sweeping floors before washing
- reusing equipment rinse water to wash floors
- using high-pressure nozzles for washing
- installing water-saving sinks
- using the first rinse water from milk lines to water calves.

Table 6.2. Water use in a milking parlour

Milking equipment	71%
Milk house floor	3%
Cold storage tank	9%
Milking parlour floor	17%

Adapted from Kienholz et al. (2000); original source: Cuthbertson et al. (1995)

6.2.3 Rural Domestic Use

Canadians are next only to Americans in the average amount of water each person uses daily. Farmers and other rural residents, unlike their urban cousins, are directly responsible for their own water supply (Kienholz et al. 2000). The source of water may include one, or a combination, of

- a shallow or deep groundwater well
- lake, stream, river, or on-farm storage pond or dugout
- a cistern filled by rainwater or by hauling water from a distant source
- a regional water supply pipeline.

In emergencies, rural families may also resort to buying bottled water for drinking and cooking. Different water sources may be used for different purposes, such as drinking, watering lawn and garden, and watering livestock. Some type of treatment is often needed to ensure that the water used for household purposes is of suitable quality. Developing a dependable supply of water often involves considerable initial costs, ranging from $5,000 to $25,000. Rural residents must also pay for in-home treatment systems and ongoing operation and maintenance costs (Kienholz et al. 2000).

How domestic use in rural areas compares with the high average amount of water used by Canadians in urban settings is difficult to say, because there are no provincial requirements for licensing domestic on-farm water use, and water used by rural residents is rarely metered. An exception to this occurs in the Prairies for recently constructed regional water supply pipelines. Meters on individual connections to pipelines in south-east Saskatchewan show that the average daily water use per person, assuming little water is used for outdoor purposes, ranges from 225 to 373 litres. The age of family members is a factor in water use, with the highest consumption occurring in households with babies and young children. In comparison, metered water use in southern Alberta for a family of five with a private water supply and treatment system was 155 litres per person daily, assuming little outdoor use. These figures indicate that rural residents use slightly less water for domestic purposes than the average Canadian residing in a community with a population of more than 1000 (Kienholz et al. 2000).

To some extent, domestic water use by rural residents is influenced by the same factors that affect urban residents. Adults generally use less water than teenagers and young children. The growing use of dishwashers and automatic clothes washers can also increase the amount of water used in both urban and rural areas, depending on how efficiently they are used. Rural residents are adopting a more urban lifestyle, and expectations for the same level of basic services are growing. This change is expected to lead to greater water use. In contrast, rural residents are still more directly affected by drought conditions and water shortages. They usually respond more rapidly by reducing water use, especially outdoors (Kienholz et al. 2000).

6.3 Impact of Agriculture on Water Quality

6.3.1 Surface Water

Surface waters in Canada range from poor to good quality. Agricultural impacts on surface waters are common, particularly after rainfall events. In general, the problem of soil erosion by water and wind is decreasing in Canada, mainly because of the use of conservation farming practices such as reduced tillage and no-till and residue management. However, cropland used to grow row crops, especially rolling land with vulnerable soils (e.g., south central British Columbia, New Brunswick, and Prince Edward Island), will continue to be at risk of erosion unless more care is taken to keep the soil covered and to adopt conservation practices such as contour strip cropping and terracing. Soil erosion figures can give only an estimate of the actual sedimentation of surface waters, for which better data are needed in many agricultural areas. Curbing the rates of erosion and sedimentation also helps to reduce the amount of nutrients, pesticides, and other substances that reach water attached to soil particles.

Eutrophication caused by nutrient inputs from agricultural sources has been observed in the Fraser Valley of British Columbia, the southern Prairie provinces, watersheds draining into lakes Erie and Ontario, and the south shore of the St. Lawrence River. Even without knowing exactly how much nutrient loading of surface waters can be attributed to agriculture, it makes sense to manage agricultural nutrients carefully, especially in these areas of intensive agriculture.

Pesticide concentrations in surface waters vary considerably across Canada. Although pesticides are detected in surface waters in many agricultural watersheds, concentrations are typically below Canadian water quality guidelines for drinking water but not for the protection of aquatic life. Wind-borne pesticides can be deposited on surface waters some distance from the source. Higher pesticide concentrations in water are associated with intensive crop production or poor management practices by some growers.

The intensity of agricultural activities continues to increase at the same time that urban populations are moving more into rural areas. Rural residents will continue to demand good quality surface water, while sometimes contributing to water quality problems themselves. Farmers too are more and more aware of the quality of the waters around them and are generally anxious to minimise any harmful effects of their practices. Improved management techniques are helping to improve surface water quality, but farmers and rural residents must remain committed to this improvement if it is to be sustained.

6.3.2 Groundwater

In Canada, groundwater quality is generally within Canadian Water Quality Guidelines in most areas of the country, but nitrate levels are a continuing concern. Research and surveys have shown that intensive agricultural practices may increase both the risk and incidence of nitrate contaminating groundwater. Nitrate leaching results mainly from the mismatch between crop demand for nitrate and microbial activity in the soil and is associated with all agricultural practice.

Bacterial contamination of groundwater is also observed, particularly in areas where large quantities of manure are applied. Pesticides have been detected in some groundwater, but concentrations that exceed water quality guidelines are uncommon and are usually associated with point sources, such as pesticide spills.

The prospects for improving this situation depend largely on producers who apply environmentally sound management practices (recognising the limitations of individual farms related to climate, topography, soil, equipment, finances, and time). Although many of these practices are easy to adopt, others require an investment of time and money.

7. Water and Industry

7.1 Mining Industries

Mining has played an important part in the history of Canada's development, from the coal mines of Nova Scotia to the Yukon gold rush. The discovery of nickel, copper, and other minerals was a key factor in the emergence of the Great Lakes Region as the economic engine of both Canada and the United States. Value-added by the mining industry continues to be a significant part of Canada's national economy. Export dollars are particularly important. All employment provided by the mining industry and the communities that flourish as a result are of direct importance to a large number of Canadians. The mining industry continues to generate these benefits for Canadians and will remain a large part of Canada's future economy. As of January 1995, there were 558 producing mining and quarrying establishments operating in Canada (Figure 7.1).

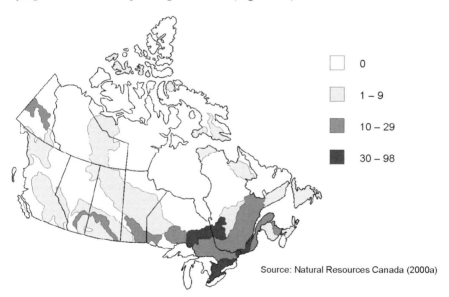

Source: Natural Resources Canada (2000a)

Fig. 7.1. Mining Sites in Canada

Exploration, mining and milling, smelting and refining, and post-operational waste management are the main phases of the mining activity, each of which has a potential to cause environmental impacts and which has historically caused significant damage to Canadian aquatic ecosystems in particular in the Yukon, Northwest Territories and BC. The major environmental impacts of mineral exploration are caused by the construction of exploration roads in previously isolated areas. Erosion, which may cause disruption to habitats of fish is one impact created by these roads. During mine operation, the main potential environmental issue is the type of effluent discharged from the tailings and waste rock to surface water and groundwater. If the effluents are not treated or acid generation is not prevented or controlled, these waters can be acidic in nature and contain concentrations of heavy metals. Depending on the ores mined and processed, these waters may also contain organic compounds, cyanide from gold milling, or ammonia (NH_3) compounds. Smelting and refining operations are used to produce most base metals (copper, lead, zinc, and nickel), all ferrous metals (iron and steel), and aluminium. Examples of, environmental impacts caused by iron and steel operations relate primarily to the discharge of toxic emissions from coke ovens and furnaces and to waterborne acidic wastes or sludge (Environment Canada 1996a). Organic compounds including ammonia, benzene, cyanide, phenols, and polycyclic aromatic hydrocarbons (PAHs) and metals including chromium, lead, nickel, and zinc (see Table 7.1) are the major contaminants released by the steel industry into Canadian waters (Environment Canada 1996a). These environmental concerns are especially problematic in the province of Ontario where four integrated facilities produce approximately 65% of the steel in Canada. About 70% of the estimated 1.3 million cubic metres per day of process effluent and the 1.6 million cubic metres per day of cooling water discharged into Ontario waterways is from plants in Sault Ste.Marie and Hamilton (Ontario Ministry of Environment and Energy 1994). As of 1996, there were over 6000 active, abandoned (i.e., no longer operating, owners known), and orphaned (i.e., no longer operating, owners unknown) tailings sites in Canada (Environment Canada 1996a). The environmental hazards posed by abandoned mine sites are potentially serious, even if a mine site does not contain any harmful materials. Acidic drainage, and the resulting contamination by metals, is the most common problem.

Table 7.1. Canadian Mining and Smelting Industry Voluntary Emission Reductions to Water, ARET Program

ARET substance	Releases to water		
	Base year - 1988 [tonnes]	1996 [tonnes]	Change from base year [%]
Arsenic	34.4	5.3	0.85
Cadmium	13.4	1.9	0.86
Copper	68.0	16.9	0.75
Cyanide	103	6.1	0.94
Lead	191	42	0.78
Mercury	1.35	0.09	0.93
Nickel	53.15	6.27	0.88
Zinc	698	90	0.87

Taken from Environment Canada, 1998a. Source: The Mining Association of Canada

The mining and smelting industry has considerably reduced emissions to water under the ARET (Accelerated Reduction/Elimination of Toxics) program (Table 7.1). ARET is a voluntary pollution prevention initiative that has been endorsed by 31 of 34 member companies of the Mining Association of Canada. These 31 companies account for 85 percent of the value of Canada's base metal production.

7.2 Pulp and Paper Industry

The production of pulp and paper generates organic and toxic wastes, which are regulated under the revised Pulp and Paper Effluent Regulations pursuant to the Fisheries Act and under the Canadian Environmental Protection Act (CEPA) dioxin and furan regulations. Because measuring and regulating all chemicals in pulp mill effluent is practicality impossible, regulatory authorities have selected a few indicators (based on ease of measurement and known impacts, for monitoring pollution). These indicators include total suspended solids, biochemical oxygen demand, and dioxins and furans. In 1992, the federal government introduced new nation-wide regulations under the Fisheries Act, including minimum discharge standards for these parameters and for non-toxic effluents. These Pulp and Paper Effluent Regulations outline minimum standards for all pulp and paper mills. In 1994, it has been stipulated that all mills must undertake an Environmental Effects Monitoring Program, pursuant to the revised Pulp and Paper Effluent Regulations. The objective is to ensure the adequacy of effluent regulatory standards for protecting fish, fish habitats , and human use of fish (Environment Canada 1996a).

To meet the federal standards, mills using chlorine bleaching had to modify their processes to prevent dioxin and furan formation. All mills were required to discharge effluents that were non-toxic to fish. Mills that did not already include secondary treatment had to install such facilities. Mills requiring time to do this were granted Transitional Authorisation Extensions. which expired 31 December 1995. As of 1996, 157 mills were subject to the requirements of the Fisheries Act. Of these mills, one or two were not expected to be in compliance by summer of 1996. Environment Canada has developed an enforcement strategy that includes inspection and investigation plans concentrating on mills that were not in compliance. Routine inspections will continue for all others (Environment Canada 1996a).

Impacts from pulp and paper mill discharges include the build-up of wood fibre in the receiving waters. The dissolved oxygen available to fish can be used up by the decomposition of organic materials in the effluent, and benthic habitats can be smothered. Depending on the production process, pulp and paper mills may also be major sources of toxic chemicals, including dioxins and furans . These chemicals have been found in the fatty tissue and muscle of crab and other shellfish in the Pacific region, with the result that some commercial and non-commercial fisheries were closed, beginning in the late 1980s. However, in February and August 1995, the Government of Canada announced re-openings of certain closed fisheries because of reduced dioxin and furan levels in shellfish tissues (e.g., the crab fishery at Howe Sound in the Strait of Georgia). These improvements were the result of significant process changes by industry in response to the federal dioxin and furan regulations (Environment Canada 1996). The pulp and paper industry has made significant efforts in the area of pollution abatement, resulting in a downward trend in total suspended solids and biochemical oxygen demand in mill effluent over the past several years. Between 1980 and 1997, total suspended solids (kilograms per tonne of pulp produced) dropped by 80%, and biochemical oxygen demand (kilograms per tonne of pulp) fell by 95 percent. Between 1988 and 1994, there was a 99.4 percent reduction in the release of dioxins and furans, reaching non-measurable levels in 1995 (Environment Canada 1998a). Between 1980 and 1993, total suspended solids (kilograms per tonne of pulp produced) dropped by 68%, and biochemical oxygen demand fell by 65% (Figure 7.2) (Environment Canada 1996a, 1998a).

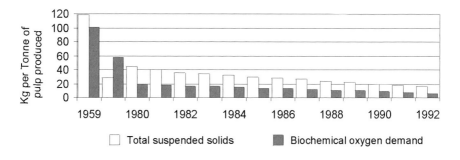

Source: Environment Canada (1996)

Fig. 7.2. Total suspended solids and biochemical oxygen demand in mill effluent, 1959-1994

8. Water and Energy[6]

Energy plays a major role in the Canadian economy and the Canadian way of living. The rigorous climate, an energy-intensive industrial base, relatively low energy prices, a low population density and large distances between population as well as the high standard of living of Canadians explain the high level of energy consumption (Environment Canada 1996a). Electricity accounts for an significant portion of the energy production in Canada: 1,692 PJ (PetaJoule or 1×10^{15} Joule) of electricity was produced in 1996 of a total energy production of 7,801 PJ (World Energy Council 1998a). However useful for Canadians, electricity production and distribution have deleterious effects on the environment and aquatic ecosystems. The recent trends in electricity production, especially the use of water in energy production and the significant impacts of electricity generation on Canada's environment are examined in this chapter.

8.1 Electricity Production Sources in Canada

Globally, Canada ranks fifth in total electricity generating capacity and electric energy production (World Energy Council 1998a). The total Canadian electricity production rose from 373 to 573 TW.h (terawatt hour or 1×10^{12} Watts hour; 1 TW.h contains 3.6 PJ of energy) between 1980 to 1996 with the relative contribution of natural gas and nuclear powered stations increasing in importance, oil use decreasing in importance and hydroelectric and coal-powered stations remaining stable (Figure 8.1; World Energy Council (WCA) 1998a). Presently, hydroelectric power is the largest domestic source of electric energy, representing 62% of the total electricity production, followed by fossil-fuelled and nuclear plants, representing 24% and 14% respectively, and alternative energy sources being only marginal (Figure 8.1; Canadian Electricity Association (CEA) 1999).

8.1.1 Hydroelectricity

Canada possesses enormous hydropower potential, with an economically exploitable capability second only to that of Brazil in the Western Hemisphere (World Energy Council 1998b). However, the fact that this potential is economically ex-

---

[6] This Chapter has been provided by Janick Lalonde, PhD candidate at INRS-Eau

ploitable does not mean that its development is environmentally or socially acceptable, as explained later in this chapter. In 1991, Canada had 650 dams with a storage capacity of at least 100×10^6 m³ (Figure 8.2; CNC ICOLD 1991). Of these dams, over 70% were used for electricity generation, with 9 out of every 10 dams constructed between 1970 and 1991 constructed for purposes of hydroelectric power generation (Canadian National Committee of the International Commission on Large Dams -ICOLD- 1991). Similarly, the 54 largest water diversion projects in Canada were undertaken predominantly for energy purposes, with about 96% of the total diverted water flow being used for the generation of electric power (Statistics Canada 1994).

	1980	1990	1995	1996
Coal	59.8	83.0	90.3	92.4
Oil	13.8	16.0	10.4	12.2
Natural gas	9.2	10.6	24.4	16.1
Nuclear	38.0	72.9	97.8	92.8
Hydro	251.1	296.8	335.5	356.1
Other	1.4	2.4	3.7	3.8
Total (TWh)	373.3	481.1	562.1	573.4

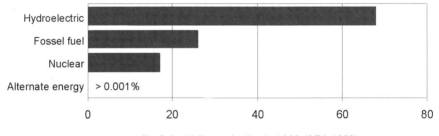

% of electricity production in 1999 (CEA 1999)

Source: Environment Canada (1996)

Fig. 8.1. Canadian electricity production by source

There is still considerable undeveloped hydro potential at numerous existing and new sites. At the end of 1996, 870 Mega Watts (1×10^6) of hydroelectric generating capacity was under construction and 12.4 Giga Watts (1×10^9) additional capacity was planned for future development (World Energy Council 1998b). Of the potential hydroelectric power, 2% is located in sites suitable for small-scale developments (World Energy Council 1998b). The present small hydro plants (less than 25 Mega Watts) had a capacity of 1450 Mega Watts at end of 1996 and plans for future installation would account for an additional 100 Mega Watts (World Energy Council 1998b). Under the Renewable Energy Strategy released by the Department of Natural Resources in 1996, an accelerated tax write-off is provided for certain classes of equipment, including hydroelectric installations with a planned average annual generating capacity not exceeding 15 Mega Watts.

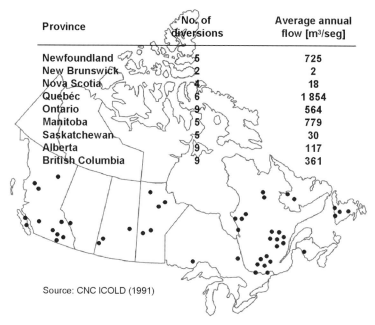

Province	No. of diversions	Average annual flow [m³/seg]
Newfoundland	5	725
New Brunswick	2	2
Nova Scotia	4	18
Quebéc	6	1 854
Ontario	9	564
Manitoba	5	779
Saskatchewan	5	30
Alberta	9	117
British Columbia	9	361

Source: CNC ICOLD (1991)

Fig. 8.2. Large hydroelectric dams and water diversion in Canada

8.1.2 Fossil Fuel

Canada possesses extensive fossil fuel resources. Almost all of the oil production in Canada comes from the Western Canada Sedimentary Basin (WCSB) (National Energy Board 1999a). In addition to the presently exploited sedimentary deposits in western Canada, there are significant unexploited oil resources in Canada's north and offshore regions (National Energy Board 1999a). The western provinces of Alberta, British Columbia and Saskatchewan supply almost all of Canada's natural gas, with approximately 83% coming from Alberta (National Energy Board 1999b). In 1999, natural gas output rose to 5.7 trillion cubic feet (National Energy Board 1999b). A new natural gas project called Sable Island situated off Nova Scotia began production in 2000. It is projected to produce over 400,000 cubic feet a day (World Energy Council 1998a). Canada has extensive coal resources, principally of the bituminous variety. It is extracted mainly by surface mining (National Energy Board 1999a). The provinces of Alberta, British Columbia and Saskatchewan account for over 94% of the 76.8 million tonnes of coal produced in 1997 with Nova Scotia and New Brunswick responsible for most of the remainder (National Energy Board 1999a).

8.1.3 Nuclear Energy

High-grade uranium occurs in Canada, particularly in northern Saskatchewan, which possesses the richest uranium ores ever discovered (Figure 8.3; INSC 1999). Canada has an indigenous nuclear power industry based on the CANDU reactor technology (Atomic Energy of Canada Limited 2000). CANDU (a registered trademark standing for *Canada Deuterium Uranium*) is a pressurized-heavy-water, natural-uranium power reactor.

In Canada, there are 14 operable nuclear power reactors utilizing uranium while 8 have temporarily suspended operations and 2 are permanently shutdown. All of these nuclear power plants are located at five sites in three provinces: Ontario, New Brunswick, and Quebec (Figure 8.3; INSC, 1999). The suspended nuclear plants are scheduled to return to service between 2000 and 2009 (World Energy Council 1998a). In the province of Ontario about 48% of the electricity supply is nuclear (Whitlock 2000). The other two provinces with nuclear power, New Brunswick and Québec, receive respectively 21% and 3% of their supply from nuclear. Total Canadian nuclear electricity production was 74 TW.h in 1999 and the net total of about 1.5 million TW.h from June 1962 to December 1999 (Whitlock 2000).

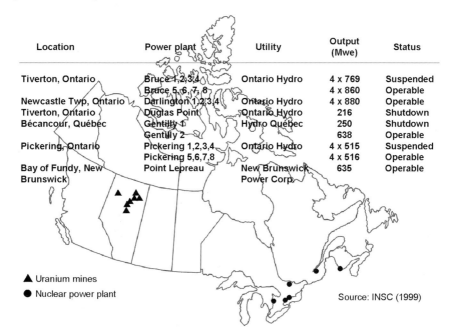

Location	Power plant	Utility	Output (Mwe)	Status
Tiverton, Ontario	Bruce 1,2,3,4	Ontario Hydro	4 x 769	Suspended
	Bruce 5,6,7,8		4 x 860	Operable
Newcastle Twp, Ontario	Darlington 1,2,3,4	Ontario Hydro	4 x 880	Operable
Tiverton, Ontario	Duglas Point	Ontario Hydro	216	Shutdown
Bécancour, Québec	Gentilly 1	Hydro Québec	250	Shutdown
	Gentilly 2		638	Operable
Pickering, Ontario	Pickering 1,2,3,4	Ontario Hydro	4 x 515	Suspended
	Pickering 5,6,7,8		4 x 516	Operable
Bay of Fundy, New Brunswick	Point Lepreau	New Brunswick Power Corp.	635	Operable

▲ Uranium mines
● Nuclear power plant

Source: INSC (1999)

Fig. 8.3. Uranium mines and nuclear power plants in Canada

8.2 Water Use for Thermal Energy Plants Cooling

Conventional thermal and nuclear power generating plants withdrew slightly more than 63% of the total water intake for all purposes in 1991. Next to fuels, water is the most important resource used in large-scale thermal power production. Production of one kilowatt-hour of electricity requires 140 litres of water for fossil fuel plants and 205 litres for nuclear power plants. Some of the water is converted to the steam which drives the turbine generating the electricity. Most of the water, however, is used for condenser cooling (Environment Canada 1993b).

Cooling is necessary because today's processes can only convert 40% of the fuel's energy into usable electricity. The rest is wasted. This shows the double cost of inefficient energy use: first, in the wasted energy, and then in the water required to cool the wasted heat to the temperature where it can be released safely into the environment. This requires a continuous flow of cooling water circulating through the condenser. All the cooling water is therefore returned to the environment much warmer. However, the temperature can be reduced using cooling towers and other such devices (see Chap. IV. Water Uses in Canada).

8.3 Environmental Impacts

Electricity is considered by some to be the foremost vector for growth and democratisation (Lafrance 2000). Paradoxically, many aspects of electricity production and use, from its generation to its final consumption, affect the environment. Some effects are temporary and easily reversed, but others like global warming are long lasting and irreversible.

8.3.1 Hydroelectricity

Hydroelectric power provides a much cleaner source of energy than the burning of fossil fuels, at least in terms of air pollution (see Table 8.1 taken from International Energy Association 2000). However, the constructions of dams and reservoirs can seriously alter aquatic ecosystem.

Table 8.1. Synthesis of environmental parameters for electricity options

Electricity generation options	Greenhouse gas emissions [kteqCO$_2$/TW.h]	Land requirements [km^2/TW.h]	SO$_2$ Emissions [t/TW.h]	NO$_x$ Emissions [t/TW.h]
Coal	790-1,182	4	700-32,321	700-5,273
Oil	555-883		8,013-9,595	316-12,300
Natural gas	290-520		4-15,000	0.3-1,500
Nuclear	2-59	0.5	3-50	2-100
Hydropower	2-48	2-152	5-60	3-42

Table 8.1. (cont.)

Electricity generation options	NMVOC Emissions [t/TW.h]	Particulate matter emissions [t/TW.h]	Mercury emissions [kg/TW.h]
Coal	18-29	30-63	1-360
Oil	22-1,570	122-213	2-13
Natural gas	65-164	2-10	0.3-1
Nuclear		2	
Hydropower		5	0.07

CO_2 carbon dioxide, SO_2 sulphur dioxide, NO_x nitrous oxide, VOC and volatile organic compounds.
Modified from International Energy Association, 2000

The restructuring of water systems alters the river flow rates, exposes riverbeds to erosion, and contributes to the loss of land resources and the destruction of wildlife and fish habitat (Statistics Canada 1994). There are therefore concerns with the preservation of productivity, biodiversity, and loss of rare or protected aquatic species, and habitat fragmentation or loss of ecologically valuable habitats. Such impacts are minimised when the flow regimes downstream of the reservoir are optimised to take into account the power generation requirements and the needs of aquatic or riverine habitats that require seasonal flooding (International Energy Association 2000). In some cases, fishways and fish ladders have permitted the passage of migratory and anadromous species at hydropower dam site (International Energy Association 2000) but in general the effectiveness of such measures has been less than satisfactory in terms of ensuring fish passage and maintaining essential migration for spawning and recruitment.

The flooding of forest and peat lands also contributes to the release of carbon dioxide and methane (International Energy Association 2000). When uncleared or partially cleared land is flooded, a large amount of biomass and soil is submerged which decays through bacterial degradation. The decaying organic matter contribute to the production and emission of carbon dioxide and methane to the atmosphere (so-called *greenhouse gases*) and can lead to oxygen depletion and to anoxic conditions in the reservoir. The anoxic conditions promote the formation of toxic substances such as hydrogen sulphide (H_2S), the mobilisation of heavy metals in the anoxic layer of the reservoir, increased water acidity levels, and subsequently problems in the downstream area (methane emissions, toxicity, off flavour and odour affecting use as drinking water, etc.).

Flooding also releases mercury (Hg) present in the soil and the flooded vegetation, which can then work its way into the aquatic food web and into the flesh of sport fish (Lucotte et al. 1999). Part of this released mercury is of natural origin and part is from Hg deposited onto the land prior to flooding. Hg can be released from thermal power plants and factories in the industrial mid-west of the North American continent and be transported by air currents and deposited before the land was flooded (Lucotte et al. 1999). The relative contribution of natural versus anthropogenic Hg is an ongoing debate and may have significant bearing upon the problem due to the relative volatility and capacity for uptake and incorporation into animal tissues.

Nonetheless, the mobilisation of Hg through the flooding of reservoirs is a serious concern because the increased Hg levels of reservoir fish can be assimilated to humans eating fish. Because fish are a major component of their diet, First Nation People living in watersheds affected by hydro development are most at risk. Exposure tends to be seasonal but chronic and is highest in the late summer and early fall. This seasonal exposure pattern may, in some cases, have prevented the mercury problem from becoming acute, as the body is able to rid itself of accumulated mercury during the low-exposure season (Environment Canada 1996a).

Problems of mercury exposure are currently most serious among the Cree of northern Quebec, who inhabit the area affected by the giant James Bay development (Environment Canada 1996a). Average mercury concentrations in whitefish taken from hydroelectric reservoirs in the area range from 0.06 to 0.21 parts per million (ppm), whereas those in Northern Pike are 3.0 ppm or higher (Bodaly and Johnston 1992). These Hg concentrations in fish far exceed the consumption guideline of Canada of 0.5 ppm. Above that level, fish consumption needs to be restricted and the fish cannot be sold in Canada (Sport Fish Contaminant Monitoring Program 1999).

8.3.2 Fossil Fuel

Fossil fuels (petroleum, natural gas, and coal) are major contributors to the three most important pollution problems of today: climate change induced by greenhouse gases (carbon dioxide, methane and nitrous oxide), acidic deposition created by nitrogen oxides and sulphur dioxide, and urban smog induced by nitrous oxide, particulates, hydrocarbon and volatile organic compounds, or VOCs (Environment Canada 1996a). Among the different electricity producing industries, the use of fossil fuels contributes the most to greenhouse gases, nitrogen oxide, sulphur dioxide and VOCs (Table 8.1; International Energy Association – IEA, 2000).

Energy use in Canada emits sulphur dioxide and nitrogen oxides, the major contributors of acid rain. The emissions are associated mainly with fossil fuel combustion by the industrial, commercial, and residential sectors as well as the electricity-generating industry and the upstream oil and natural gas industries. Eastern Canada is mostly sensitive to acidic pollutants since it has little ability to neutralise with its thin, coarsely textured soil and granitic bedrock, characteristic of the Canadian Shield. Acidic deposition can contribute to declining growth rates and increased death rates in trees. In aquatic ecosystems, it can cause metals to leach from surrounding soils into the water system. Increased metals concentrations impair the ability of water bodies to support life, resulting in a decline in species diversity. Also, recent studies suggest that when lakes acidify their water becomes more transparent due to the decrease in plankton and more harmful sunlight radiation (usually filtered by coloured water) can penetrate the water and affect other aquatic organisms. Between 1970 and 1990, sulphur dioxide emissions declined for most energy applications, except for emissions from thermal electric power generating plants, which increased by 25% (Environment Canada 1999b).

Land disturbance occurs through the extraction of oil and gas generally recovered by means of wells drilled into subsurface reservoirs (National Energy Board 1999a). The extraction generally involves injecting steam through wells into the subsurface oil deposits. This lowers the viscosity of the bitumen so that it can be pumped more easily to the surface. Oil sands projects disturb more land per unit of oil produced than other methods since the bitumen has to be mined by surface mining techniques, leading to large quantities of overburden and oil sands tailings for disposal (Environment Canada 1996a). The disposal of large quantities of tailings, if not properly managed, can increase the acidity and introduce metals in nearby water systems (Yanful and Simms 1997).

Similarly, coal mining generates acid tailings, especially in eastern Canada where coal deposits contain significant amounts of sulphides, the main cause of acidification (Environment Canada 1999b). Once mined, coal consumes large quantities of water to be processed. Raw coal needs to be cleaned of non-combustible inorganic material, which is removed or washed out prior to use (Environment Canada 1996a). The inorganic material and the large quantities of cleaning water have to be collected, controlled, and treated (Environment Canada 1996a). For every 10 tonnes of coal produced, 2 tonnes of coal waste are generated. In 1991, close to 20 million tonnes of coal waste were generated in Canada (Statistics Canada 1994).

Oil extraction from deposits under water bodies can lead to oil spills. These accidents can be very damaging to the environment by coating shorelines, affecting wildlife, especially seabirds, damaging fish spawning areas and killing micro-organisms that form the base of the food chain (Newman 1998). Based on current tanker traffic levels, it has been estimated that Canada can expect over 100 small oil spills (<1 tonnes), 10 moderate spills (1-100 tonnes), and at least one major spill (100-10,000 tonnes) every year (Statistics Canada 1994). Close to 10,000 m^3 of oil per year are released in oil-producing areas of Canada through spillage or well blowouts (Statistics Canada 1994). Although legislation exists to cover clean up costs and compensate fishermen and others affected economically by such spills, inevitably the costs, both environmental and economic, exceed the recompense available.

After its extraction, petroleum's refining process generates wastewaters containing a wide range of organic and metallic pollutants (Environment Canada 1996a). These contaminants may originate in the crude oil itself, be produced during refining, originate from the chemicals employed for the refinery process, or be produced by corrosion of the processing equipment (Environment Canada 1996a). Contaminants in oil refinery effluents are controlled by both federal and provincial legislation (Environment Canada 1996a). While most of the contaminants have been significantly reduced since 1980, concerns still persist regarding discharges of carcinogenic substances like polycyclic aromatic hydrocarbons (PAHs) and other potentially toxic organic and metallic contaminants (Environment Canada 1996a).

8.3.3 Nuclear

The advantage of nuclear power is that it does not produce atmospheric emissions associated with climate change like fossil fuel-fired facilities (Table 8.1; IAE 2000). However, human and environmental concerns exist with the short-term safety of operating nuclear power plants (especially since the 1986 Chernobyl accident) and with the long-term safety of stored radioactive wastes. The federal government established the Canadian Nuclear Safety Commission (CNSC) to regulate the development, application, and use of nuclear energy in Canada. The general approach to safety, including radioactive waste management, is that the producer/owner is responsible for complying with regulatory criteria (Natural Resources Canada 2000c). The CNSC's policies span issues related to management of radioactive wastes produced during the fuel cycle associated with three specific types of nuclear wastes: low-level radioactive waste generated by uranium mining and milling operations; low-level radioactive waste associated with other phases of the nuclear cycle (e.g., contaminated garbage from operations and maintenance activities, filters, process residues, and irradiated equipment); and nuclear fuel wastes, consisting mainly of spent nuclear fuel rods (Natural Resources Canada 2000c).

The on-site storage of nuclear fuel waste may be adequate for short to medium term but not for the long run because some of these wastes will remain radioactive for thousands of years. A method to dispose of these and contain their radioactivity permanently will have to be found. Atomic Energy of Canada Limited developed a concept for the permanent disposal of nuclear waste and recommended disposal in stable geologic formations, called plutons, which are found in the Canadian Shield (Atomic Energy of Canada Limited 2000). After eight years of study, the Canadian Environmental Assessment Agency Panel concluded that the concept was technically safe, but it lacked the required level of public acceptability to be adopted as Canada's approach for managing nuclear fuel wastes (Atomic Energy of Canada Limited 2000).

The outlook for the construction of new nuclear power plants in Canada is uncertain. However, most of the existing nuclear power reactors are expected to remain in operation. As well, the production, mining, and processing of uranium in Canada are likely to continue because of the relatively high-grade uranium found in Saskatchewan (Figure 8.3; INCS 1999). This means that inventories of radioactive wastes will continue to increase with time. The inventory of low-level radio-active waste could rise by 33% between 1991 and 2025 (Statistics Canada 1994).

8.4 Energy Outlook

The outlook shows that Canada will continue to rely on conventional sources of energy supply to meet its growing demand, such as hydro, coal, natural gas and nuclear, with natural gas being preferred for the new power plants after 2010 (Natural Resources Canada 1997). Despite a move towards relatively less polluting electricity with an increased use of natural gas, greenhouse gas emissions are expected to increase past the recommendations of the Kyoto Protocol (Bérubé 2000).

9. Fisheries and Aquaculture[7]

This chapter draws extensively on the State of Canada's Environment Report (Environment Canada 1996a) and data assembled by the Department of Fisheries and Oceans, and Statistics Canada (http://www.dfo-mpo.gc.ca/communic/Statistics). Results are presented for freshwater commercial and recreational fisheries and for that component of the emerging Canadian aquaculture industry which is, in whole or in part, related to freshwater.

9.1 Fisheries Statistics

Information on volume and value of fish landings in Canada are collected by both federal and provincial agencies. Data on freshwater fisheries is provided by a number of sources including regional offices of DFO, provincial governments, the Freshwater Fish Marketing Corporation and fish processing companies. Information contained in this Chapter covers volume and value of landings in sea fisheries by species-groups, by main species, by province and, for Atlantic Canada by DFO regions. These tables cover the period from 1989 to 1998. Due to the nature of the fishing industry, it should be noted that data provided may be modified later should new information be provided to DFO. Any such changes will generally be minor

9.2 Freshwater Commercial Fisheries

Canada has about 9% of the world's renewable supply of fresh water. The species of greatest economic importance for commercial fishing include trout, walleye, yellow perch, rainbow smelt, whitefish, Northern pike, and various bass species. Catches of these species have remained relatively unchanged since 1990. The annual commercial catch of freshwater fish in Canada averaged 50,800 tonnes from 1955 to 1982 and totalled 44,718 tonnes in 1990, representing a landed value of $66.4 million. In 1993, the total catch was lower (38,000 t), but the landed value was higher ($76.2 million) (see Table 9.1).

---

[7] This Chapter has been provided by Chris Morry, Programme officer at IUCN-Canada.

Table 9.1. Profile of Canadian commercial fisheries, 1989-1993

Species	Regions	1989 Q [t]	1989 V [$.000s]	1990 Q [t]	1990 V [$.000s]	1991 Q [t]	1991 V [$.000s]
Groundfish	Atl. coast	384,504	359,161	646,161	386,456	623,221	395,896
	Pac. coast	132,084	72,946	139,538	85,550	161,538	100,925
	Canada	816,508	432,107	785,699	472,006	784,759	496,781
Pelagic and other finfish	Atl. coast	359,939	85,414	423,407	84,495	304,540	69,493
	Pac. coast	130,074	326,850	142,263	340,074	125,886	250,265
	Canada	489,467	412,264	565,670	424,569	430,426	319,758
Shellfish	Atl. coast	227,816	503,531	227,116	460,412	225,258	533,004
	Pac. coast	21,237	45,157	21,521	46,051	24,538	46,835
	Canada	249,053	548,688	248,637	506,427	249,796	579,839
Total see fisheries	Atl. coast	217,713	959,775	1,296,684	936,020	1,153,019	1,007,781
	Pac. coast	283,395	453,664	303,322	479,943	311,962	406,888
	Canada	1,555,108	1,413,439	1,600,006	1,415,963	1,464,981	1,414,669
Inland Fisheries	Canada	51,199	82,690	44,178	66,413	77,184	73,402
Grand Total	Canada	1,606,307	1,496,129	1,644,724	1,482,376	1,542,165	1,488,071

Table 9.1. (cont.)

Species	Regions	1992 Q [t]	1992 V [$.000s]	1993 Q [t]	1993 V [$.000s]
Groundfish	Atl. coast	464,157	315,553	286,634	187,221
	Pac. coast	162,280	98,275	126,568	88,988
	Canada	626,437	413,828	413,202	276,209
Pelagic and other finfish	Atl. coast	286,637	69,844	281,031	74,157
	Pac. coast	102,370	246,587	126,627	259,078
	Canada	389,007	316,431	407,658	333,235
Shellfish	Atl. coast	238,214	587,136	260,030	634,595
	Pac. coast	31,201	62,350	26,789	73,485
	Canada	269,415	649,513	286,819	708,080
Total see fisheries	Atl. coast	989,008	984,139	827,695	902,244
	Pac. coast	295,851	416,127	279,984	422,611
	Canada	1,284,859	1,400,266	1,107,679	1,324,855
Inland Fisheries	Canada	38,697	70,354	38,000	76,200
Grand Total	Canada	1,323,556	1,470,620	1,145,670	1,401,055

Q Quantity, *V* live-weight-value, *Atl. Coast.* Atlantic coast, *Pac. Coast.* Pacific coast.
Note: 1993 data were preliminary estimates
Taken from Environment Canada, 1996a; original source: Department of Fisheries and Ocean (1994a, 1995c)

Ontario, Manitoba, Saskatchewan, Alberta, and the Northwest Territories contribute about 95% of the Canadian commercial freshwater fishery catch. Table 9.2, shows the landings and values for these areas from 1987 to 1992. In Ontario, the catch, which was mainly from the Great Lakes, declined somewhat every year throughout this period. Pollution, overexploitation, and the introduction of exotic species are among the factors influencing the declines of some Great

Lakes species (Government of Canada 1991). To counter the declines, efforts are continuing to reintroduce species such as Atlantic Salmon and Lake Trout. The other provinces and the territories show fluctuations in landings and values over the six-year period (Table 9.2) but a downward trend is evident in both weight and value of catch for all fisheries continuing into 1998, as shown in Table 9.3.

Table 9.2. Landings and value trends of major inland commercial fisheries, 1987-1992

Province/ Territory	1987		1988		1989	
	Landings [t]	Value [$.000s]	Landings [t]	Value [$.000s]	Landings [t]	Value [$.000s]
Ontario	27,759	48,340	27,591	54,710	25,610	48,123
Manitoba	12,151	25,313	14,094	25,196	14,699	21,538
Sakatchewan	3,860	5,153	3,680	4,672	3,904	4,165
Alberta	1,866	2,263	2,185	2,842	1,594	1,912
Nortwest Territories	1,572	2,268	1,747	2,763	1,954	2,730
Total	47,208	88,697	49,297	90,183	47,761	78,468

Table 9.2. (cont.)

Province/ Territory	1990		1991		1992	
	Landings [t]	Value [$.000s]	Landings [t]	Value [$.000s]	Landings [t]	Value [$.000s]
Ontario	24,853	42,370	23,140	38,140	19,639	41,523
Manitoba	12,273	15,650	12,693	19,973	13,624	24,960
Sakatchewan	2,964	2,704	3,890	3,527	3,797	4,223
Alberta	1,525	1,395	1,773	1,537	1,875	1,951
Nortwest Territories	1,769	2,052	1,860	1,860	1,538	1,537
Total	43,384	64,171	43,350	65,031	40,473	74,194

Note: Landings are in live equivalent tonnes.
Taken from Environment Canada, 1996a original source: Department of Fisheries and Ocean (1994a, 1995c).

9.3 Freshwater Recreational Fisheries

The 1995 Survey of Recreational Fishing in Canada collected information about recreational fishing in Canada in 1995 which indicates the economic and social importance of recreational fisheries to the country. This survey was conducted by the Statistical Services unit of the federal Department of Fisheries and Oceans in Ottawa, in cooperation with provincial and territorial governments.

The component provincial and territorial surveys formed part of a nationally co-ordinated study which provides data on topics of interest to the public and government fisheries managers. The results of these surveys provide the most authoritative and up-to-date information on activity and harvest in Canadian recreational fisheries. As well, they provide the only published detailed source of information on the economic dimensions of recreational fisheries across all regions of the country. Once aggregated, the results of the surveys are used to provide managers and researchers with nation-wide data. These data are used to assess changes with respect to the level of fishing effort as well as to determine the economic impacts of angler expenditures at both the regional and national levels.

The 1995 survey was the fifth in the series of surveys that have been undertaken once every five years since 1975. Highlights from previous surveys were also published in bulletins similar to this one.

Table 9.3. Summary of Canadian commercial catches and values, 1996-1998

	1996		1997		1998	
	Q	V	Q	V	Q	V
Atlantic - Total	686,441	1,148,884	735,324	1,214,572	778,453	1,294,431
Pacific - Total	240,715	428,069	250,771	406,501	217,080	290,875
Seafisheries - Total	927,156	1,576,953	986,095	1,621,073	995,533	1,585,306
Freshwater fish - Total	38,295	69,249	38,798	70,505	37,638	48,194
New Brunswick	1,072	657	1,432	982		
Quebec	1,429	4,178	1,515	4,872		
Notario	17,003	41,249	19,463	43,151	20,074	48,194
Manitoba	11,593	15,966	10,125	14,955	11,044	
Saskatchewan	3,615	3,546	3,157	3,375	3,388	
Alberta	1,716	1,642	1,695	1,702	1,811	
NWT	1,867	2,011	1,411	1,468	1,321	
Canada - Total	965,451	1,646,202	1,024,893	1,691,578	1,033,171	1,633,500

Note: Q Quantity in tonnes live weight, V Value in thousand dollars
(Last Updated on 01/19/2000 By Statistical Services, DFO)

9.3.1 1995 Recreational Fisheries Survey Highlights

Recreational fishing is an important economic activity in the natural resources sector. Over 4.2 million adult anglers fished in Canada in 1995. Of this total, almost 749,000 were anglers visiting this country, many for the sole purpose of recreational fishing. The estimated 4.2 million anglers covered in the survey fished for 55.5 million days and caught over 254 million fish of all species (see Table 9.4). Of the fish caught, just over 113 million (or 44.6%), were kept. Although fishing provided the focus of activity for anglers, the most important factors associated with their enjoyment of the sport are the lack of pollutants in the fish they caught and clean water.

In total, the angling population spent $7.4 billion in Canada in 1995 of which $4.9 billion was directly associated with their sport. Anglers spent over $2.5 billion on package fishing trips, accommodation, food, transportation, fishing supplies and other services directly related to their angling activities. These expenditures were further augmented by monies they spent on durable goods purchased to improve their access to the resource and their enjoyment of it. Investments in 1995 totalled close to $4.9 billion for such durables as fishing equipment, boats, motors, camping equipment and special vehicles. Anglers estimated that almost $2.4 billion of these investment expenditures were wholly attributable to their participation in angling.

Table 9.4. Fish caught by species, 1995

Species	Resident	Non-resident Canadian	Non-Canadian	Total
Trout	52,943,547	1,036,953	2,074,767	56,055,267
Walleye	25,315,886	1,463,410	19,493,513	46,272,809
Perch	37,106,747	736,554	4,547,213	42,390,514
Northern Pike	15,808,454	964,295	11,524,695	28,297,444
Bass	16,514,013	429,333	5,353,748	22,297,094
Smelt	13,000,874	3,518	57,158	13,061,550
Shellfish	5,395,760	249,787	622,170	6,267,717
Salmon	4,005,132	281,094	569,841	4,856,067
Whitefish	1,788,173	52,978	125,384	1,966,535
Kokanee	741,158	66,258	37,428	844,844
Charr	397,707	48,668	38,854	485,229
Mackerel	443,437	1,500	612	445,549
Grayling	410,258	75,369	126,656	612,283
Flounder	107,856	0	0	107,856
Cod	104,734	1,084	325	106,143
Other	21,799,349	326,925	8,142,842	30,269,116
Total	195,883,086	5,737,725	52,715,206	254,336,017

Note: Last Updated on 03/27/98 by Statistical Services, DFO

Anglers further indicated that they were willing to spend on average $17.37 per day over and above their current expenditure levels for costs associated with fishing. The total additional amount anglers indicated they were willing to pay in 1995 was $964 million. Resident anglers accounted for 83% of this total. There were notable differences when comparing resident and non-resident anglers. Resident anglers indicated they were willing to pay approximately $16.35 per day compared to $21.52 per day for Canadian non-residents and $25.77 per day for visitors to Canada.

Non-resident anglers made over three million trips for fishing and other reasons. Just over half these trips were made by visitors to Canada with the other half representing trips by Canadians to other provinces/territories. Overall, non-resident anglers fished on 58% of the trips they made, with visitors to Canada indicating fishing activity on almost 80% of their trips.

The framework of socio-economic and biological information provided by this survey will help managers of this resource better assess and manage fish resources. For example, the impressive catch and release results achieved over the past five years indicate that anglers understand better the need to conserve this fragile and important natural resource.

9.3.2 Great Lakes Recreational Fisheries

As mentioned above, the Great Lakes recreational fishery is by far the most important geographic component of overall recreational fisheries in Canada. Since 1980, the Great Lakes Fishery Commission has requested that information specific to the Great Lakes system be developed in order that they could determine the economic importance of this fishery. This request was made of both the Department of Fisheries and Oceans and the Fish and Wildlife Service of the United States Department of the Interior since national surveys were being conducted by both agencies. The surveys are no longer carried out in the same year. Canada has maintained the original five-year interval whereas in the U.S., the survey is conducted to coincide with census years. Data has been generated to reflect activity by both residents of the province and visitors who fished in Ontario and, more specifically, on the Great Lakes.

Highlights

- an estimated 552,710 anglers fished on the Great Lakes in 1995; 485,096 residents of Ontario and 67,614 visitors to Ontario, primarily from the United States,
- Great Lake anglers were primarily male (85%), between 35 and 44 years of age (26.8%), with a college education (36.9%), employed full-time (70.1%) and with personal income between $20,000-49 999 (51.8%) and household incomes between $40,000-59,999 (30.9 %),
- Great Lake anglers fished for over 7 million days, with residents averaging 12.0 days and non-residents 7.1 days,
- most days were spent fishing on Lake Huron (2.6 million) followed by Lake Ontario (2.1 million)
- Great Lakes anglers caught a total of 47.3 million fish of all species of which 32.6 million were in the Great Lakes System,
- of all fish retained on the Great Lakes, perch ranked first in terms of numbers at 13.5 million, followed in order by smelt (7.7 million), walleye (5.3 million) and smallmouth bass (5.2 million)
- on average anglers retained 22 fish each from the Great Lakes of which 62% were smelt and perch. This amounts to 41% of all fish caught,
- Great Lakes anglers kept over 50% of only three species: smelt (77%), splake (72%) and Coho salmon (65%),
- Great Lakes anglers spent an estimated $284 million on durable goods used in whole or in part for recreational fishing, allocated on the basis of their Great Lakes fishing activity,

- of this amount, they attributed $142 million (50%) wholly to their recreational fishing activities
- Great Lakes anglers spent $220 million on goods and services directly related to their fishing activities, averaging almost $397 per angler,
- these anglers further indicated they were willing to pay an additional $131 million, had costs been higher for service-related expenses, before they would have stopped fishing,
- these additional costs averaged $238 per year, or $19 per day.

9.4 AQUACULTURE

9.4.1 Socio-economic Aspect

Aquaculture, the cultivation of aquatic life, is an established practice in many parts of the world. In Canada, aquaculture was first used to enhance natural stocks of freshwater and anadromous[8] finfish. However, it is now a large-scale commercial industry across the country providing direct and indirect economic benefits to many local and regional economies. All ten provinces and the Yukon Territory currently have a stake in commercial aquaculture and interest is increasing in the Northwest Territories. Aquaculture production (both mariculture and freshwater aquaculture) in 1995 accounted for 7% of total fish production in Canada.

Although some attempts at freshwater aquaculture were attempted as long ago as the late 1800s in Newfoundland, commercial aquaculture in Canada began in earnest in the 1950s, when brook trout and rainbow trout were the principle freshwater species of interest, and remained in a developmental stage until the early 1980s. During that period, production of trout was mostly centred in Ontario and British Columbia. Over the past 20 years, commercial production has expanded to include several salmon species as well. The information presented here covers the years 1984 to 1999 and provides information on volume and value of aquaculture production by species and by province/territory. The analysis focuses on freshwater and anadromous finfish production, since these are the species most greatly affected by changes in the freshwater environment. During this period, the culture of these finfish species gravitated more and more into the marine coastal environment, due to the improved growth conditions afforded there. For the purpose of this study, no attempt is made to separate strictly freshwater aquaculture from that conducted in the coastal zone or involving production on land near the coast utilis-

---

[8] *Anadromous* fish are those that spawn in freshwater but spend at least part of their life-cycle in the marine environment. All salmon, charr and trout species are naturally anadromous, though many have adapted to landlocked conditions over the millennia. *Finfish* is a generic term which distinguishes all free-swimming (finned) fish from *shellfish*, a term which generally includes both crustaceans (like lobsters and shrimp), bivalves (like mussels and oysters), and univalves (like whelks and other marine snails).

ing seawater, because these segments of the industry are equally susceptible to water quality problems associated with adjacent freshwater rivers and streams that affect coastal waters.

Similarly, some discussion of the culture of mussels, oysters and other marine species (mariculture) is included because of the indirect effects of freshwater quality on this segment of the industry, especially the impact of pollution and silt from freshwater sources in the coastal zone where most shellfish mariculture is conducted. The first wave of aquaculture development took place between 1984 and 1991. During this period, the production value for the four main marine and freshwater species increased more than 36 fold, from $7 million to more than $250 million, recording an extraordinary average growth of 67% per year (Department of Fisheries and Oceans 1995b). The industry expanded to every province and the Yukon, whereas the species base broadened to include salmon, mussels, clams, scallops, Arctic charr, and marine plants. In 1991, the commercial aquaculture industry provided more than 5,200 jobs, about 2,800 in the production sector and 2,400 in the supply and services sector. About 80% of aquaculture production is exported (Department of Fisheries and Oceans 1993b).

In 1993, Canadian aquaculture producers generated more than $289 million. Salmon and trout are the principal farm-raised finfish species. Along with oysters and mussels, they had a production value of more than $284 million in 1993, with the finfish constituting more than 95% of that total (Table 9.5). In 1993, the output from these species was 49,493 t, or 4.3% of the total Canadian fisheries production. However, aquaculture accounted for more than 17% of the total fisheries landed value (Department of Fisheries and Oceans 1995b). Additionally, the supply and services sector of the aquaculture industry generated more than $266 million annually, including more than $53 million in exports. Table 9.5 also indicates the projected production of finfish in aquaculture up to the year 2000. Table 9.6 which provides the actual production figures for 1999, indicates that these estimates were actually quite accurate, though the anticipated level of growth was in fact achieved a year earlier than projected.

Table 9.5. Aquaculture tonnage and value produced, 1991, 1993 and outlook for 2000

Main species	1991		1993		Outlook to year 2000	
	Quantity [t]	Value [$ million]	Quantity [t]	Value [$ million]	Quantity [t][a]	Value [$ million][a]
Salmon	29,099	220.0	32,523	245	89,400	502.6
Trout	4,808	24	5,267	26.8		
Oyster	6,830	6.2	6,528	6.8	18,500	20.5
Mussel	4,046	0.5	5,175	5.7	35,000	32
Total	44,783	250.7	49,493	284.3	142,900	555.1

[a] Numbers for salmon and trout are presented together.
Taken from Environment Canada, 1996a.

Table 9.6. Aquaculture tonnage and value produced, 1999

1999 Canadian Aquaculture Production Statistics (tonnes)

	Nfld	PEI	NS	NB	Que	Ont	Man	Sask	Alta	BC	CANADA
Finfish											
Salmon	399	x	791	22,000	-	-	-	-	-	49,100	72,290[2]
Trout	10	x	-	550	1,084	4,000	4	875	x	100	6,623[2]
Steel-head	2,078	-	3,924	-	-	-	-	-	-	-	6,002[2]
Other[1]											488[1]
Total Finfish[3]	2,487	82	4,715	22,550	1,084	4,000	4	875	x	49,200	85,485
Clams	-	-	-	-	-	-	-	-	-	900	900
Oysters	-	2,423	776	286	1	-	-	-	-	5,800	9,286
Mussels	1,700	13,890	945	665	139	-	-	-	-	-	17,339[2]
Scallops	-	-	25	-	-	-	-	-	-	30	55[2]
Other	-	-	16	-	2	-	-	-	-	-	18
Total Shellfish	1,700	16,313	1,762	951	142	-	-	-	-	6,730	27,598
Total	4,187	16,395	6,477	23,501	1,226	4,000	4	875	x	55,930	113,083

1999 Canadian Aquaculture Production Statistics (' $000)

	Nfld	PEI	NS	NB	Que	Ont	Man	Sask	Alta	BC	CANADA
Finfish											
Salmon	2,462	x	7,022	150,000	-	-	-	-	-	292,200	451,684[2]
Trout	80	x	-	6,100	4,303	15,880	16	3,859	x	500	30,738[2]
Steel-head	11,402	-	17,352	-	-	-	-	-	-	-	28,754[2]
Other[1]											4,711[1]
Total Finfish[3]	13,944	786	24,374	156,100	4,303	15,880	16	3,859	x	292,700	516,673
Shellfish											
Clams	-	-	-	-	-	-	-	-	-	3,800	3,800
Oysters	-	5,075	1,815	788	3	-	-	-	-	6,000	13,681
Mussels	3,800	16,845	1,485	798	167	-	-	-	-	-	23,095[2]
Scallops	-	-	166	-	-	-	-	-	-	200	366[2]
Other	-	-	43	-	4	-	-	-	-	-	47
Total Shellfish	3,800	21,920	3,509	1,586	74	-	-	-	-	10,000	40,989
Total	17,744	22,706	27,883	157,686	4,477	15,880	16	3,859	x	302,700	557,662

[1] Includes Char, Other Finfish and Total Alberta Finfish; [2] Excludes Confidential Data.
[3] Excludes *Other* for provinces

The production and value of Aquaculture include the amount and value produced on sites and exclude hatcheries or value added products. Statistics Canada - Agriculture Division.

9.4.2 Impact to the Aquatic Ecosystem

Although there have been significant socio-economic benefits from the expansion of the aquaculture industry in Canada, especially in coastal regions, there is concern regarding the potential impact of this development on the aquatic ecosystem, both freshwater and coastal, wherever this industry is prosecuted. These concerns centre on visual impacts, fouling of the benthic environment under pens and the water around them, pollution of surrounding waters with wastes and chemicals (e.g. antibiotics and hormones) used to treat the aquaculture organisms, interference with recreational or commercial fishing and boating, introduction or transportation of fish disease and its impacts on native wild stocks, reduction in genetic diversity due to wild and cultured stocks interbreeding, and competition between native wild stock and escaped cultured fish for habitat and food resources.

Many of these concerns are being addressed through legislation and industry self-policing. For example, in British Columbia, the impacts of waste feed and faecal material on the benthos, which are primarily local in nature, can be largely avoided by proper siting of the culture operation (Department of Fisheries and Oceans 1993c). The Department of Fisheries and Oceans (1993c) contends that there is no evidence that wild fish in British Columbia are at serious risk from disease in farmed fish, and reports that the potential for reducing genetic diversity of wild stock through breeding with cultured fish is low. Similar concerns have been noted for the Atlantic coast and are addressed through application of regulations, policies, and guidelines.

On the Atlantic coast, the impacts of deposition are a greater concern because of the shallower locations in which marine cages are sited. In New Brunswick in 1992, a study of the seafloor around 48 cage sites in about 12 m of water indicated that 8 sites had high impact on the bottom, 29 had moderate impact, and the remaining 11 had low impact (Thonney and Garnier 1992). The industry has responded to many of these concerns by developing diets designed to enhance survival and growth and reduce disease. These measures are linked with feeding regimes, such as demand feeding, that reduce waste and deposition (Stewart 1994). The Province of New Brunswick, through its Water Quality Regulation permit process, stipulates general and specific environmental requirements that must be met by each inland aquaculture site. Other Atlantic provinces have similar regulatory controls. Generally speaking, stakeholders in aquaculture recognise the importance of harmony between the industry and the environment if a sustainable fishery is to be realised.

10. Water and Nature

10.1 Wetlands in Abundance?

Canada has more than 1.2 million square kilometres of wetlands or 14% of the country's total land area. The distribution of wetlands varies greatly across Canada. As shown in Table 10.1 and Figure 10.1 most of Canada's wetlands are located in Manitoba, Ontario, and the Northwest Territories. In Manitoba and Ontario, wetlands represent respectively 41% and 33% of the land area in the province. With 4,000 hectares Prince Edward Island is the *poorest* province in terms of wetlands.

Table 10.1. Total wetland area in Canada

Province and Territory	Total Wetland Area		
	In hectares	% of Land Area in Province or Territory	% of Total Canadian Wetlands
British Columbia	3,120,000	3	2
Alberta	13,704,000	21	11
Saskatchewan	9,687,000	17	8
Manitoba	22,470,000	41	18
Ontario	29,241,000	33	23
Quebec	12,151,000	9	10
New Brunswick	544,000	8	<1
Nova Scotia	177,000	3	<1
Prince Edward Island	4,000	<1	<
Newfoundland	6,792,000	18	5
Northwest Territories (incl. Nunavut)	27,794,000	9	1
Yukon Territory	1,510,000	3	22
CANADA	127,199,000	14	100%

Source: Environment Canada, 1993c

10.2 Importance of Wetlands for Migratory Birds

Wetlands in Canada are essential to the survival of migratory bird populations in the Western Hemisphere and polar regions. Figure 10.2 shows the routes and sanctuaries used by migratory birds.

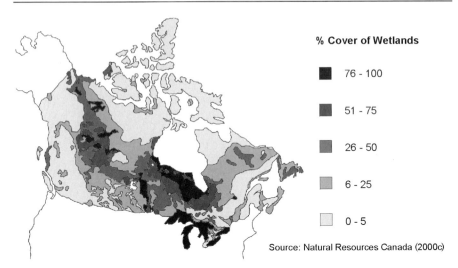

% Cover of Wetlands

■ 76 - 100

■ 51 - 75

■ 26 - 50

□ 6 - 25

□ 0 - 5

Source: Natural Resources Canada (2000c)

Fig. 10.1. Wetlands in Canada

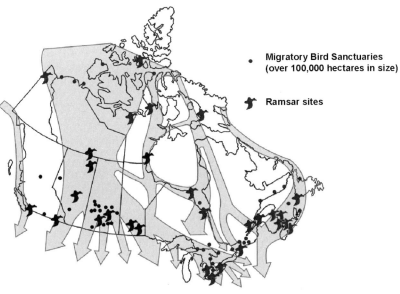

• **Migratory Bird Sanctuaries (over 100,000 hectares in size)**

🦆 **Ramsar sites**

Source: Natural Resources Canada (2000c)

Fig. 10.2. Routes and sanctuaries used by migratory birds in Canada

To preserve the major migratory flyways in Canada a network of 49 National Wildlife Areas and 98 Migratory Bird Sanctuaries managed by the Canadian Wildlife Service has been established across the country. Migratory Bird Sanctuaries in Canada protect about 11.3 million hectares of both dry land and wetland areas. The Migratory Bird Sanctuary Regulations prohibit all disturbance, areas. The Migratory Bird Sanctuary Regulations prohibit all disturbance, hunting and collecting of migratory birds and their eggs within the sanctuary. Visitors are also prohibited from carrying firearms and or allowing their pets to run at large. The first Migratory Bird Sanctuaries were established in 1917 to protect migratory birds from physical disturbances and hunting. By the middle of the century, however, the loss and degradation of habitat had become the greatest threat to migratory birds and other wildlife. The creation of National Wildlife Areas with the primary focus on habitat protection was a major response to this threat. These sites protect about 300,000 hectares for wildlife and contain a wide diversity of habitat of both national and international importance.

Ramsar sites (sites declared for protection under the *Convention on Wetlands*, adopted in Ramsar, Iran in 1971) are another form of migratory bird sanctuary. There are 32 sites in Canada designated under the Ramsar Convention as wetlands of international importance in the promotion of intercontinental flyway habitat conservation for migratory birds. Seventeen of the Migratory Bird Sanctuaries or National Wildlife Areas are recognised as Ramsar sites or have been included within Ramsar sites.

In addition, there are two sites designated as Western Hemisphere Shorebird Reserve Network sites. This network links together sites in different countries of North, Central and South America that are essential to large number of shorebirds as they complete their annual migration. Some of these migrations routes are 10,000 kilometres or more in length.

Single flocks of 250,000 Semipalmated Sandpipers (10% of the world's population) move as one body and occupy the beaches and mudflats of Shepody and Chignecto National Wildlife Areas at migration time. The Great Plains of the Koukdjuak on Baffin Island (part of Dewey Soper MBS) shelters the largest goose colony in the world. The leafy woodlands, marshes and beaches of Long Point (part is a NWA) contain over 90 species of boreal and Carolinean plants that are rare in Ontario. It also includes several birds, reptiles and amphibian species. More than 10% of North America's population of Redheads (diving ducks) stop at Long Point during the Spring and Fall migration. Clouds of Monarch butterflies also alight at Long Point during their migration southward. Over 100,000 Snow Geese use the Cap-Tourmente National Wildlife Area for resting and feeding during the Spring and Fall migration across North America. The Alaksen National Wildlife Areas in the Fraser Delta of British Columbia is a central area for waterfowl and shorebird population of the Pacific Coast of North America.

10.3 Threats to Wetlands in Canada

Despite the legal protection afforded wetlands in Canada today, much of the original base of wetlands has already been lost and many threats still remain. For instance, conflicts between wetland conservation and wetland utilisation are still important, especially in southern Ontario where population, agriculture and development pressures are the most intense. Agricultural expansion is the major cause of 85% of Canada's wetland losses. Since European settlement, wetland conversion to agriculture is estimated at over 20 million hectares (65% of the coastal marshes of Atlantic Canada, 70% in southern Ontario, 71% in the Prairie Provinces, 80% of the Fraser River Delta in British Columbia). Despite the fact that less than 0.2% of Canada's wetlands lie within 40 kilometres of major urban centres, urbanisation is a major threat for this wetlands. Indeed, over 80% of the wetlands near major urban centres have been converted to agricultural use or urban expansion.

10.4 Freshwater Biodiversity Loss in Canada

The authors of the first estimate of extinction rates of North America's freshwater fauna came recently to the conclusion that "A silent mass extinction is occurring in our lakes and rivers." (Ricciardi and Rasmussen 1999). According to this study Freshwater species groups are the most endangered on the North American continent and are dying out as fast as rainforest species. Ricciardi and Rasmussen (1999) predict that about 4% of freshwater species will be lost each decade if more is not done to conserve them. Since 1900, at least 123 freshwater animal species have been recorded as extinct in North America. Common freshwater species are dying out five times faster than land species in North America, and three times faster than coastal marine mammals. The modern extinction rate in North America, is about one extinction every 2600 years – about 1000 times higher than the background rate. Ricciardi and Rasmussen (1999) predict that many species considered at risk will disappear within the next century. At risk species account for 49% of the 262 remaining mussel species, 33% of the 336 crayfish species, 26% of the 243 amphibian species, and 21% of the 1,021 fish species.

Causes of this high extinction rate are diverse. For example, non-native species pose a serious threat to indigenous freshwater fauna. European zebra mussels are out-competing native mussels in Canadian and North American lakes and rivers. Sea lampreys invade lakes and attach themselves to native fish, killing them. Even sport fish transplanted from one lake to another can take over an ecosystem, driving less aggressive native fish toward extinction. Dams that obstruct river flow are also threats causing habitat deterioration. For example, in the lower 48 US states, only 40 rivers longer than 200 kilometres remain free flowing. "Such massive habitat deterioration threatens some of the world's richest freshwater faunal assemblages," the study says. Excess sediment, toxic contaminants and organic pollutants from agriculture threaten also most waterways causing water quality contamination and habitat deterioration.

11. Public Health Issues

This Chapter gathers together the information related to human health problems that have already been presented in preceding Chapters. It gives the opportunity to present in the study a general portrait of Canadian Public Health problems related to water issues. The main health problems associated with contaminants found in Canadian drinking waters are presented as well as a brief examination of hazards associated with recreational use of water. Finally, the problems caused by mercury contamination in fish are presented as a special case because of the extreme importance of this issue to the Aboriginal people of Canada in particular. This Chapter is largely inspired by the Health Canada (1997) report *Health and Environment* that describes the current understanding of the relationship between human health and the Canadian environment.

11.1 Drinking Water Issues

Although there is need for improvement in some areas of Canada, the overall quality of drinking water is very high. For example, surveys have shown that when chemicals are detected in municipal tap water, their levels are generally many times lower than the maximum acceptable levels set by federal-provincial-territorial guidelines. With a few exceptions, such as chlorination by-products (like Trihalomethanes – THMs –), the most potentially serious contamination problems in Canada generally involve elevated levels of natural contaminants in tap water from inadequately treated surface water or groundwater sources, such as private wells and water sources used by First Nations communities (Health Canada 1997).

11.1.1 Water-borne diseases

In Canada, municipally treated tap water is generally safe to drink. This does not mean, however, that the water is absolutely safe. Indeed, even treated water may contain some micro-organisms, although at very low levels. But water-borne diseases caused by bacteria (such as *Campylobacter*, *Escherichia coli*[E. coli], *Salmonella*, *Shigella*, *Staphylococcus aureus* and *Pseudomonas aeruginosa*) , viruses, protozoa and certain classes of phytoplankton are the most common health hazards associated with drinking water (and recreational waters – see section below) in Canada.

Generally speaking, the true incidence of water-borne diseases is likely much higher than what is usually reported, because the majority of cases involve mild, flu-like symptoms that do not require medical treatment. Groups that are at high risk of exposure to microbial contaminants include members of First Nations communities and rural Canadians who depend on private well water, since private well water is generally untreated.

Lately (in spring 2000) in Walkerton, Ontario, the illness caused by contamination of drinking water with *Esherichia coli* bacteria caused the death of seven people and the hospitalisation of 2,000. Since this accident water authorities are very prudent. For example, Saskatchewan authorities very recently (December 2000) issued an order to about 1,500 people in 28 small rural communities to boil their drinking water, on the suspicion that it might be contaminated, though in the end this did not prove to be the case. No bacterial contamination was found in the water but many of these towns have no proper filtering or disinfection systems and the communities' water treatment systems failed to meet provincial standards.

In a 1991 study funded by Health and Welfare Canada, scientists at the University of Quebec evaluated the risk of gastrointestinal illness due to consumption of treated water that was prepared from a sewage-contaminated water source and that met provincial water quality standards after treatment. The study involved two groups of suburban Montreal residents. Over a 15-month period, one group drank municipally treated tap water, while the second group drank treated water that was further purified using reverse osmosis water filters installed inside their homes. Throughout the study, the incidence of gastrointestinal illness was significantly higher in every age group of those who drank plain tap water. The authors concluded that about 35% of the reported gastrointestinal illnesses among the tap water drinkers were water-related and preventable.

11.1.2 Persistent Pollutants

In recent decades, concern over the presence of toxic chemicals in Canadian drinking water has surpassed other water quality issues. Hundreds of different chemical compounds have been found in Canadian drinking water supplies, although generally at very low levels. Many of these substances are persistent, which means that they degrade very slowly and may remain in the environment for years or even decades. Since the 1970s, controls have been placed on the most toxic pollutants, including lead, mercury, polychlorinated biphenyls (PCBs), chlorinated dioxins and chlorinated furans. Registrations of the remaining uses of the persistent pesticide dichlorodiphenyl-trichloroethane (DDT) in Canada were discontinued in 1985 with the understanding that existing stocks would be sold, used or disposed of by the next registration renewal date, December 31, 1990. Although these substances and their degradation residues are still a health concern, their concentrations in many water bodies are declining (Health Canada 1997). Scientists have estimated that drinking water accounts for less than 1% of our total exposure to persistent pollutants, with most of our intake coming from food (Health Canada 1997). Exposure from consumption of wildlife and fish makes first nation communities especially vulnerable to these contaminants in the Northern Regions. Contaminants are found in highest concentrations in mother's milk, exposing new born babies and very young children to these substances at a period when they are most vulnerable. This situation is threatening public health and traditional cultures in these communities, as they have to choose between maintaining their traditional culture of subsistence fishing and hunting, or abandoning it for public health consideration.

11.1.3 Health Problems due to Groundwater Contamination

A 1993 survey conducted by Agriculture Canada suggests that farm families relying on private wells have a higher risk of exposure to groundwater contaminants than those who receive treated water from municipal wells. The study, which analysed the water quality from 1,300 rural wells, found that approximately 40% exceeded provincial drinking water objectives for at least one of the contaminants surveyed (Health and Welfare 1993). The results showed that 25% contained faecal coliform bacteria (these include *Escherichia*, *Klebsiella*, *Citrobacter* and *Enterobacter* bacteria, which are present in fresh faeces), 15% exceeded the Ontario maximum acceptable concentration for nitrate and 12% had detectable levels of pesticides (two wells had pesticide concentrations that exceeded Ontario interim maximum acceptable concentration values).

11.2 Recreational Water Hazards

In Canada, fresh and salt water are used for a variety of recreational purposes, such as swimming, water-skiing, windsurfing, boating and fishing. Although the overall quality of our lakes and rivers is high, people may be exposed to small amounts of water-borne pollutants during recreational activities in certain areas (see Chap. on Water Uses in Canada). Microbial contaminants such as bacteria and viruses from sewage pose the greatest potential health risk to recreational water users. Other common sources of contamination include industrial waste, agricultural runoff, urban runoff, storm water runoff, faeces, oil and gasoline spills from power boats and marinas and pollution from boaters. Swimmers can also be a source of contamination, especially if some people have infections or open wounds (Health Canada, 1997). Other hazards include exposure to phytoplankton and chemical contaminants. Swallowing water is one of the ways in which pollutants may enter the body during outdoor water activities, although people may also be exposed to contaminants through contact with their skin, eyes, ears or nose (Health Canada 1997).

11.3 Mercury Concentrations in Fish

Because fish are a major component of their diet, First Nation People living in watersheds affected by hydro development are most at risk of suffering from mercury contamination (see Chap. VIII on Water and Energy). Exposure tends to be seasonal but chronic and is highest in the late summer and early fall. This seasonal exposure pattern may have prevented the mercury problem from becoming acute, as the body is able to rid itself of accumulated mercury during the low-exposure season. Average mercury concentrations in whitefish taken from hydroelectric reservoirs in northern Quebec range from 0.06 to 0.21 ppm, whereas those in Northern Pike are 3.0 ppm or higher (Bodaly and Johnston 1992). Fish containing more than 0.5 ppm are considered unsafe for continued long-term consumption and cannot be sold in Canada.

12. Institutional and Legal Framework for Water Management[8]

12.1 Introduction

The underpinnings of institutional responsibility for water management in Canada are found in the Canadian Constitution. The general approach in the Constitution is to divide authority between the federal and provincial levels of government on the basis of individual specified heads of power, which in most cases are exclusive to one of the two levels (agriculture is a notable exception that is shared between the two levels of government). Because fresh water is not listed as one of these heads of power, provincial and federal legislative authority with respect to water must be found in a number of sections of the Constitution. Jurisdiction over water may also be conferred as the result of proprietary rights, as opposed to legislative authority. Because of the potential overlap of authority on the part of federal and provincial governments, Canadian management of freshwater resources depends in important respects on co-operation between these two levels of government.

As a practical matter, municipal governments also bear important responsibilities for water management. However, these governments exercise such powers solely as a matter of delegated authority and have no independently recognised status in the Canadian Constitution. Because many of Canada's water resources lie in basins that are shared with the United States, international agreements have also played an important role in the management of water resources, notably, but not exclusively, in the Great Lakes Basin. The transboundary management of water resources is peculiarly affected by the unique provisions of the Canadian Constitution dealing with treaties entered into on Canada's behalf by the British Empire.

12.2 Federal Responsibilities

The constitutional ability of the federal government to legislate with respect to water can be found in a number of heads of federal power. For example, the federal government is given specific legislative authority over navigation and shipping

---

[8] This Chapter has been provided by Owen Saunders, executive director of the Canadian Institute of Resources Law.

and over fisheries. The federal government can also find constitutional support for certain legislative initiatives with respect to water under other provisions in the Constitution, for example, the federal powers over trade and commerce, Indians and lands reserved for Indians, criminal law, agriculture (a power shared with the provinces), and undertakings (including canals) connecting or extending beyond the limits of a province. An even broader source of potential federal legislative authority, and one that has been relied upon by the Supreme Court of Canada in the context of water management, is the general power of the federal government to make laws for the *peace, order and good government* of Canada.

Finally, the federal government has certain authority with respect to implementation of treaties that has proven important in international water management, and which is discussed further on.

In addition to its legislative powers, the federal government also possesses some proprietary rights that bring with them the ability to manage water. This is particularly the case in the three northern territories where the federal Crown is the dominant owner of public lands (although in practice significant authority is being devolved to territorial governments and to structures set up as the result of land claims agreements with aboriginal peoples). Even in the provinces, however, there are some federal public lands and works which give rise to proprietary rights, in addition to the federal interest in aboriginal lands. In general, though, the federal proprietary interest in waters lying within the provinces pales next to that of provincial governments.

One area of potential federal jurisdiction over freshwater resources that does exist in the Constitution – but which in general has not been actively exploited by the federal government – is with respect to water resources that are shared by two or more provinces. In this regard, the Canadian federal government has been far less active than its counterpart in the United States, for example, despite what would seem to be a more explicit constitutional role for the former. While the federal government has exercised some role in the management of inter-provincial watercourses, its role has been primarily facilitative in nature rather than dirigiste. This approach is reflective of a more general tendency in Canadian federalism, under which federal governments have typically exercised significant restraint in intervening in natural resources management issues. Indeed, in those cases where the federal government has acted unilaterally in natural resources policy (especially energy policy), it has given rise to significant political opposition in the provinces.

Just as federal authority with respect to water is fragmented in the Constitution under a number of heads of power, so too there is no general piece of legislation that establishes a plenary federal role, with much of the federal legislation focussed on specific federal powers over fisheries and navigation (in addition to its responsibilities with respect to internationally shared waters, discussed further on). For example, the *Fisheries Act* confers broad powers on the federal government to protect both fish and fish habitat. Similarly, the *Navigable Waters Protection Act* (NWPA) allows the federal government to take broad measures to regulate activities and works that may affect navigation. Because of the general requirement for federal agencies to abide by the *Federal Environmental Assessment Act* (CEAA) and Regulations for matters that fall under their jurisdiction, the NWPA also confers a certain degree of environmental protection to the watersheds which fall under the legislation (i.e. all navigable waters). Neither statute, however, creates a comprehensive framework for managing water resources – and, indeed, the Supreme Court has effectively precluded this possibility in holding, for example, that the federal fisheries power cannot be used as a pretext for regulating water quality (and, presumably, water quantity) more generally.

Under the *Oceans Act*, administered by Fisheries and Oceans Canada, estuarine, coastal, and marine ecosystems are protected from the negative effects of land– and marine-based activities. This Act allows for guidelines for marine environmental quality to be developed, including those pertinent to water quality. It also emphasises integrated management, providing for the development of management plans to protect ecological resources and ecosystem integrity and productivity in co-operation with other responsible federal authorities, as well as stakeholders.

The federal government also has legislation addressing international transboundary water concerns, both in the context of implementing specific treaty obligations (discussed further on) and more generally in the context of regulating certain aspects of water flows across the boundary. In this latter respect, the *International Rivers Improvements Act* of 1955 (originally directed at possible dam construction by British Columbia on the Columbia River) requires a federal licence for any *improvements* (broadly defined under the Act as virtually any use that would affect the flow or use) of internationally-shared rivers (again, broadly defined as *water flowing from any place in Canada to any place outside Canada*). While the Act does give significant authority to the federal government for regulating some aspects of transboundary water relations, it is, however, limited in important respects – most significantly in that it does not apply to improvements in boundary waters (which would include, for example, the Great Lakes), nor to a number of important consumptive uses.

The closest the federal government has come to taking a plenary role with respect to water management is with the *Canada Water Act* of 1970. Although the Act is designed primarily to encourage and facilitate federal-provincial co-operation, it does include some provisions for unilateral federal action with respect to trans-jurisdictional waters. In practice, however, the Act has not been used to the extent that is anticipated in the legislation – particularly with respect to those provisions that set out a federal role in concluding agreements with the provinces regarding waters in which there is a significant national interest.

While, as noted earlier, the Canadian federal government has not been as aggressive as its U.S. counterpart in exercising a role in inter-provincial water management, it has facilitated and participated in a number of intergovernmental agreements designed to improve the management of inter-provincial (and provincial-territorial) watercourses. The most ambitious of these are certain interrelated agreements which together address both water quantity and (more recently) water quality issues in eastward-flowing prairie rivers, and which establish an inter-jurisdictional board (the Prairie Provinces Water Board) to oversee the implementation of the agreements. Other, less far-reaching agreements, which also create inter-jurisdictional boards, include those with respect to the Ottawa River, the Lake of the Woods, and, more recently, the Mackenzie River. These agreements are in addition to a number of federal-provincial shared funding initiatives in areas such as hydrometric research and flood damage reduction.

In recent years, the public interest in a stronger role for the federal government (as well as for provincial governments) in water management has centred on the issue of water exports. The possibility of large-scale water exports to the United States is a subject that has engaged Canadians on a number of occasions over the past several decades, beginning with a number of proposals, especially in the 1960s, for massive diversions of northern rivers southward to the U.S. market (typically the arid south-west). These proposals have waxed and waned, depending upon the vagaries of politics and weather, but in general have not been considered feasible in the foreseeable future. The debate over such potential exports nevertheless was an important feature of the federal election campaign in Canada in 1988, where it became joined with the issue of Canada-U.S. free trade. Subsequently the same debate was engaged in the context of the negotiation of the Canada-U.S.-Mexico free trade agreement (NAFTA) a few years later. Indeed, one of the requirements insisted on by Canada for ratifying the NAFTA was a joint statement clarifying that water in its natural state was not subject to trade obligations. Nevertheless the issue of trade obligations with respect to water arose again in 1998 in the context of possible tanker exports, as the result of the issuance of a licence (subsequently revoked) by the Ontario government for the export of water by tanker from Lake Superior.

Despite the questionable economics of the tanker-export proposal, and the relatively small quantity involved, the public reaction was strong (and generally negative), and gave rise to three federal initiatives: first, a joint reference with the United States to the IJC in February 1999, requesting the Commission to investigate and make recommendations on a number of issues related to consumptive uses, diversion and removal of Great Lakes water resources; second, the introduction of amendments to the *International Boundary Waters Treaty Act* (the Act implementing the Canada-U.S. Boundary Waters Treaty, discussed further on), designed to prohibit water removals from boundary water basins; and third, a strategy of working with the provinces to develop a federal-provincial accord to prohibit bulk water removals from other parts of Canada. As to the first of these initiatives, the IJC delivered its final report in February 2000. As to the second, the promised amendments were introduced in November 1999, but the legislation died on the order paper with the federal general election of November 2000. Finally, the negotiation of a federal-provincial accord has not yet proved successful. It should be noted that even if the federal legislation is subsequently introduced and passed in the new Parliament, the scope of the legislation is such that it will by no means have plenary effect, as it is closely tied to the federal authority with respect to boundary waters as set out in the Boundary Waters Treaty.

A final federal document of specific relevance to water management is the *Federal Water Policy* of 1987, which was issued as an articulation of *the federal government's philosophy and goals for the nation's freshwater resources.* While the *Policy* did not lead directly to legislation, and as such is without binding effect, it is still in many respects probably a good indication of how the federal government sees its role in water management. In particular, the document emphasises a co-operative approach with the provinces for many areas of policy – an approach that reflects not only constitutional constraints, but also political realities in Canada. Among the issues on which the *Policy* provides general policy statements are management of toxic chemicals, water quality, contamination of groundwater, municipal water and sewer infrastructure, safety of drinking water, irrigation, wetlands preservation, navigation, heritage river preservation, aboriginal water use, climate change and inter-basin transfers. A more recent attempt to update this policy seems to have been at least temporarily put on hold, perhaps as a result of the difficulty of coming to a common understanding with the provinces on the many current controversial issues in regard to water resource management described elsewhere in this analysis.

Apart from policy and legislation that is directed specifically at water resources, there are numerous federal statutes, especially dealing with environmental protection, that may play an important part in Canadian water management. One of the most important of these is the Canadian Environmental Protection Act (CEPA), the primary federal statute with respect to the environment – passed originally in 1988. Of particular relevance to water management, CEPA empowers the federal government to create and enforce regulations with respect to toxic substances, fuels, and nutrients from cleaning products. It also enables the federal government to undertake environmental research, develop guidelines and codes of practice, and conclude agreements with provinces and territories. Environment Canada administers CEPA but assesses and manages the risk of toxic substances jointly with Health Canada. Following an extensive five-year review, the Act was amended in 1999 (proclaimed in force effective March 2000), so as to make pollution prevention a priority. The Act, now requires pollution prevention planning for substances declared toxic under CEPA, and includes expanded regulatory powers. CEPA does not apply, however, to aspects of substances that are regulated for environmental and human health protection under any other federal act. (For example, Health Canada's Pest Management Regulatory Agency administers the *Pest Control Products Act,* including certain aspects of health and environmental assessment and regulatory decisions, as well as policy issues respecting pest control products. Once a pesticide has been registered federally, the provinces regulate its sale and use.).

Another important weapon in the arsenal of modern environmental protection measures is environmental impact assessment. The *Canadian Environmental Assessment Act* (CEAA) requires federal authorities, to conduct environmental assessment of any proposed project or activity that they fund or carry out. Environmental assessment is an important means of reviewing potential environmental impacts of proposed projects, so as to allow informed decisions on how to proceed in such a way that environmental concerns are addressed. CEAA and its predecessor procedures (which operated under a non-statutory *Guidelines Order*) have in fact played an important role in a number of controversial water-related projects in both northern and southern Canada (see the discussion of NWPA above). Finally, there is specific legislation – discussed below – that addresses Canada's international obligations with respect to water.

12.3 Provincial Responsibilities

While the federal interest in water management under the Constitution is based primarily on a number of legislative powers, provincial jurisdiction is rooted in both legislative authority and proprietary rights. The Constitution confers on the provinces ownership of those public lands and resources – including water resources – that the provinces brought with them when they entered Confederation. (The prairie provinces were in the anomalous situation of having the federal Crown retain ownership until the 1930s, when the lands and resources were transferred to them) These proprietary rights are subject only to certain specified exceptions. The practical result is provincial ownership of all navigable watercourses, including their beds, outside the territories.

The provincial proprietary rights with respect to water are supported by a number of legislative powers in the Constitution. For example, the provinces are given authority with respect to the management and sale of public lands, local works and undertakings, property and civil rights in the province, and generally over matters of a local or private nature. The provinces were given additional rights with respect to non-renewable natural resources in a 1982 amendment to the Constitution; however, this amendment does not increase in a significant way the practical ability of provinces to manage their water resources.

The legal regime that governs water in the provinces will of course vary from province to province. As a preliminary observation, however, it is worth noting three distinct underlying approaches to water management; these distinctions are particularly important with respect to water quantity issues. In the eastern provinces, with the exception of Québec, water management is rooted in the riparian doctrines of the common law (largely as received from English law); in Québec, the civilian legal regime possesses its own peculiarities; and in the western provinces, the doctrine of prior appropriation, originally developed in the arid American west, provides the conceptual basis for the statutory regime. While the doctrinal differences in these approaches should not be minimised, the significance of such differences has arguably declined with the growing attention paid to water quality as opposed to water quantity in all jurisdictions, and with the increasing role of legislation in this respect as opposed to the common law.

All provinces have legislation designed to protect water quality; some of this legislation is specific to water, other is more general in nature and deals with pollution of land, air and water. Whether general in nature or specific to water, there is a commonality to many of these environmental provisions across Canada. A non-exhaustive list of such controls would include provisions relating to:

- water treatment and sewage treatment facilities,
- well drilling and construction,
- testing of well water and water supply,
- protection of watersheds that supply drinking water,
- alterations to watercourses (including rivers, streams, brooks, lakes, or ponds),
- disposal of waste materials, crop wastes, and hazardous wastes,
- handling, storage, and use of manure,
- handling, storage, and use of pesticides,
- septic system construction and maintenance,
- installation, maintenance, and removal of petroleum storage tanks,
- handling and disposal of used oil,
- emergency spill procedures,
- marshland or wetland protection,
- land drainage.

In addition to more traditional types of regulatory instruments dealing with water quality, provinces also provide for environmental impact assessment. The particular nature of impact assessment processes varies from province to province, so that there are significant differences amongst jurisdictions as to what body conducts the assessment, which projects will be assessed, and the particular procedures that attach to the assessment. Despite these differences, there is a general consensus at both the federal and provincial levels (as well as in the territories) that environmental assessment is a fundamental tool in the protection of the environment. There has, however, been considerable discussion over which level of government should conduct an assessment where there are both federal and provincial interests at stake. This issue is part of a broader debate in Canada as to the appropriate role of each level of government in environmental protection. An important document in this respect is the Canadian Council of Ministers of the Environment (CCME) *Canada-Wide Accord on Environmental Harmonisation,* agreed to by the federal government and all provinces and territories except Québec in

1998. The Accord attempts to ensure that *each government will retain its existing authorities but will use them in a co-ordinated manner to achieve enhanced environmental results*. There are a number of sub-agreements under the Accord, including one on environmental assessment. Bilateral agreements providing for joint environmental assessment procedures have already been signed between the federal government and some provinces in this respect (Alberta and Saskatchewan both have signed such agreements), and more are likely to follow.

Finally, it should be noted that the same concerns with respect to bulk exports of water that have operated recently at the federal level (as discussed above) also have exercised provincial governments. In the result, a number of provinces have recently passed legislation to restrict such exports.

12.4 Municipal Responsibilities

Municipalities have no distinct recognition in the Canadian Constitution. Outside the northern territories, municipalities are purely creatures of provincial legislation, and depend on such legislation for all their powers; the same is true for other regional government entities that may be established by the provinces. The roles that attach to municipalities with respect to water management vary, then, from province to province. While, typically, municipal governments are given responsibility for the delivery of water and sewage services, there are significant differences in other aspects of water management, such as the testing of drinking water and the assurance of adequate standards of safety. As governments in Canada have increasingly downloaded responsibilities (and costs) of government services to other levels of government, or have privatised them, concern has been expressed about the ability of these smaller entities to ensure continuing standards of water quality. This concern was highlighted recently by a particularly dramatic case in Ontario, where the failure of a municipal government to adequately safeguard the quality of the drinking water led to a number of deaths and a high-profile public inquiry. Especially in eastern Canada, where water delivery and sewage systems are of an older vintage, the need to renew declining infrastructure will continue to place strains on municipal and provincial budgets at a time when there is political pressure to reduce the tax burden on citizens.

An additional problem that is increasingly facing certain municipal government structures in both western and eastern Canada is the growth of intensive livestock operations and the difficulties this causes for assuring the integrity of drinking water supplies, particularly, but not exclusively, in those areas that are heavily dependent upon groundwater.

12.5 International Water Management

Canada's management of internationally shared water resources is largely governed by a number of treaties with the United States. The most significant of these is the *Boundary Waters Treaty* of 1909 (BWT). The BWT not only establishes the legal principles governing boundary and transboundary waters, but also – and of at least equal importance – creates an institutional structure for implementing the treaty, the International Joint Commission (IJC). The treaty has added importance in a Canadian context because of a peculiarity of Canadian federalism, under which the federal government is given the authority to implement treaties concluded by the British Empire on Canada's behalf – even where this would otherwise intrude on provincial jurisdiction. However, under existing judicial precedent, the federal governments authority to implement treaties concluded by Canada on its *own* behalf (which at a minimum would include all treaties since 1931) is constrained by the division of powers in the Constitution. Since the BWT is an *Empire* treaty, the federal government retains broad authority to implement it – an authority that it arguably would lack with respect to other areas of water management not covered by the treaty.

Although the wording of the BWT very much reflects its vintage (for example, in its emphasis on water quantity as opposed to water quality issues), the treaty has proved remarkably adaptable over time, largely owing to the work of the IJC, and some of the treaty's limitations in this respect have been more apparent than real. The first and perhaps most obvious limitation is the differential treatment accorded to shared water resources depending upon whether, on the one hand, they form part of the boundary (i.e. boundary waters) or, on the other, are tributary to boundary waters or flow across the boundary. As to boundary waters, the two nations enjoy "equal and similar rights in [their] use", while for tributary or transboundary waters, each nation reserves "exclusive jurisdiction over [their] use and diversion". This distinction similarly affects the jurisdiction of the IJC, which exercises authority with respect to the approval of "uses or obstructions or diversions. . . of boundary waters on either side of the [boundary] affecting the natural level or flow of boundary waters on the other side of the line". Also with respect to the differential treatment of different waters, the treaty makes no reference to groundwater. A second limitation of the treaty is that, in exercising its jurisdiction, the IJC is required to recognise the following precedence of uses: domestic and sanitary uses, uses for navigation, and uses for power and irrigation. Clearly, a modern understanding of water management would yield different priorities more attuned to environmental requirements. Finally, the jurisdiction of the IJC with respect to approving uses of boundary waters is further constrained by the fact that such jurisdiction must be triggered by an application from either the United States or Canada. The Commission has itself noted that in the cases of minor diversions, this trigger may not be engaged (on the rationale that there is no appreciable effect on the level or flow of boundary waters), despite the potential problem of cumulative impacts of many small projects.

Despite the apparent problems noted with respect to the BWT and the associated constraints placed on the IJC, in practice the regime has proven remarkably adaptable to emerging water management issues. This is perhaps most striking with respect to the issue of water quality, an issue on which the treaty is silent except for a single sentence, which provides that boundary and transboundary waters "shall not be polluted on either side to the injury of health or property on the other". Nevertheless, largely as a result of joint references from the two governments, the Commission since its early days has taken an active role in addressing water quality concerns – especially in recent decades. A notable example of the significance of the Commission's investigatory function (which is provided for in the BWT) was the conclusion of the Great Lakes Water Quality Agreement of 1972, which followed directly from the Commission's 1970 report on pollution of the lower Great Lakes. The 1972 Agreement was subsequently supplanted by the Great Lakes Water Quality Agreement of 1978, which is notable for the adoption of an ecosystem approach to water quality management. The 1978 Agreement was further amended in 1987 so as to introduce new water management approaches, including remedial action plans and plans with respect to critical pollutants.

Just as the Commission has taken on a major role with respect to water quality despite the cryptic treatment of it in the BWT, so too it has also been relatively successful despite the other constraints noted. For example, while it is true that the treaty regime distinguishes between different types of waters depending upon whether they form part of the boundary, in practice this has been of decreasing significance, especially in the IJC's reports, where the Commission has increasingly focussed upon the needs of basins and ecosystems. This is perhaps most evident in the increasing attention that the Commission has given to groundwater (for example, in its recent reference with respect to water uses in the Great Lakes Basin), despite the silence of the treaty in this respect.

The Commission operates in two sections – the U.S. section with its office in Washington and the Canadian section with its office in Ottawa. Additionally there is now an office in Windsor, Ontario with the responsibility of overseeing the implementation of the Great Lakes Water Quality Agreement. The Commission is composed of six Commissioners, three appointed by each nation; there are similarly two co-chairs, one appointed by each government. While each section maintains what is in effect its own secretariat, the Commission relies heavily on government officials from both nations in carrying out the various duties assigned to it. As a result, the Commission has reporting to it a number of boards related to water management – including boards of control, investigative-engineering boards, pollution advisory boards, and Great Lakes Water Quality Agreement Boards.

While the BWT is the central instrument in the management of Canada-U.S. water relations, there are other agreements dealing with specific watercourses - usually in the context of some particular aspect of resource development such as hydroelectricity. These other treaties include agreements with respect to the Co-lumbia River, the Skagit River, the Lake of the Woods, the Niagara River and Rainy Lake. Also relevant are the agreements with respect to the St. Lawrence Seaway and Great Lakes water quality (discussed above). Some of these agreements have associated with them boards that report to the two governments.

An additional element in Canada-U.S. transboundary water relations that is of particular relevance to water management in the Great Lakes Basin is the emergence of sub-national co-operation - that is, between U.S. states on the one hand and Canadian provinces on the other. This co-operation recognises that states and provinces are the primary managers of water resources, and suggests that international instruments other than binding international agreements may have a practical role to play in transboundary water management. The most important document illustrative of such co-operation is the Great Lakes Charter of 1985, a non-binding agreement between the Great Lakes states and the provinces of Québec and Ontario. The Charter arose out of an initiative of the Council of Great Lakes Governors, and is designed to protect the waters and ecosystems of the Great Lakes Basin. It does this through agreement on five principles fundamental to the management of the Basin: integrity of the Basin; inter-jurisdictional co-operation; protection of the Lakes' waters; prior notice and consultation for new or increased diversions or consumptive uses above a certain threshold; and co-operative programmes and practices. While the Charter and its implementation have not been free from criticism, it is generally agreed that this regional approach to transboundary water management holds considerable promise.

12.6 Summary of Institutional Legal Framework in Canada

In summary, both the federal government and the provinces are given significant powers under the Constitution with respect to water management. The boundary between provincial and federal authority, however, is not always clear – nor, given the reluctance of Canadian governments to engage in inter-jurisdictional litigation, is there a mechanism that can readily answer the question as to which level of government should exercise control in a particular matter. In the face of this imprecise understanding of where federal authority ends and provincial authority begins, the effective management of water resources in Canada depends to a significant degree on federal-provincial (and inter-provincial) co-operation. Beyond this generality, however, one can observe that, in practice, most of the important water management issues in southern Canada – with the possible exception of international concerns – are dealt with at the provincial level, and even with respect to transboundary water issues involving the United States there is a significant degree of provincial input. In the absence of a strong federal interest, the clear tendency of the federal government is to exercise restraint in asserting a role in water management and to defer as much as possible to the provinces as resource owners. This is true even with respect to matters such as the management of interprovincial watercourses, where the federal government has preferred to take a facilitating role – even though this has often resulted in very slow progress.

13. Partnerships and Community Involvement[9]

Across Canada, individuals, communities, and governments are taking up the challenge of sustainable development by working together to ensure a healthy environment to the future generation and the full richness of the country's natural legacy (Environment Canada 1998a).

Environment Canada has developed, in partnership with these other interests, a number of *ecosystem initiatives* that respond to the unique problems of targeted areas and communities and address environmental, economic, and social concerns. They are characterized by a number of principles, including:

1. an ecosystem approach – recognising the interrelationships between land, air, water, wildlife, and human activities;
2. decisions based on sound science – including natural and social sciences combined with local and traditional knowledge;
3. federal–provincial–territorial partnerships – governments working together to achieve the highest level of environmental quality for all Canadians;
4. a citizen/community base – working with individuals, communities, Aboriginal peoples, industry, and governments in the design and implementation of initiatives;
5. pollution prevention – promoting a precautionary approach.

Ecosystem initiatives help to achieve environmental results through partnerships, pooling resources, focusing science, co-ordinating efforts, sharing information and experiences, and generating a broad basis of support. Moreover, they help build the capacity of all the players involved to make better decisions and to effect change.

Five Ecosystem Initiatives are currently inplace to address issues in specific ecosystems (Figure 13.1):

- the Atlantic Coastal Action Program
- the St. Lawrence Action Plan - Vision 2000
- Great Lakes 2000
- the Northern River Basins Study/Northern Rivers Ecosystem Initiative

---

[9] This Chapter has been written with the support of contributions from Jean Burton, St. Lawrence Centre, Environment Canada and André Beauchamp Chair of the commission on water management in Quebec.

• the Fraser River Action Plan/Georgia Basin Ecosystem Initiative.

Atlantic Coastal
Action Program

St. Lawrence Action Plan
Vision 2000

Fraser River Action Plan/
Georgia Basin Ecosystem Initiative

Northern River Basins Study/
Northern Rivers Ecosystem Initiative

Great Lakes 2000

Source: Environment Canada (1998a)

Fig. 13.1. Five ecosystem initiatives in Canada

Discussions have begun on a sixth Initiative, addressing the unique environmental circumstances of Northern ecosystems. The Canadian approach to community involvement has evolved over the last twenty years. Originally centred on habitat and endangered species conservation, its attention has shifted progressively to a broader range of issues, in a newly defined relation between governments and citizens. The Ecosystem Initiatives which have been successfully field tested in the country can be described as a model of the Canadian approach to community involvement. The 6 following sections are drawing a general description of each Ecosystem Initiative and summarises the major community involvement issues the section named *Concluding remarks on Community involvement in Canada* discusses the common features of these programs and their future challenges. Finally, the last section presents as an example of public involvement, the 1999 public hearings undertaken in Quebec to direct future provincial water policy.

13.1 Great Lakes 2000

Great Lakes 2000 (GL2000) was formally launched in 1989 as the Great Lakes Action Plan to fulfil Canada's commitments under the 1972 Canada–U.S. Great Lakes Water Quality Agreement to protect and sustain the world's largest freshwater ecosystem. GL2000 integrates diverse programs across seven federal departments and involves close partnerships with community organisations, individual citizens, industry groups, academics, municipal governments, and the province of Ontario. Eighty percent of the pollution in the Great Lakes originates in the United States, making successful Canada-U.S. partnerships essential to deal with bi-national problems.

Through a Canada-Ontario Agreement, targets for improving the environment focus on restoring degraded areas, reducing and preventing pollution, and protecting human and ecosystem health. Community-based action is a central element in developing and implementing Remedial Action Plans (RAPs) for Canadian *Areas of Concern*. Citizens are directly involved in Public Advisory Committees; they take part in all stages of the RAP process, namely the problem definition stage, the selection of remedial and regulatory measures stage, and the restoration stage. A RAP program exists in each of 17 Canadian Areas of Concern, and to date, more than 50 percent of the necessary actions to restore these areas have been implemented. Collingwood Harbour has been fully restored.

Progress in the Great Lakes Basin ecosystem has included reductions in environmental levels of targeted pollutants, leading to a decline in contaminant levels in humans; the recovery of wildlife populations such as the peregrine falcon and the bald eagle; the development and implementation of recovery plans for 14 threatened species; the protection of 3,000 hectares of natural areas; and a reduction by 4,500 tonnes of emissions of toxic substances.

GL2000 has also contributed to leading edge environmental research on toxic substances and to the advancement of the concept of *virtual elimination* in national policy development. GL2000 research indicates that there is a continuing need to measure the effects of pollution on wildlife and humans and to further reduce contaminant levels.

13.2 Atlantic Coastal Action Programme (ACAP)

The Atlantic Coastal Action Programme (ACAP) is a network of 13 community-driven, watershed-based ecosystem initiatives located across the four Atlantic provinces. Since 1991, citizens, community organisations, private sector organisations, municipalities, universities, First Nation representatives, and a number of federal and provincial government departments have been collaborating to develop broadly supported strategies for the restoration and sustainable use of their coastal environments and the watersheds that form a part of them. For each of the 13 identified sites, an accreditation process for participating representatives is put into place. An overall environmental management plan is developed through a participatory process. Here again, both the diagnostic and the remedial actions are identified through consensus developed by all parties involved. This is a shared responsibility, and everyone contributes, learns, and benefits.

To date, over 400 projects have been undertaken involving hundreds of organisations and thousands of volunteers. Science and monitoring projects have supported informed decision making, linking scientists with communities and combining scientifically derived information with other forms of traditional information and knowledge. Results have included pollution prevention, restored habitats, reduced waste, upgraded sewage treatment facilities, improved energy efficiency, the establishment of new parks, the creation of artificial wetlands, reduced risks to human health, and increased employment.

Phase II of ACAP is focusing on implementing individual site strategies, expanding the ACAP network, and collaborating with others to better understand the science and achieve measurable ecosystem goals.

13.3 St. Lawrence Action Plan - Vision 2000

The St. Lawrence Action Plan was launched in 1988 and is now in its third phase. Its objectives are to achieve a healthy St. Lawrence ecosystem, healthy communities, and greater public access to the St. Lawrence River and Estuary. The St. Lawrence Action Plan openly recognises public participation as a key factor for governmental actions at the local scale. The third phase of the St. Lawrence Action Plan emphasises a preventative approach in biodiversity conservation, focusing on agriculture, industry, human health, and navigation. Community organisations will also play an increasingly active role in improving the St. Lawrence ecosystem.

Results from the first 10 years of the plan include a 96 percent reduction in toxic effluent discharges by 50 priority industrial plants; the creation of the first federal–provincial marine park in the mouth of the Saguenay River; the protection of 12,000 hectares of wildlife habitat; and an increase in the population of beluga whales, one of the most well-known symbols of the St. Lawrence, from 500 to approximately 800.

For several years, public hearings and round tables have been standard public involvement procedures in Quebec. With the creation of the St. Lawrence Action Plan in 1988, the ground was ready for a full scale use of these well honed processes. Since the very beginning, the St. Lawrence Action Plan has had a strong commitment to the publication of information held by federal and provincial departments. This effort constitutes a common base for problem identification and a shared understanding of issues by all parties involved.

The success of the plan is due, in large part, to a close working relationship between the Government of Canada and the Government of Quebec. Both governments have responsibilities in the basin and recognise that co-operation on environmental problems is essential. Other partners involved in this initiative include the private sector, universities, environmental groups, research centres, and local community organisations. Ten community groups, or *ZIP* committees (*Zones d'Intervention Prioritaire*, or Zones of Priority Intervention), have been set up along the St. Lawrence River and Estuary.

Current partners in the ZIP Programme are the federal and provincial governments and Strategies Saint-Laurent, an NGO acting as an umbrella organisation to co-ordinate the activities of the 14 ZIP Committees. The ZIP Committees are made up of representatives from the community: mayors, industry representatives, environmental groups, teachers, and ordinary individuals from the community.

The ZIP Committee is created on the sole initiative of the community, with technical assistance from Strategies Saint-Laurent. Following an information and familiarisation period led by a local NGO, interested citizens are invited to join in and assist in the creation of a ZIP Committee . The overall goals are quite broad; to conserve, restore and promote the sustainable use of the resources of a section of the river. The representation rules are quite simple: all interested parties are welcome, without predetermined number of seats; but, at no time can any given sector of society hold a majority of seats. They define their own territory based on their membership and their capacity to manage public participation within these boundaries. The ZIP Committee is then formally recognised by Strategies Saint-Laurent and the government agencies participating under the St. Lawrence Action Plan.

What is typical of the ZIP Committee is its operational mode: it provides a neutral forum to all interested parties. Everyone is welcome as long as they share the overall objectives of the program. Priority issues are identified through a public hearing; then remedial actions are defined by the citizens themselves and incorporated in a local action plan. Given the composition of each ZIP Committee and the specific aspects of their environment, each Committee takes on its own character and mode of operation. Within their own territory, the ZIP Committee also assists other NGOs with the preparation of project proposals for funding through the St. Lawrence Community Interaction Funding Program.

The government's role is two fold; it provides funds, both for core-funding of the ZIP Committee and for projects; and it provides technical and scientific assistance at all stages of the program. Funds are provided to support the basic needs of the ZIP Committee: the committee does not have to compete with other NGOs for funds, so they can play a support role within their territory. Funds are also provided for projects selected by a multilateral committee. The information support role of governments is also quite important; over a five year period, federal and provincial departments have collected and published together the ZIP Reports covering the entire St. Lawrence River; a four-report series was published for each of the 14 sectors of the river, dealing with the topics of physics and chemistry, biology, sociology and economics, and human health. Government representatives also provide information and technical assistance, both at the public hearings and through the development and implementation phases of the local action plans.

13.4 Northern River Basins Study / Northern Rivers Ecosystem Initiative

The Northern River Basins Study was launched in 1991. Its findings improved our understanding of the impacts of the growing number of industrial developments in these northern watersheds, particularly pulp and paper and oil sands projects, as well as the effects of human activity on these ecosystems. The study focused on the Peace, Athabasca, and Slave river systems.

The Northern River Basin Study has developed an interesting participatory model well adapted to these northern regions. A 25 member committee has been set-up to manage this program; federal, provincial and territorial representatives, along with First Nations, industry, education and health community groups, environmental NGOs and the public, all joined efforts to share knowledge and experience. The inclusion of traditional knowledge within a scientific study was given special attention.

The Northwest Territories and the Federal Ministry of the Environment are working together through the Northern Rivers Ecosystem Initiative (NREI) to address the recommendations of the Northern River Basins Study. Building on the success of the study and working with individuals, industry, Aboriginal peoples, communities, and others, the NREI will focus on priority issues and areas of concern, including promoting pollution prevention to maintain the long-term quality of the ecosystems within these river basins.

13.5 Fraser River Action Plan / Georgia Basin Ecosystem Initiative

The Fraser River Action Plan was also launched in 1991. The uniqueness in this case in the creation of the Fraser Basin Council a long-term and broadly representative non-governmental organisation which is made up of 36 members representing federal and provincial governments, First Nations, industry, and representatives from the 5 regional districts. The overall goal is the sustainable development of the Basin as described in the Fraser River Basin Charter, developed through a participatory process and to promote a balance between social, economic, and environmental well-being within the Fraser Basin.

Environment Canada, with a wide range of partners including Fisheries and Oceans Canada, completed the Fraser River Action Plan (FRAP) in March 1998. Results included the protection of almost 65 000 hectares of waterfowl habitat; a reduction in the release of toxic wood preservatives by 90 percent; and the implementation of best management practices and pollution prevention plans in many business and industry sectors.

The successes, knowledge, and lessons learned from FRAP have helped shape the Georgia Basin Ecosystem Initiative (GBEI), which was launched in 1998. The geographic focus is the Georgia Basin (which includes Puget Sound in the State of Washington), where the human population is projected to double in the next 20 years. Growth pressures are already imposing high levels of physical, chemical, and biological stress on the ecosystem. Left unchecked, these stresses will affect not only the state of the environment but also the foundation of the region's economy, the health of individuals, and the overall quality of life. The vision for the GBEI is *managing population growth to achieve healthy, productive, and sustainable ecosystems and communities.*

The GBEI is a results-based federal-provincial program founded on sound scientific knowledge. It focuses on priority environmental issues for British Columbia residents: air and water quality, including the reduction of urban smog, the clean-up and prevention of pollution from non-point sources, and the identification of impacts of toxic substances; and the conservation and protection of nature. It strives for a co-operative programming approach to improve the efficiency and effectiveness of government and depends upon partnering with others to enhance the ability of individuals, communities, and the private sector to make decisions that promote sustainable development. The GBEI also provides a context for working with the United States to address shared challenges, including the effective management of airsheds, watersheds, and wildlife populations within the transboundary Georgia Basin/Puget Sound region.

13.6 Northern Ecosystem Initiative

Northern Canada is a geographically and politically diverse area. The northern coastline and oceans represent about 80 percent of the marine area in Canada. The North is rich in biodiversity and natural resources, sensitive to environmental impacts, and sparsely populated. Predominantly Aboriginal, the population in the North is the youngest and fastest growing in Canada.

The North is undergoing significant economic, socio-political, and environmental change. Resource development is rapidly accelerating and will have an increasing impact on Canada's economy. The North is also affected by global environmental problems including the long-range transport of toxic substances and climate change. These challenges are global and require international co-operation to find solutions.

Scientific knowledge of northern ecosystems must be improved to promote sustainable development in the North. Discussions on the Northern Ecosystem Initiative (NEI), engages a number of partners, including northern populations and communities, in setting priorities for action in the North.

13.7 Summary of Community Involvement in Canada

Several programmes building on community involvement have been developed in Canada over the past 20 years. These programmes share some common features. They result from a rapid change in the way governments conceive and deliver their programs in the field; from a top-down centralised attitude, governments are moving to a more decentralised approach recognising the importance of community participation and empowerment of local partners. Governments, within this new scenario, become one of many partners in action, bringing expertise and resources, without imposing their views.

Nevertheless, in spite of a solid volunteer base throughout the country, community involvement programmes are quite vulnerable. It has become very difficult for governments to provide core-funding to these organisations; without the support for a permanent staff in the field, volunteer organisations depend on administration fees derived from projects and are competing amongst themselves for limited resources. Moreover, it is very difficult to maintain public involvement in the long term as some problems may take several years to be resolved. Finally, there is a serious gap between the level of action desired by the community and the technical and financial resources available to implement them.

13.8 The Public Hearings in Québec (BAPE Project)

In recent years, water has become a hot topic of debate the world over. There is fear of drinking water shortages affecting much of the world's population. There is also fear of serious droughts in some nations, which could threaten both food security and economic development. Water management methods are being examined. Many observers use the term *water war* to describe the increasing scarcity of water, a resource which is becoming a major geopolitical issue.

The Province of Québec possesses enormous amounts of relatively good quality water. There is no threat of its losing this water or seeing its quality deteriorate significantly, at least in the short term. However, this issue is perceived differently within Quebec's borders. For many years now, various topics have fuelled public opinion.

For thirty years now, a potential *Grand Canal* (Great Recycling and Northern Development) project has been the topic of discussion. It consists in channelling into Lake Superior water flowing towards James Bay, and diverting water from Lake Superior to the southern United States. While this project has not been scientifically studied or assessed, it fuels the public imagination and constantly reappears in the headlines, as a fantasy of US ownership of Canadian water.

This idea of bulk water exports from Canada to the United States has also been hotly debated by Canadians from sea to sea. Many observers think that projects to export bulk water via diversion, pipeline or containers would make water a commercial commodity governed by the specifications of NAFTA (North America Free Trade Agreement) and would result in Canada's losing sovereignty over its waters. Hence, the demand that water exports be prohibited by law matters (see Chap. XII on Institutional and Legal Issues above).

The fear that water would become a merchantable resource has also increased in Québec in recent years, following numerous local conflicts involving groundwater collection projects. A number of groundwater bottling companies, affiliated primarily with French water multinationals, wanted to tap water tables in the rural environment. The resulting conflicts were intense and received extensive media coverage.

Finally, there has also been keen controversy over municipal management of drinking water filtration and wastewater purification plants. Privatisation and/or partnership projects met with enormous resistance, primarily from the union and popular sectors. In the context of grappling with the deficit and questioning the welfare state situation characterising the Canadian and Québec political context, a large portion of the population seems to fear that the state will withdraw from essential services, with drinking water heading the list. The private partners envisaged are, once again, water multinationals.

It was in this rather turbulent context that the Québec government launched its consultation on water management in Québec. As a natural resource, water is, first and foremost, under provincial jurisdiction, although the federal government has jurisdiction over fishing, navigation and international matters (see Chap. XII on Institutional and Legal Issues above).

The Québec government mandated the Institut national de recherche scientifique sur l'eau (INRS-EAU), a water research institute within the decentralised framework of the University of Québec, to prepare a symposium on water management in Québec, which was held in Montréal on December 10-12, 1997. The symposium resulted in an awareness of the state of scientific expertise as concerns water in Québec. The symposium proceedings were published in three thick volumes (INRS-EAU, 1998). However, far from achieving unanimity, the symposium rekindled public debate. Consequently, the government decided to mandate the Minister of the Environment to carry out a far-reaching public consultation on water management.

The Minister of the Environment at then time, asked the *Bureau d'audiences publiques sur l'environnement* (BAPE), Québec's environmental hearings board, to conduct an inquiry and hold public hearings on water management in Québec. The three-member commission (André Beauchamp, chair, Gisèle Gallichan, commissioner, Camille Genest, commissioner) began its consultation on March 15, 1999 and submitted its report to the Minister of the Environment on May 1, 2000.

To facilitate the consultation, the government published *Water Management in Québec: Public Consultation Document*, a document summarising the government's overall aims and objectives, and proposing four themes for discussion: groundwater; surface waters; municipal infrastructure and water services management; and water as a strategic world issue (Gouvernement du Québec 1999). It is important to note that the terms of reference assigned to this investigation by BAPE expressly excluded examination of issues pertaining to hydropower, one of the most contentious areas of water use in Québec. While this in no way negates the importance and value of the results of these public consultations as published, it is necessary to remember this important omission in reviewing and making use of these results. In keeping with the BAPE's procedure, the hearings were held in two parts: Part One was devoted to questions from the public and providing information, and Part Two was devoted to hearing briefs. During Part One of the hearings, participants were able to ask questions of representatives of the government departments concerned (ten provincial government departments, five federal government departments) on issues they considered important. Québec's procedure, which is easy to apply and not subject to legal control, allows anyone to attend the hearing and raise points of personal interest. The Commission toured Québec's 17 administrative regions and, from the different government departments and agencies concerned, obtained approximately 800 documents and studies produced by government machinery. The Commission also organised 11 workshops with various specialists on a number of more specific themes.

The James Bay and Northern Québec territories are governed by agreements between the Cree and Inuit nations and the Canadian and Québec governments. To respect the framework of these agreements, the Commission on water management entered into agreements with the James Bay Advisory Committee on the Environment and the Kativik Environmental Advisory Committee for consultation on the lands governed by agreements. In Part Two of the hearings, participants were invited to express their opinion before the Commission in the form of briefs. The Commission made a second round of Québec's 17 regions and heard 379 briefs.

Sum total, the Commission held 143 public hearings and heard 379 briefs. Combined with the 800 government documents released and the transcripts of the public hearings, the mass of documentation resulting from the exercise represents an impressive volume of highly variable scientific quality. The Commission's report (approximately 750 pages), the hearing transcripts and the briefs and documents transmitted in electronic form were collated onto a CD-ROM, thereby enabling the public to access both the Commission's report and analysts' and researchers' comments in order to continue its deliberations (Bureau d'audiences publiques sur l'environnement 2000).

The Commission's report is not scientific in nature. It gives no opinion on the scientific controversies and does not touch on the highly technical aspects of certain issues. Its objective is twofold. It has a political goal, proposing the broad lines of an integrated water and aquatic environment management policy to government, and a teaching goal, giving militants and organisations concerned about water management a general frame of reference.

With regard to the issues mentioned above (bulk water export, groundwater withdrawal for commercialisation and municipal water management privatisation), the Commission's opinions are:

- The Commission recommends that it be prohibited to export bulk surface water or groundwater by diversion, pipeline or container. Despite its apparent abundance, the water of the Great Lakes – St. Lawrence basin is a vulnerable resource and potential climate changes would result in a significant drop in St. Lawrence River water levels with a foreseeable impact on biodiversity. These conclusions are quite similar to those of the International Joint Commission, a joint Canada-U.S. agency for international and transboundary water management, which simultaneously issued an opinion on the same subject. According to the Commission, water exports by ship to Middle Eastern countries are unlikely, since seawater de-salinisation is 2-3 times less expensive than water transport. As for sending water south toward the United States, no serious economic study is currently available. A request to this effect could come from rural communities. Rather than devising a strategy for the bulk export of Canadian water, it would be preferable to develop demand side management strategies within an overall strategy of integrated resource management.
- The Commission recommends increased groundwater use, provided that all uses requiring more than 75 m^3/day undergo the environmental impact assessment and review procedure. While bottled groundwater exports do not, in themselves, threaten Québec water reserves, they may conflict with local uses. Consequently, projects must undergo public assessment and review.
- The Commission does not recommend privatising municipal drinking water and wastewater purification facilities, but is favourable to certain limited, well-supervised partnerships.

The report's principal recommendations target:

- implementation of management by drainage basin and establishment of a system of royalties for major users for water collection and for waste;
- comprehensive reform of the Watercourses Act to put groundwater and surface water on a common footing (*res communis*) and to ensure the implementation of integrated water management by drainage basin;
- designation of a Minister of State for Water responsible for ensuring joint efforts between the ten Québec government department with sector-specific water-related mandates;

- creation of an agency called Bassin Versant Québec (drainage basin Québec) mandated to implement management by drainage basin;
- creation of a *Conseil de l'eau et des milieux aquatiques* (council for water and aquatic environments) responsible for issuing opinions to the Minister of State and initiating certain consultations.

The Commission affirms that the following three principles underpin its recommendations: the principle of integrating government decisions to overcome the current contradictions inherent in sector-based water management; the principle of articulated efforts throughout the territory by drainage basin, a form of management which, through the world, increasingly seems to be the appropriate management method for the water sector; the principle of democratic participation and citizen involvement in water management.

Operationally speaking, the Commission recommends a review of the drinking water regulation, and its now-obsolete standards; reform of the regulation governing the disposal of wastewater in isolated dwellings; revival of a federal-provincial infrastructure renovation program to help municipalities renovate their system; and intensification of joint efforts related to measures concerning the St. Lawrence River. Above all, it recommends a fundamental review of the agricultural clean-up program. In the past 20 years, Québec has seen considerable progress in urban and industrial clean-up but subsequent phases may be jeopardised unless a substantial, systematic effort is made in the farm sector. In particular, the Commission stresses issues related to non-point source pollution: harmful cropping practices, intensive use of chemical inputs, over-fertilisation, improper management of manure, farm drainage and forests, etc.

It is important to point out that, throughout the consultation process, participants insisted extensively on water's symbolic dimension. At the end of Part One of the public hearings, a coalition of artists came to testify. It would be fair to say that a large part of the population is very concerned about water, since it sees it as a source of life. The possibility of state withdrawal and water becoming a commercial commodity raised fear and indignation among participants and the media.

The Commission's report has been received very positively both by government and hearing participants. No group or sector has decried its orientations. It is as if the Commission had voiced others' thoughts. In order to gauge the report's real impact in the future the Commission recommended producing a summary of the report's implementation five years after its release.

14. Prevalent Water Issues

Because of the complexity of Canada's fresh water system and because many issues are highly variable at the regional or community levels, it is difficult to present a national portrait of Canada's water issues. However, this report has brought up a certain number of prevalent issues related to Canada's fresh water that are briefly presented below.

The most challenging issues in Canada's water sector include surface water quality and groundwater pollution from agriculture (livestock operations), conflict between rural and urban land uses as Canada's population density (in narrow areas) grows, insufficient water quality monitoring, insufficient training of water plant operators, and fragmentation of governance. In addition to these overarching issues, the report addresses climate change and the hydrological cycle, floods and drought management, ageing infrastructure and urban water management, water exports and biodiversity loss as some of the current topical and controversial water issues in Canada. Although the big dam and diversion era is probably over in Canada, inter-basin water transfers are still occurring and diversions that took place years ago still result in significant ongoing impacts. The need to continue to assess these ecological effects is the reason that this topic is highlighted here.

14.1 Water Quality and Pollution

Though Canada is blessed with an abundance of clean water, several rivers, lakes, aquifers and coastal areas have been polluted by industrial and municipal discharges, runoff, spills and deposition of airborne contaminants. Acid rain is still affecting some lakes in eastern Canada. Some Canadian cities which are lacking stormwater management facilities have still to deal with untreated urban sewage effluents. Quite a few large Canadian cities and most small communities in coastal areas provide little or no treatment of sanitary sewage before its entry into coastal waters. In rural areas non point sources of pollution are greatly affecting Canadian freshwater bodies causing bacterial and nutrient contamination.

There is a need to make sure that all users of water resources in Canada are aware that they are guardians of some of largest freshwater bodies in the world and therefore they are responsible for its present and future quality. Indeed, even if these large freshwater bodies (e.g. Great Lakes) are capable of assimilating waste to some extent, there are limits to what they can absorb.

14.2 Groundwater

In Canada the pressure on groundwater is high. Reliance on groundwater for most domestic and industrial uses is especially high in Prince Edward Island and in rural areas elsewhere in Canada. Roughly one in every five Canadians (six million in total) rely on groundwater for their domestic water supplies (100% of the population of Prince Edward Island and over 60% of the population of New Brunswick and the Yukon). Most of the people who rely on groundwater in Canada live in rural areas and draw their water from private wells. A number of groundwater bottling companies already tap water tables in the rural environment and other new ventures are pressuring governments for approval. Although the quality of the groundwater is generally very high, locally problems due to contamination remain. Pollutants resulting from human activities may come from a range of sources, including septic systems, leaky storage tanks, municipal landfills, industrial discharges, and agricultural practices, including improper handling of livestock wastes and spraying with pesticides and fertiliser (causing, for example, high levels of nitrate concentration). Due to its inaccessibility, groundwater is difficult if not impossible to clean up once polluted.

In Canada, a better scientific understanding, including groundwater hydrological studies and a cartography of the resource, is needed as well as a better knowledge of the biogeochemical processes that act on groundwater and either diminish or restore its quality.

14.3 Climate Change and the Hydrological Cycle

As in other parts of the world, the potential effects of climate change on precipitation and the hydrological cycle in Canada are strictly speculative at this time. Many theories abound but consensus is far from being achieved. There is a strong likelihood, however, that areas of Canada presently subjected to sporadic periods of drought will see the frequency and severity of such events increase, while those areas currently subject to flood risk can expect to be more critically and frequently at risk. Improvements in monitoring capacity and research in dealing with both cases are essential.

14.4 Inter-basin Water Transfers

Inter-basin transfers, primarily for hydroelectric power generation and irrigation, are more common in Canada than in any other part of the world and are the subject of many environmental concerns. Aquatic habitat and productivity losses in the watersheds from which the water is extracted are frequently matched or exceeded by the impacts due to increased flooding, erosion and instream flows in the receiving watershed. In addition, these transfers of water are accompanied by a transfer of pollutants, alien species and occasionally disease organisms into recipient system causing great impact on the ecosystem. Consequently, these transfers cause profound impacts on biological diversity. Though environmental impact assessments that have addressed these issues and shed some light on these concerns, there is still a great deal more to learn about the individual and cumulative effects of these changes over time on the Canadian aquatic environment and resources.

14.5 Floods and Droughts

In the past, Canada has had in place a flood prediction, prevention and relief programme that has been the envy of many other parts of the world. But through decreasing investment by senior levels of government, division of jurisdictional responsibilities and a failure to capture the public concern and create a culture of flood control and management, this situation has deteriorated in recent years to the point where significant improvements are needed.

Canada is not as severely affected by drought as many other parts of the world. But regionally the problem can be severe and it is likely to become even more severe in future due to climate change and increasing demand on limited resources. The need for a national strategy to cope with these problems is indicated.

14.6 Allocation

The majority of Canada's population lives in a narrow band within 300 kilometres of the Canada-US southern border. This concentration of population leads to environmental degradation and damage to the drainage basins and to the surrounding ecosystem and it places an enormous demand on local water supplies and the treatment of wastewater. In these areas where water scarcity, water quality deterioration, economic, environmental, social and cultural factors conflict, tensions between users are commonly arising and future conflicts can be expected. In seeking to draw a balance between affordability, accessibility and the demands of multiple user groups, absence of a formal water resource management mechanism has led to conflicts between user groups. Therefore, the definition of new mechanisms to allocate water to its users and the criteria for determining the priority of need between user groups need to be established. Allocation based on economic criteria alone can further compromise allocation to non-commercial water users and the environment. Demand management, especially through metering of individual consumption and appropriate pro-rata pricing, is an important element of any such strategy.

14.7 Ageing Infrastructure and Urban Water Management

In Canada, the state of urban infrastructure has become of increasing concern in recent years to municipalities, and the provincial and federal governments. Many studies have been undertaken to assess the state of urban infrastructure, which generally appears to be in critically poor condition in many instances. Projections indicate that an investment of 36 billion dollars is needed in Canada over the next 15 years to bring municipal infrastructures up to acceptable North-American standards. Moreover, the studies show as well the lack of information available to assess the state of urban infrastructure and the need for better knowledge base on which to make decisions.

More generally speaking, although measures have been taken in specific locations, governments at all levels must give adequate attention to the issue of urban sprawl. Integrated approaches that address loadings from treatment plants, stormwater sewers, combined sewer overflows and other wastewater sources are in urgent need of strengthening.

14.8 Monitoring and Information

Freshwater monitoring programs are spread among many governmental levels (federal, provincial, regional and municipal). No single agency operates a general, nationally integrated monitoring program. National coverage is therefore incomplete and inconsistent. Moreover the recent download of responsibilities by provincial governments to the municipalities without provision for ensuring appropriate resourcing or training has led to dramatic consequences, as in the case of Walkerton, Ontario, where seven people died due to *Escherichia coli* contamination in the municipal water supply. The need for strengthening intergovernmental (federal, provincial, regional and municipal) collaboration and partnership in freshwater monitoring and for tightening the training of operators is very real.

14.9 Biodiversity Loss

In Canada, despite the legal protection afforded to wetlands, much of the original base of wetlands has already been lost and many threats to remaining areas still remain. The authors of the first estimate of extinction rates of North America's freshwater fauna recently came to the conclusion that *A silent mass extinction is occurring in our lakes and rivers*.

Many feel that wetlands and freshwater biodiversity loss in Canada should be as much of concern to Canadians as the loss of tropical rainforests, but public ignorance of these concerns prevents the kind of urgent attention by governments that will stem these losses and lead to efforts to restore damaged ecosystems.

14.10 Fragmentation of Governance

In Canada, both the federal government and the provinces are given significant powers under the Constitution with respect to water management. The boundary between provincial and federal authority, however, is not always clear – nor, given the reluctance of Canadian governments to engage in inter-jurisdictional litigation, is there a mechanism that can readily answer the question as to which level of government should exercise control in a particular matter. In the face of this imprecise understanding of where federal authority ends and provincial authority begins, the effective management of water resources in Canada depends to a significant degree on federal-provincial (and inter-provincial) co-operation, and on the willingness to work towards a common set of principles at the federal and provincial levels. There is a compelling and urgent need for national strategies and policies to address issues that are common to most provinces or that are national in scope.

However, one can observe that, in practice, most of the important water management issues in southern Canada – with the possible exception of international concerns – are dealt with at the provincial level, and even with respect to transboundary water issues involving the United States there is a significant degree of provincial input. The federal presence in water management in southern Canada is strongly reflected in Great Lakes 2000 and the St. Lawrence Vision 2000. In the absence of a strong federal interest, the clear tendency of the federal government is to exercise restraint in asserting a role in water management and to defer as much as possible to the provinces as resource owners. This is true even with respect to matters such as the management of inter-provincial watercourses, where the federal government has preferred to take a facilitating role – even though this has often resulted in very slow progress.

14.11 Water Exports

Given the large volume of renewable surface water available in Canada the question is: With so much water available, should Canada allow the export of freshwater resources?

Recently, in response to this question the public reaction was strong and generally negative. A concrete response arose in November 1999 when an amendment to the International Boundary Waters Treaty Act, designed to prohibit water removal from boundary water basins, was introduced. But the legislation died on the order paper with the federal general election of November 2000 and a strategy of working with the provinces to develop a federal-provincial accord to prohibit bulk water removal from other parts of Canada has still to materialise. In the province of Quebec, no application for authorisation to export water in bulk or to divert watercourses to the U.S. has yet been made. To prevent such export of water, the Québec Provincial Government promulgated on November 24, 1999 Law 73 (*Loi visant la préservation des ressources en eau*). This law forbids groundwater and surface water transfers outside the Province of Québec. Similar laws have been adopted in British Columbia and Ontario. Most provinces have, or are in the process of developing, legislation or regulations to prohibit the bulk removal of water in Canada.

14.12 The Shift to Integrated Water Resource Management

In Canada, the *ecosystem approach* for water management has been successfully implemented with the ecosystem initiatives which have been adopted in the largest river basins. However, the concept needs to be more actively promoted throughout the country, especially in the smaller river basins. An holistic ecosystem approach takes into account the inter-relationships between the air, water, land, fish, wildlife and people, and this ought to be stressed as the core of integrated water resource management programs.

There is in Canada a real need to infuse economic, social, environmental and cultural factors into water management plans. Present practices in some basins have failed to respect natural hydrological characteristics, and the flora and fauna of the surrounding areas. In some cases, the natural boundaries of hydrological units are incompatible with the administrative and jurisdictional boundaries. These shortcomings have impeded the development of long term plans in the conservation and preservation of natural resources.

References

Alberta Environment (1999) Alberta's Water Resource, http://www.gov.ab.ca/env/water/reports/index.html

Atomic Energy of Canada Limited. (2000). Annual Report 1999-2000. Supply & Services Canada, Canada

Bérubé G (2000) New co-operation dawning for public and private sectors. Forces 126: 73-78

Boyd D (1997) Saguenay Storm applied to the Grand River Watershed, http://www.cwra.org/hydrology/arts/dwight.html

Brown DW, Moin SMA, Nicolson ML (1997) A Comparison of Flooding in Michigan and Ontario: 'Soft' Data to support 'Soft' Water Management Approaches. Canadian Water Resources Journal 22:125-139

Bruce JP (1976) National Flood Damage Reduction Program. Canadian Water Resources Journal 1: 5-14

Bruce J, Mitchell B (1995) Broadening Perspectives on Water Issues. Royal Society of Canada, Ottawa

Bureau d'audiences publiques sur l'environnement (2000) L'eau, ressource à protéger, à partager et à mettre en valeur. Bureau d'audiences publiques sur l'environnement, Québec

Canadian Council of Forest Ministers (1995) Canadian Council of Forest Ministers. 1995. Canadian Council of Forest Ministers, Ottawa

Canadian Council of Ministers of the Environment (1999) Canadian Environmental Quality Guidelines, http://www.ccme.ca/ceqg_rcqe/

Canadian Council of Resource and Environment Ministers (1987) Canadian Water Quality Guidelines, http://www.ccme.ca/ceqg_rcqe/

Canadian Electricity Association (1999) 1999 Industry Annual Report. Canadian Electricity Association, Ottawa

CNC ICLD (1991) Canadian National Committee of the International Commission on Large Dams. CNC ICLD, Canada

Canadian Pulp and Paper Association (1995) Reference Tables: Economics and Statistics Section. Canadian Pulp and Paper Association, Montreal

Cardy WFG (1976) Flood Management in New Brunswick. Canadian Water Resources Journal 1: 40-46.

Chambers PA, Anderson C, Bernard LJ, Gregorich B, McConkey PH, Milburn J, Painchaud NK, Patni RR, Simard, van Vliet LJP (2000) Surface Water Quality. In: Coote DR, Gregorich LJ (eds) The health of our water - toward sustainable agriculture in Canada. Research Planning and Coordination Directorate, Research Branch, Agriculture and Agri-Food Canada, Ottawa, pp 43-60

Chambers PA, Allard M, Walker SL, Marsalek J, Lawrence J, Servos M, Busnarda J, Munger KS, Adare K, Jefferson C, Kent RA, Wong MP (1997) Impacts of Municipal Wastewater Effluents on Canadian Waters: a Review. Water Quality Research Journal of Canada 32: 659-713

Coalition pour le renouvellement des infrastructures du Québec (2000) Un projet de société pour le Québec de l'an 2000, http://www.bape.gouv.qc.ca

Comfort L, Wisner B, Cutter S, Pulwarty R, Hewitt K, Oliver-Smith A, Wiener J, Fordham M, Peacock W, Krimgold F (1999) Reframing Disaster Policy: The Global Evolution of Vulnerable Communities. Environmental Hazards 1: 39-44

Communauté urbaine de Québec (1999) Mémoire présenté par la Communauté urbaine de Québec (CUQ) en septembre 1999 lors de l'audience publique du BAPE sur la Gestion de l'eau au Québec. Bureau d'audience publique sur l'environnement, Québec

Connor R (1999) North America's Freshwater Resources: Emerging Trends and Issues. Commission for Environmental Co-operation, Montreal

Coote DR, Gregorich LJ (eds) (2000) The health of our water - toward sustainable agriculture in Canada. Research Planning and Coordination Directorate, Research Branch, Agriculture and Agri-Food Canada, Ottawa

Cuthbertson EH, Senyshyn L, Koppen-Train S (1995) Milking Centre Waste Management in Ontario. Can. Soc. Agric. Engr., Ontario

Day JC, Quinn F (1992) Water diversion and export: learning from Canadian experience. University of Waterloo, Ontario

Day JC (1999) Adjusting to Floods in the Lower Fraser Basin, British Columbia: Toward an Integrated Approach?. Environments 27: 49-66

Department of Fisheries and Oceans (1993b) Canadian fishing industry overview. DFC, Ottawa

Department of Fisheries and Oceans (1993c) Wild/farmed salmonid interactions: review of potential impacts and recommended actions. DFC, Ottawa

Department of Fisheries and Oceans. (1994a). Canadian fisheries statistical highlights 1992. DFO, Ottawa

Department of Fisheries and Oceans (1994d) Annual summary of commercial freshwater fish harvesting activities 1992: central and Arctic region freshwater fisheries. DFO, Winnipeg

Department of Fisheries and Oceans (1995b) Federal aquaculture development strategy. DFO, Ottawa

Department of Fisheries and Oceans (1995c) Canadian fisheries statistical highlights 1993. DFO, Ottawa

Department of Indian Affairs and Northern Development (1993) Basic departmental data – 1993. DIAND, Ottawa

Doern GB, Conway T (1994) The Greening of Canada: Federal Institutions and Decisions.: University of Toronto Press, Toronto

Eaton PB, Gray AG, Johnson PW, Hundert E (1994) State of the environment in the Atlantic region. Dartmouth

Emergency Preparedness Canada (1997) Emergency Preparedness Canada, http://epc-pcc.gc.ca/pub/factsheets/en_epc.html

Emergency Preparedness Canada (1998a) A National Mitigation Policy, http://www.epc-pcc.gc.ca/hottopics/what_hot/old_mitiga.html

Emergency Preparedness Canada (1998b) The First Nation Experience, http://www.epc-pcc.gc.ca/epfn/the_first_nation_experience.html

Emergency Preparedness Canada (1999) Federal Disaster Assistance Arrangements (DFAA), http://www:epc-pcc.gc.ca/publicinfo/fact_sheets/disas_fin_ass.html

Emergency Preparedness Canada (2000) Federal Payments under DFAA for Floods occurring Since 1970. Environment Canada, Canada

Environment Canada (1993a) Water works, http://www2.ec.gc.ca/water/index.htm

Environment Canada (1993b) Water use in Canada, 1991, http://www2.ec.gc.ca/water/index.htm

Environment Canada (1993c) Wetlands A Celebration of Life, http://www.atlas.gc.ca/english/quick_maps/index_issues.htm

Environment Canada (1994) Urban water: municipal water use and wastewater treatment. Environment Canada, Ottawa

Environment Canada (1995) Climate change impacts: an Ontario perspective. Environment Canada, Ottawa

Environment Canada (1996a) The State of Canada's Environment – 1996, http://www.ec.gc.ca/soer-ree/

Environment Canada (1996b) The Canadian Climate and Water Information and Data Site, http://www.cmc.ec.gc.ca/climate/

Environment Canada (1998a) Canada and Freshwater: Experience and Practices, http://www2.ec.gc.ca/agenda21/98/Default.htm

Environment Canada (1998b) Municipal water use and wastewater treatment, http://www.ec.gc.ca/Ind/English/home/default1.htm

Environment Canada (1999a) The Canada Water Act Annual Report 1998-1999. Minister of Public Works and Government Services, Ottawa

Environment Canada (1999b) Acid Rain, http://www.ec.gc.ca/ind/English/AcidRain/Bulletin/arind6_e.cfm

Environment Canada (1999c) Canadian Environmental Quality Guidelines, http://www.ec.gc.ca/ceqg-rcqe/water.htm

Environment Canada (2000) The Water Survey of Canada, http://www.msc-smc.ec.gc.ca/index_e.cfm

Epp D. Haque CE, Peters B (1998) Emergency Preparedness and First Nation Communities in Manitoba, http://www.epc-pcc.gc.ca/epfn/the_first_nation_experience.html

Fairchild GL, Barry DAJ, Goss MJ, Hamill AS, Lafrance P, Milburn PH, Simard RR, Zebarth BJ (2000) Groundwater Quality. In: Coote DR, Gregorich LJ (eds) The health of our water—toward sustainable agriculture in Canada. Research Planning and Coordination Directorate, Research Branch, Agriculture and Agri-Food Canada, Ottawa, pp 61-73

Federation of Canadian Municipalities (1996) Report of the State of Municipal Infrastructure in Canada. FCM, Canada

Forget S, Robert B, Rouselle J (1999) The Effectiveness of Flood Damage Reduction Measures in the Montreal Region. University of Toronto, Toronto

Francis GR, Regier HA (1995) Barriers and bridges to the restoration of the Great Lakes Basin ecosystem. In: Holling CS, Light SS (eds) Barriers and bridges to the renewal of ecosystems and institutions. Columbia University Press, New York, pp 239–291

Gouvernement du Québec (1999) La gestion de l'eau au Québec - Document de consultation publique. Ministère de l'environnement, Québec

Government of Canada (1991) The state of Canada's environment. Government of Canada, Canada

Government of Manitoba (1997) State of the Environment Report for Manitoba, 1997, http://www.gov.mb.ca/environ/pages/soe97/soe97.html

Government of Saskatchewan (1999) State of the Environment Report for Saskatchewan, http://www.serm.gov.sk.ca/publications.php3

Governments of Canada and the United States (1995) State of the Great Lakes 1995, http://www.on.ec.gc.ca/glimr/data/sogl-final-report/intro.html

Handmer JW (1980) Flood Hazard Maps as Public Information: An Assessment within the Context of the Canadian Flood Damage Reduction Program. Canadian Water Resources Journal 5: 113-122

Handmer JW (1996) Policy Design and Local Attributes for Flood Hazard Management. Journal of Contingencies and Crisis Management 4: 189-197

Handmer JW, Parker DJ (1992) Hazard Management in Britain: Another Disastrous Decade?. Area 24: 113-122

Haque CE (2000) Risk Assessment, Emergency Preparedness and Response to Hazards: The Case of the 1997 Red River Flood, Canada. Natural Hazards 21: 225-245

Harker DB, Chambers PA, Crowe AS, Fairchild GL, Kienholz E (2000) Understanding Water Quality. In: Coote DR, Gregorich LJ (eds) The health of our water—toward sustainable agriculture in Canada. Research Planning and Coordination Directorate, Research Branch, Agriculture and Agri-Food Canada, Ottawa, pp 27-42

Harrison K (1996) Passing the Buck: Federalism and Canadian Environmental Policy. UBC Press, Vancouver

Health Canada (1996) Guidelines for Canadian Drinking Water Quality, http://www.hc-sc.gc.ca/ehp/ehd/catalogue/bch_pubs/dwsixth.htm

Health Canada (1997) Health and Environment - Partners for Life, http://www.hc-sc.gc.ca/ehp/ehd/catalogue/general/97ehd215.htm

Health and Welfare Canada (1993) Well Water Quality in Ontario. Cardio-Respiratory Diseases. 1: 7

Hofmann N, Mortsch L, Donner S, Duncan D, Kreutzwiser R, Kulshreshtha S, Piggott A, Schellenberg S, Schertzerand B, Slivitzky M (1998) Climate Change and Variability: Impacts on Canadian Water. In : Koshida G, Avis W (eds) Canada Country Study Volume VII-National Sectoral Volume. Environment Canada, Toronto

Hunt C (1999) A Twenty-First Century Approach to Managing Floods. Environments 27: 97-114.

INRS-EAU (1998) Actes du symposium sur la gestion de l'eau au Québec Palais des Congrès de Montréal 10-11-12 décembre 1997. Édité par : Jean-Pierre Villeneuve, Alain N. Rousseau et Sophie Duchesne. Publié par l'INRS-Eau

INRS-Eau et INRS-Urbanisation (1998) Synthèse des rapports INRS-Urbanisation et INRS-Eau sur les besoins des municipalités québécoises en réfection et construction d'infrastructures d'eaux, Rapport de recherche No R-517, Sainte-Foy, février

Insurance Bureau of Canada (1999) A National Mitigation Strategy, http//www.ibc.ca/English/articles/nat_mit_strategy.htm

Insurance Council of Canada (1998) Fact Book, http://www.ibc.ca/English/facts/factpageb.htm

International Energy Association (2000) Hydropower and the Environment: Present context and guidelines for future action. Technical Report, Paris

International Joint Commission (1997) Red River Flooding Short-Term Measures, http://www.ijc.org/boards/rrb/taskforce.html

International Joint Commission (2000a) Living with the Red, http://www.ijc.org

International Joint Commission (2000b) Focus, http://www.ijc.org/focus/newsletter.html

International Nuclear Safety Center (1999) Maps of nuclear reactors: Canada. International Nuclear Safety Center, USA

Kienholz E, Croteau F, Fairchild GL, Guzzwell GK, Massé DI, Van der Gulik TW (2000) Water Use. In: Coote DR, Gregorich LJ (eds) The health of our water—toward sustainable agriculture in Canada. Research Planning and Coordination Directorate, Research Branch, Agriculture and Agri-Food Canada, Ottawa, pp 15-25

Koshida G, Avis W (1998) Canada Country Study Volume VII-National Sectoral Volume. In: Koshida G, Avis W (eds) Canada Country Study Volume VII-National Sectoral Volume. Environment Canada, Toronto

Kreutzwiser R, Woodley I, Shrubsole D (1994) Perceptions of Flood Hazard and Development Regulations in Glen Williams, Ontario. Canadian Water Resources Journal, 19: 115-124

Kuban R (1996) The Role of Government in Emergency Preparedness. Canadian Public Administration, 39: 239-244

Lafrance G (2000). Forces, 126: 73-78

Leith RM (1991) Patterns in snowcourse and annual mean flow data in British Columbia and the Yukon. In: National Hydrology Research Institute (eds) Hydrometric data to detect and monitor climate change. NHRI, Saskatoon, pp 225–231

Liebscher H, Trew DO, Bowen GS, Painchaud J, Mutch JP (2000) Provincial and local programs to monitor water quality in Canada. In: Coote DR, Gregorich LJ (eds) The health of our water—toward sustainable agriculture in Canada. Research Planning and Coordination Directorate, Research Branch, Agriculture and Agri-Food Canada, Ottawa, p 33

Linton J (1997) Beneath the Surface : The state of water in Canada. Canadian Wildlife Federation, Canada

Lucotte M. Schetagne R, Thérien N, Langlois C, Tremblay A (1999) Mercury in Biogeochemical Cycle: Natural Environments and Hydroelectric Reservoirs of Northern Québec. Springer, Berlin Heidelberg New York

Marsalek and Kok (2000) Urban Stormwater Management for Ecosystem Protection. Water Quality Research Journal of Canada 35: 313-314

Ministry of Environment, Lands and Parks of British Columbia and Environment Canada. (1996) British Columbia Water Quality Status Report, http://www.elp.gov.bc.ca/wat/wq/public/bcwqsr/bcwqsr1.html

Ministry of Environment, Lands and Parks of British Columbia and Environment Canada. (2000) Water quality trends in selected British Columbia waterbodies, http://www.env.gov.bc.ca/wat/wq/wqhome.html

Morrison IN, Kraft DF (1994) Sustainability of Canada's agri-food system: a Prairies perspective. International Institute for Sustainable Development, Winnipeg

Morris-Oswald T, Simonovic SP, Sinclair J (1998) Case Study: Efforts in Flood Damage Reduction in the Red River Basin-Practical Considerations, http://www.epc-pcc.gc.ca/research/scie_tech/ef_flooding/casestudy/index.html.

National Energy Board (1999a) Canadian Energy Supply and Demand to 2025. National Energy Board, Calgary, Alberta

National Energy Board (1999b) 1999 Annual Report. National Energy Board, Calgary, Alberta

Natural Hazards Centre and the Disaster Research Institute (1999) An Assessment of Recovery Assistance Provided after the 1997 Floods in the Red River Basin: Impacts on Basin-wide Resilience. Environment Canada, Toronto

Natural Resources Canada (1997) Electricity generation in Canada's Energy Outlook 1996-2020. Environment Canada, Toronto

Natural   Resources   Canada   (2000a)   National   Atlas   of   Canada, http://atlas.gc.ca/english/index.html.

Natural Resources Canada (2000b) Uranium, nuclear energy & wastes management. National Resources Canada, Canada

Natural Resources Canada (2000c). National Atlas of Canada: Canadian Issues and Themes Wetlands, http://www.atlas.gc.ca/english/index.html

Newman MC (1998) Landscape to global effects in Fundamentals of Ecotoxicolgy. Sleeping Bear, Chelsea, MI

Ontario Ministry of Agriculture, Food and Rural Affairs (1994) Water Management. Guelph, Ontario

Ontario Ministry of Environment and Energy (1994) Clean water regulations proposed for the iron and steel manufacturing and the electric power generation sectors. Minister of Environment and Energy, Toronto

Pears PH (1988) Rising the to the Challenge: A New Policy for Canada's Freshwater Fisheries. Canadian Wildlife Federation, Ottawa

Phillips D (1990) The Climates of Canada, http://www.ns.ec.gc.ca/climate/cofc.html

Platt RH (1999) From Changing Control to Flood Insurance: Changing Approaches to Floods in the United States. Environments, 27: 67-78.

Rahman MM (1998) Roseau River Anishinabe First Nations, http://www.epc-pcc.gc.ca/epfn/the_first_nation_experience.html

Ricciardi A., Rasmussen JB (1999) Extinction rates of North American freshwater fauna. Conservation Biology 13: 1220-1222.

Rochfort Q, Grapentine L, Marsalek J, Brownlee B, Reyoldson T, Thompson S, Milani D, Logan C (2000). Using Benthic Assessement techniques to determine Combined Sewer Overflows and Stormwater Impacts in the Aquatic Ecosystem. Water Quality Research Journal of Canada, 35: 365-397.

Rogers P (1994) Assessing the Socioeconomic Consequences of Climate Change on Water Resources. Climatic Change, 28:179-208

Roy E, Rouselle J, Lacroix J (1997) Flood Damage Reduction Program in Quebec: Case Study   of   the   Chaudiere   River.   http://www.epc-pcc.gc.ca/research/scie_tech/ef_floding/fdrp_chaud/index.html

Sandra K, Shaw J, Seto P, Weatherbe D (2000) The Urban Drainage Program of Canada's Great Lakes 2000 Cleanup Fund. Water Quality Research Journal of Canada 35: 315-330

Shrubsole D, Hammond VJ, Green M (1995) Floodplain Regulation in London, Ontario, Canada: Assessing the Implementation of Section 28 of the Conservation Authorities Act. Environmental Management, 19: 703-717

Shrubsole D, Scherer J (1996) Floodplain Regulations and the Perceptions of the Real Estate Sector: The Cases of Brantford and Cambridge, Ontario, Canada. Geoforum. 27: 509-525

Shrubsole D, Hammond VJ, Green M (1996) Floodplain Regulation in London, Ontario: Status and Prospects. Canadian Water Resources Journal, 21: 367-386

Shrubsole, D, Hammond VJ, Kreutzwiser R, Woodley I (1997) Assessing Floodplain Regulation in Glen Williams, Ontario, Canada. Journal of Environmental Management, 50: 301-320

Sport Fish Contaminant Monitoring Program (1999) 1999-2000 Guide to eating Ontario sport fish. Ministry of the Environment, Etobicoke, Canada

Statistics Canada (1994) Human activity and the environment. National Accounts and Environment Division, Ottawa

Statistics   Canada   (1996)   Census   of   Agriculture, http://www.statcan.ca/english/censusag/tables.htm

Statistics Canada (1997) Historical Overview of Canadian Agriculture (data and analytical products : 1996 Census of Agriculture). National Accounts and Environment Division, Ottawa

Statistics Canada (2000) Canada at a glance, http://www.statcan.ca

Stewart JE (1994) Aquaculture in Atlantic Canada and the research requirements related to environmental interactions with finfish culture. In: Ervik A, Kupka Hansen P, Wennevik V (eds) Canada–Norway workshop on environmental impacts of aquaculture. Institute of Marine Research, Havbruk, Norway, pp 1–18

Thonney JP, Garnier E (1992) Bay of Fundy salmon aquaculture monitoring program 1991–1992. Department of the Environment, New Brunswick

Topp E (2000) Endocrine-disrupting chemicals. In Coote DR, Gregorich LJ (eds) The health of our water—toward sustainable agriculture in Canada. Research Planning and Coordination Directorate, Research Branch, Agriculture and Agri-Food Canada, Ottawa, pp 38

Trépanier M (1998) Les infrastructures d'eau dans les municipalités québécoises : Évaluation de leur état et des coûts de réfection. In: Jean-Pierre Villeneuve, Alain N. Rousseau et Sophie Duchesne. Publié l'INRS-Eau (eds) Symposium sur la gestion de l'eau. Montréal

Union des Municipalités du Québec (1994) L'état actuel et les nouveaux modes de gestion et de financement des infrastructures municipales. Union des Municipalités du Québec, Québec

Watt WE (1995) The National Flood Damage Reduction Program: 1976-1995. Canadian Water Resources Journal, 20: 237-247.

Whitlock J (2000) The Canadian Nuclear FAQ, http://www.freenet.carleton.ca

World Energy Council (1998a) International Energy Data Report. World Energy Council, UK

World Energy Council (1998b) Survey of Energy Resources. World Energy Council, UK

Woo MK (1993) Northern hydrology. In: French HM, Slaymaker O (eds) Canada's cold environments. McGill-Queen's University Press, Canada, pp 117–142

Yanful EK, Simms P (1997) Review on water cover sites and research projects. MEND, Canada

15. Physical Context

The United States of America (U.SA.) is located in the central portion of the North American continent, between 25 and 48 degrees north latitude, and 68 to 125 degrees west longitude (Figure 15.1). The total area of the United States is 9,629,091 square kilometers, 470,131 square kilometers of which are water. The nation contains 50 states, several associated territories, the District of Columbia, and approximately 275 Native American land areas administered as Indian reservations. Forty-eight contiguous states comprise the mainland. Non-contiguous states include the continental state of Alaska to the north and the islands of Hawaii in the South Pacific (Figure 15.2).

The mainland is bordered by Canada to the north, Mexico to the south, the Pacific Ocean to the west, and the Atlantic Ocean to the east. The border with Canada is 8,893 kilometers long, including 2,477 kilometers between Alaska and Canada. The border with Mexico is 3,326 kilometers long.

Fig. 15.1 The U.S.A. is located in the western hemisphere. Forty-eight conterminous states comprise the mainland. The state of Alaska lies to the northwest and the state of Hawaii lies in the Pacific Ocean to the southwest.

Fig. 15.2. The 50 states of the United States of America.

The United States has kilometers of coastline, and claims a territorial strip extending 12 nautical miles and an exclusive economic zone extending 200 nautical miles (CIA, 2000). The country exhibits a broad diversity of landforms, climates, and ecological regions. The topography varies from mountainous regions in the western third of the country, to a large central plain, to lower elevation mountains and Piedmont terrain in the eastern portion (Figure 15.3). Rugged mountain ranges and broad river valleys characterize the northern state of Alaska, and the Hawaiian Islands are composed of mountainous volcanic terrain. The western continental U.S. has the highest degree of relief, with elevations ranging from below sea level to above 4,000 meters in many areas. The overall elevation of the country ranges from 86 meters below sea level at Death Valley, California, to 6,194 meters above sea level at Mount McKinley, Alaska.

15.1 Landforms

The mainland of the U.S. can be divided into 24 physiographic provinces, as shown in Figure 15.4. In the eastern part of the country are the Appalachian Mountains, Piedmont, and plateaus (regions 1, 2, 4, 14, 23, and 19). A broad coastal plain lies along the Atlantic Coast and the Gulf of Mexico (region 7).

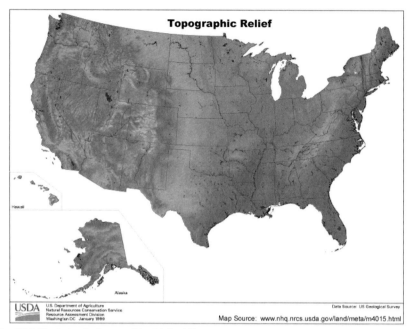

Fig. 15.3 Shaded relief map of the United States.

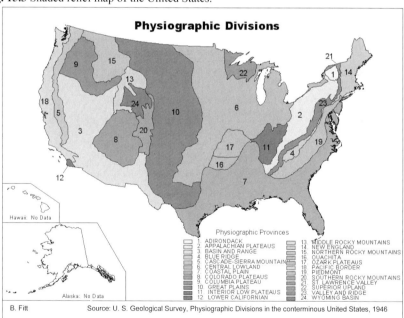

Fig. 15.4. USGS divides the conterminous U.S. into 24 physiographic provinces. These are defined by physical features such as landforms, elevation, and geology

The central U.S. is composed of the Great Plains (region 10), interior lowlands (regions 6 and 11), and some isolated upland areas (regions 22, 16 and 17). The western United States contains the Rocky Mountains, high upland plateaus (regions 8, 9, 13, 15, and 20), and the Great Basin, a large region characterized by a series of mountain ranges and closed basins (region 3). The Pacific Border province (region 18) is located along the Pacific Coast, to the west of the high mountains that comprise the Cascade and Sierra ranges of the Pacific Cordillera (region 5).

15.2 Waterways and Aquifers

The United States contains 5.8 million kilometers of intermittent and perennial rivers and streams, 41,009 kilometers of which are designated as navigable inland channels, exclusive of the Great Lakes. Major river systems are shown in Figure 15.5.

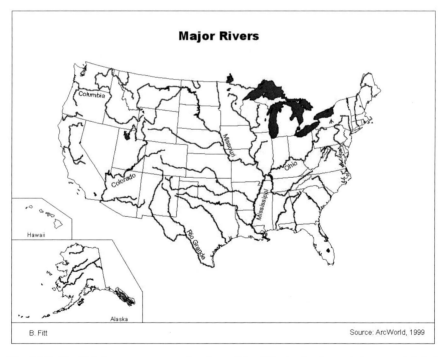

Fig. 15.5. The U.S. has numerous major rivers. Six of the twelve largest rivers are labeled here.

The eastern seaboard contains a number of large rivers that drain the eastern slopes of the Appalachian Mountains and empty into the Atlantic. These rivers include the Penobscot, Connecticut, Hudson, Susquehanna, Potomac, Pee Dee, Savannah, and St. Johns. The largest river in the country is the Mississippi, which empties into the Gulf of Mexico. The Mississippi River drains water from the central portion of the country. It is fed by the Ohio River, which drains the western slope of the Appalachian Mountains, and the Missouri and Arkansas River systems, which drain the eastern slope of the Rocky Mountains. Other major rivers that drain to the Gulf of Mexico include the Chattahoochee, Alabama, and Brazos. The Rio Grande also enters the Gulf of Mexico, draining the southern Rocky Mountains and forming the border between Texas and Mexico. The Colorado River system originates on the western slope of the Rocky Mountains and drains to the Gulf of California. The Columbia River originates in Canada, drains the northern Rocky Mountains and enters the Pacific Ocean in the Pacific Northwest.

Principal aquifer systems in the United States are shown in Figure 15.6. Major aquifers in the eastern portion of the country occur along the Atlantic and southeastern coastal plains (aquifers 12, 17, and 18), and along the western slope of the Appalachian Mountains (aquifers 14 and 20).

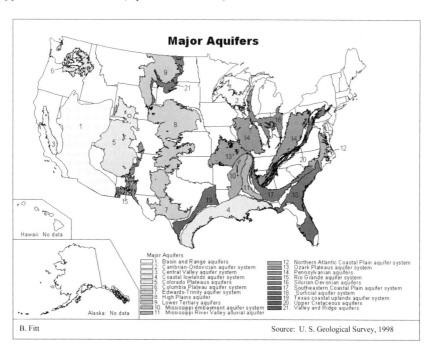

Fig. 15.6. Principal aquifers of the United States. The U.S. Geological Survey has identified more aquifers than are pictured here, but only the 21 largest or most spatially contiguous are shown

A series of aquifers occurs throughout the Midwest (aquifers 2, 13, 14, and 16) and in the lower portions of the Mississippi Valley (aquifers 10 and 11). Several large aquifers are located along the Texas Coastal (aquifer 19) and Coastal Lowlands (aquifer 4). Beneath the Great Plains lies an extensive system of aquifers stretching from Canada to the Mexican border (aquifers 7, 8, 9, and 21). The Colorado Plateau contains the principal aquifers in the Rocky Mountain region (aquifer 5). Other important aquifers are the Columbia Plateau system (aquifer 6) in the Northwest, the Central Valley aquifer system (aquifer 3) in the agricultural region of California, and the Basin and Range aquifers (aquifer 1) occurring throughout most of Nevada and southern Arizona.

15.3 Climate and Precipitation

Climate in the United States is mostly temperate, but is tropical in Florida and Hawaii, and is arctic in Alaska. The mean annual temperature in the United States is controlled by both latitude and topography (Figure 15.7). Annual mean temperatures in the contiguous 48 states range from less than 0°C in high alpine regions to approximately 25°C in areas of the desert Southwest. The climate is humid in areas east of the Mississippi River and in the Pacific Northwest, semi-arid in the Great Plains and Rocky Mountains in the West, and arid in the Great Basin and other areas of the Southwest (Figure 15.8).

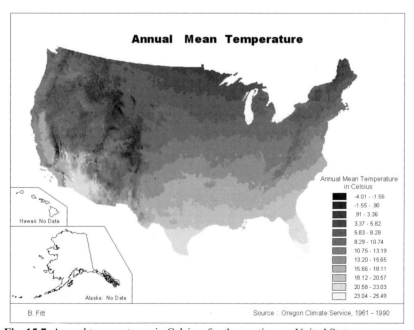

Fig. 15.7. Annual temperatures, in Celsius, for the contiguous United States.

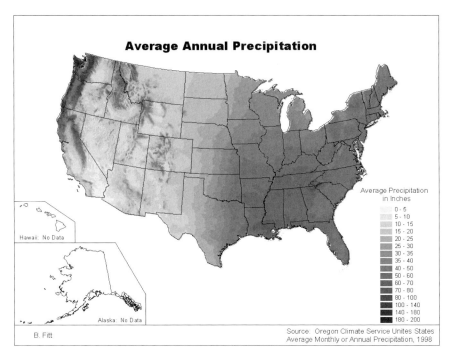

Fig. 15.8. Average annual precipitation, in inches, for the contiguous United States.

The highest annual precipitation occurs in the mountainous regions of the Pacific Northwest, and is strongly correlated with elevation in most of the western regions. East of the Rocky Mountains, annual precipitation generally increases to the east, as shown in Figure 15.8.

Average annual runoff, or stream discharge from small basins is closely related to average annual precipitation; runoff is generally higher in the Pacific Northwest and in mountainous areas (Figure 15.9). Very little runoff is present in the Great Basin and in smaller inland basins in the West. The annual runoff values illustrate the relative abundance of available water in mountainous and eastern regions, and the lack of water in the interior western basins and Great Plains. The amount of snow that accumulates at high elevations each winter (snowpack) mainly determines the amount of annual runoff in western regions, whereas the amount of rain and snow throughout the year largely controls runoff in eastern regions.

The distribution of precipitation and the potential rate of surface evaporation, called evaporative demand, control the susceptibility of regions to droughts (Figure 15.10). The population in the humid region east of the Mississippi River primarily relies on rivers and streams for water supply; therefore, this region is vulnerable mainly to short droughts.

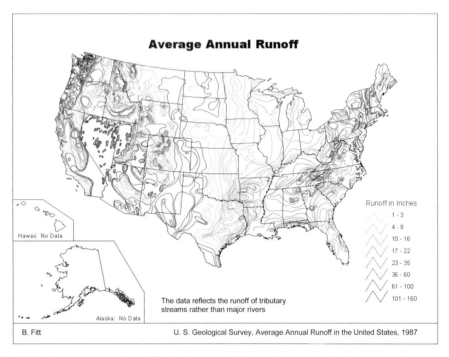

Fig. 15.9. Average annual runoff of tributary or low-order streams in the conterminous United States.

In the West, groundwater and water stored in reservoirs comprise a greater proportion of water resources; therefore, the arid regions west of the Great Plains are vulnerable to droughts lasting several years in duration. A marginal region, chiefly comprising the Great Plains, is vulnerable to droughts of both short and long duration. Finally, several humid mountainous areas in the Pacific Northwest and in the East maintain water surpluses, even during periods of less-than-average precipitation (Moreland, 1993).

Over the past 100 years, 0% to 45% of the land area in the United States has been affected by severe to extreme drought conditions each year. During this same 100-year-period, 0% to 35% of the land area has also been affected by severe to extreme wetness. Drought and wetness cycles range from approximately 2 to 6 years and are generally negatively correlated (Figure 15.11).

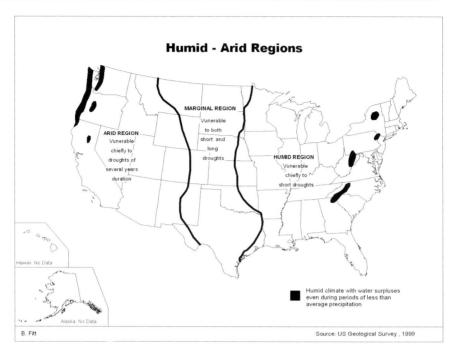

Fig. 15.10. The conterminous United States ranges from humid to arid from East to West. Elevation changes create more complex local conditions. For example, the high peaks of the Sierra Nevada in northern California capture and retain winter precipitation in the form of snow and snowpack.

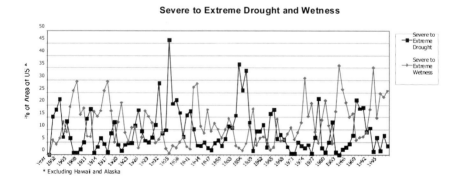

Fig. 15.11. Percentage of land in the conterminous United States affected by severe to extreme drought and wetness for most of the 20th century, as measured by the Palmer Drought Severity Index. The index is based on a hydrologic accounting model of water supply and demand. (Source: Source: US Department of Commerce, National Oceanic and Atmospheric Administration, National Climatic Data Center, 1997.)

15.4 Land Use and Cover

Arable land comprises 19% of the total land area in the U.S., and 207,000 square kilometers of land are under irrigation (CIA, 2000). Permanent pastures comprise 25% of the total land surface, forests and woodlands make up 30%, and other categories comprise the remaining 26%.

Dominant land covers in the eastern United States are forests and woodlands interspersed with croplands and urban areas (Figure 15.12). Many coastal areas along the southern Atlantic Ocean and Gulf of Mexico are mainly composed of wetlands. The Midwestern portion of the country is primarily cropland, with forests and woodlands in the south and semi-arid grazing lands to the west. The western United States is composed of a diverse mix of land covers, including croplands in areas of low relief, forests in mountainous regions, and large expanses of desert. The distribution of land cover is a particularly important parameter for the analysis of water resource issues because water quality and demand are strongly affected by land use and hydrology.

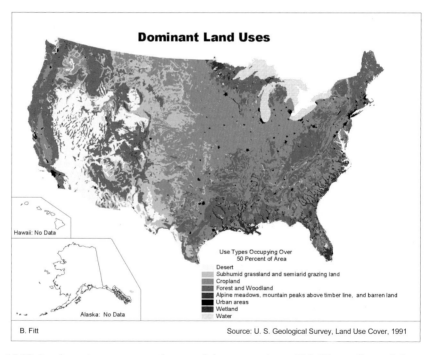

Fig. 15.12. Land-use types vary greatly around the conterminous U.S. The quality and timing of runoff are affected by type and intensity of land use. Thus, while urban areas may comprise a relatively small area of land use, they have a significant effect upon their local drainage basins.

15.5 Land Ownership

The distribution of public and private land ownership throughout the United States also is important to the analysis of water resources because land ownership affects land management (Figure 15.13). For example, although federal, state, or local laws may regulate private lands, the individual landowners largely control the dominant uses of those lands.

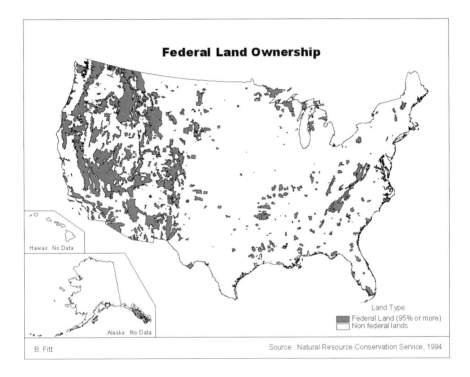

Fig. 15.13. Federally owned (public) lands throughout the conterminous United States. Public lands comprise a large proportion of many western states.

Fig. 15.14. Tribal lands in the conterminous United States (Alaska and Hawaii not included). There are approximately 275 reservations and more than 550 federally recognized tribes. (Source: Mapped by L. Porter, 5/98. USGS Digital Line Graph-Administrative Boundaries. http://www.epa.gov/ceisweb1/ceishome/atlas/nationalatlas/triballands.htm.)

Federal lands are generally managed for maximum public benefit, which often means managing for many competing uses. These multiple uses may include resource exploitation, ecological protection, recreational interests, and aesthetic values. There are large tracts of public lands in the western United States composed predominantly of range, forests, mountains, and wilderness. These lands are managed by federal agencies, such as the National Park Service and the Bureau of Land Management, both of which are bureaus within the U.S. Department of the Interior, and the USDA Forest Service, which oversees the National Forests and Grasslands.

The eastern part of the country contains a smaller proportion of federal lands, which are mainly held along the crest of the Appalachian Mountains. Privately held agricultural lands make up much of the central portion of the United States. In addition, approximately 56.2million acres are held in trust by the federal government for Indian tribes and individuals (Figure 15.14). Most Indian reservations are in the western United States; they vary greatly in size from less than 100 acres to about 16 million acres (see Figure 15.14).

16. Socioeconomic Aspects

The population of the United States is approximately 272.6 million people (CIA, 2000), and is composed of 51% females and 49% males (Figure 16.1). Population has grown steadily over the past century (Figure 16.2, Table 16.1), at annual rates ranging from 0.95% to 2.10%. Currently the rate of population growth has slowed to an annual rate of 0.85%. The birth rate is 14.3 births per 1000 people, the death rate is 8.8 deaths per 1,000 people, and the migration rate is 3.0 migrants per 1,000 people. The total fertility rate is 2.07 children born per woman. The infant mortality rate is 6.33 deaths per 1,000 live births.

Life expectancy has increased over the past century for both sexes (Figure 16.3), and is currently 72.95 years for males, and 79.67 years for females, with a life expectancy for the total population of 76.23 years (CIA, 2000). The U.S. Census Bureau projects high, middle, and low levels of population increase based on assumptions regarding life expectancy, fertility, and immigration. The middle series of population projections indicates that the U.S. population will increase 41%, from 263 million in 1995 to 370 million in 2040. The low and high projections indicate that the population may increase by as little as 9% and by as much as 74% between 1995 and 2040 (US Census Bureau, 1996).

Population by Gender

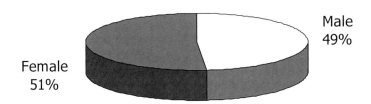

Fig. 16.1. U.S. Population by gender. (Source: US Census Bureau, *Census of Population and Housing,* 1990.)

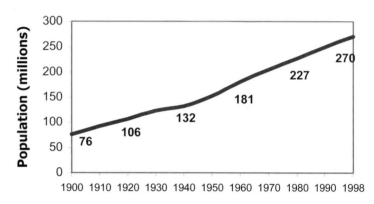

Fig. 16.2. Population growth steadily increased over the past century, but has slowed to an annual growth rate of 0.85%. (Source: Population Estimates Program, Population Division, US Census Bureau, Internet Release Date: June 4, 1999.)

Table 16.1. Growth rate of U.S. population since 1900.

Decade	% Change
1900	(no data)
1910	21.4
1920	15.2
1930	15.6
1940	7.3
1950	15.3
1960	18.7
1970	13.5
1980	10.8
1990	9.8
1998	8.4

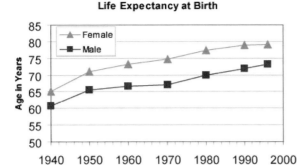

Fig. 16.3. Life expectancy has increased for both men and women over the past century. (Source: National Center for Health Statistics, *Deaths: Final Data for 1996*, 1998.)

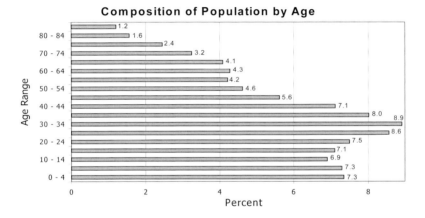

Fig. 16. 4. Percent of the U.S. population by different age groups. (Source: US Census Bureau, *Census of Population and Housing*, 1990.)

The composition of the U.S. population by age is shown in Figure 16.4. Twenty-two percent of the population is below 14 years of age, 66% is in the 15–64 years range, and 12% is above 65 years of age. The largest population group is in the 25–34 years range, comprising 17.5 % of the population.

The number of persons per household is relatively low in the United States. Over 56% of the population lives in households with two or fewer persons, 33% of the households in the U.S. have from three to four members, and 11% have five or more people (Figure 16.5). The distribution of number of persons per household is an important determinant of domestic water use. A certain volume of water is necessary to support the infrastructure of each household (e.g., lawn and garden watering and cleaning), independent of the number of residents.

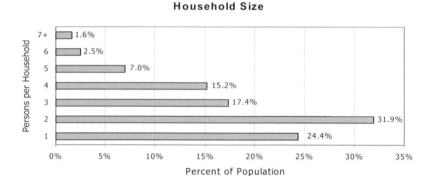

Household Size

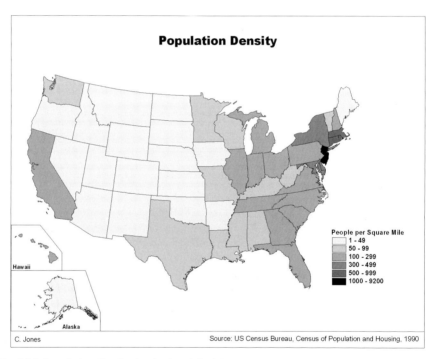

Fig. 16.5. More than 56% of the U.S. population lives in households with two or fewer persons. (Source: US Census Bureau, *Census of Population and Housing*, 1990.)

Fig. 16.6. Population distribution in the United States.

The U.S. population is unevenly distributed, with the highest population densities occurring along the Atlantic and Pacific Coasts, and along the Great Lakes (Figure 16.6). Population density is lowest in the interior states west of the Mississippi River. Similarly, most of the large U.S. cities are concentrated in the eastern part of the nation and along the Pacific Coast (Figure 16.7). Three-quarters of the

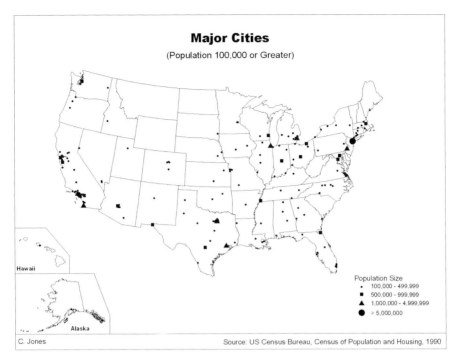

Fig. 16.7. Most major U.S. cities are concentrated along the mid-Atlantic Coast, the southern Pacific Coast, and along the Great Lakes.

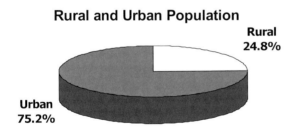

Fig. 16.8. Most residents of the United States live in urban areas. (Source: US Census Bureau, *Census of Population and Housing*, 1990.)

U.S. population is concentrated in urban areas, with the remaining quarter in rural regions (Figure 16.8). Only a small percentage of the rural population lives on farms (Figure 16.9).

Educational attainment in the United States is typically high, with 74% of the population completing secondary education (through 12 years), and 24.8% of the population receiving a college degree (Figure 16.10). Less than one-tenth (9.2%) of the population has not completed primary education, leaving school prior to the 10th year of education. The literacy rate for the U.S. population is 97%.

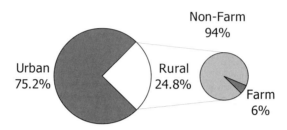

Fig. 16.9. Of the minority of Americans living in rural areas, most do not live on or operate farms. (Source: US Census Bureau, *Census of Population and Housing,* 1990.).

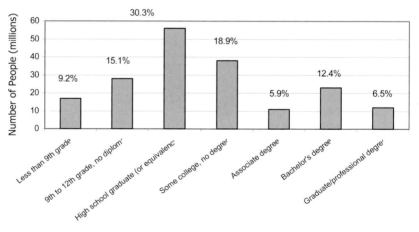

Fig. 16.10. Educational attainment in the U.S. is typically high, with 30.3% of the population completing secondary education.

Education is important to natural resource management because it will lead to more effective understanding of the complex issues associated with natural and human systems. It will also enhance possibilities for the development of effective policies and strategies to sustain limited resources.

Post-primary educational attainment in the United States is also comparatively high, with 30.3%of the population completing secondary education and 24.8% receiving a college degree.

The Economy

The U.S. currently maintains a strong economy marked by a steady growth rate of 3.9%, low unemployment rate of 4.5%, low inflation rate of 1.3%, and a high degree of technological advancement. The U.S. gross domestic product (GDP) has grown consistently, from $750 billion in 1960, to over $9 trillion in 1999 (Figure 16.11). The per capita GDP is $31,500, and is the highest among the major industrial nations. By sector, the GDP is composed of 75% services, 23% industry, and 2% agriculture. GDP consumption by income class is unequally distributed in the United States, with 28.5% going to the highest 10% of household incomes, and 1.5% going to the lowest 10%. Employment in the United States is dominated by the managerial, professional, technical, sales, and administrative support categories, which comprise 58% of all employment (Figure 16.12). One-quarter of jobs in the United States are located in the manufacturing, transportation and mining categories. Services (14%) and farming, forestry and fishing (3%) jobs comprise a minor fraction of the total employment.

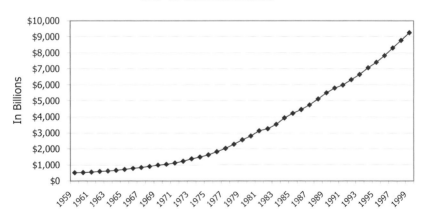

GDP in Current Dollars

Fig. 16.11. The U.S. gross domestic product (GDP) has grown consistently from $750 billion in 1960 to over $9 trillion in 1999. (Source: Bureau of Economic Analysis, National Income and Product Accounts, 2000.)

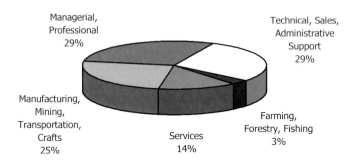

Fig. 16.12. Employment in the United States is dominated by the managerial, professional, technical, sales, and administrative support sectors.

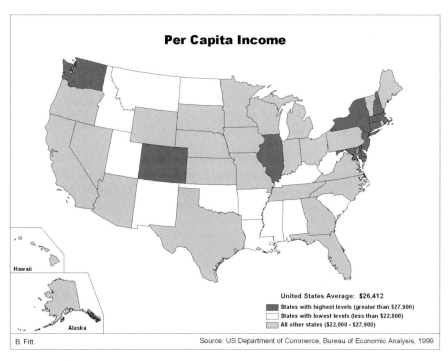

Fig. 16.13. A few states, particularly in the Northeast, exceed the annual average per capita income of $26,412. Several states, particularly in the Southeast and North, fall below $22,000 per capita. Most are near the average, ranging between $22,000 and $27,900.

In 1998, the mean per-capita income in the United States was $26,412 (BEA, 1999). The spatial distribution of per-capita income is generally correlated with areas of high urbanization, such as the northeastern, mid-Atlantic, and West Coast states (Figure 16.13). The lowest per-capita incomes occur in the South and in the more rural western and Midwestern states where populations are sparse.

Overall, 13% of U.S. residents live below the poverty level (Figure 16.14). The age distribution of individuals below poverty level (Figure 16.15) is similar to the distribution of population by age (Figure 16.4). The notable exception is that a large proportion of the elderly population exceeding 65 years of age lives below the poverty level. The highest poverty rates in the United States are in the District of Columbia, New Mexico, Louisiana, and Mississippi. Over a 3-year period (1996-1998), average poverty rates in those areas ranged from 18% to 23% of the population (U.S. Census Bureau, 1998).

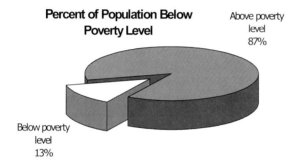

Fig. 16.14. Although most of the U.S. population lives above the poverty line, 13% falls below it. (Source: US Census Bureau, *Census of Population and Housing*, 1990.)

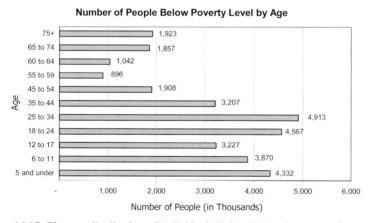

Fig. 16.15. The age distribution of individuals living below the poverty level. (Source: US Census Bureau, *Census of Population and Housing*, 1990.)

Rapid technological developments since World War II are partially responsible for the development of the diverse and powerful U.S. economy, as well as for the widening gap between economic classes.

The increase in technology has also led the development of a two-tiered labor market, where those at the bottom lack the professional and technical skills of those at the top. As a result, those at the bottom increasingly fail to obtain income increases, health insurance coverage, and other benefits, further increasing the economic disparity between the two groups. Primary economic problems include inadequate investment in economic infrastructure, increasing medical costs of the aging population, large trade deficits, and stagnation of family incomes in the lower economic groups (CIA, 2000).

17. Institutional Framework For Water Management

Water resources in the United States are managed by a number of federal, state, and local agencies, with a variety of missions and sometimes overlapping jurisdictions. Resources are frequently managed through cooperation between many federal and state agencies, or by delegation of authority by federal agencies to state entities. This chapter provides brief overviews of the legal framework governing water use and the functions of the primary federal agencies responsible for various aspects of water management in the United States.

17.1 Surface Water

Within the United States, individual states reserve the power to allocate all surface waters, lakes, rivers, and streams within their borders that are not encumbered by federal law or interstate compact. Each state has a distinct legal system for establishing water rights, changing approved uses of water rights, and administering uses. In almost all eastern states, where water is fairly abundant, laws governing surface-water use are based on the *riparian doctrine,* which is derived from English common law. In the more arid western states (Arizona, California, Colorado, Idaho, Kansas, Montana, Nebraska, Nevada, New Mexico, North Dakota, Oklahoma, Oregon, South Dakota, Texas, Utah, Washington, and Wyoming), water laws generally are based on the doctrine of *prior appropriation.* There is no distinct geographical division between areas where laws are based on one or the other doctrine, and some states employ a combination of the two.

The riparian doctrine grants the owner of land immediately adjacent to a body of water the rights to use that water, as long as use is reasonable and does not diminish the flow or impact other users. Riparian rights are inseparable from the land and may not be transferred or sold independently. And because no user has priority over others, all riparian users must reduce their water use proportionally in times of water shortage.

The principle of *'first in time, first in right'* is the foundation of the prior appropriation doctrine. In this system, the holders of water rights with the earliest date of appropriation (senior water rights) have priority over all holders of water rights appropriated at a later date (junior water rights). As a result, junior users may be denied water during shortages. Furthermore, ownership of water rights is not limited to owners of property adjacent to a body of water. The nature of the prior appropriation system can allow for the transfer of water rights, as well as for the sale of water where permissible under state law.

Under the prior appropriation system, diverting water and putting it to beneficial use can secure water rights. Beneficial use is the measure and the limit of the water right. Only that amount of water that can be put to beneficial use may be diverted. If the water is not put to beneficial use, it is subject to forfeiture for nonuse. In addition, twelve of the western states specifically rank preference of use (e.g., domestic supercedes agriculture) during periods of water shortages (Tarlock et al., 1993).

Regardless of the legal system governing water use and allocation, however, federal law takes priority over state law and individual user rights when and where local governance contradicts a federal directive. For example, the Endangered Species Act (ESA) can serve to protect in-stream water needs at the expense of senior water rights holders during times of drought.

17.2 Groundwater

The allocation of *groundwater*—underground aquifers—has evolved differently from that of surface water and varies widely between states. Where the riparian doctrine has been applied to groundwater, two variants have emerged. The *English* or *absolute use* doctrine states that a landowner may withdraw as much water as desired for any use. The only consideration of other users is that water may not be extracted with the intent of harming others. The *American* or *reasonable use* doctrine states that groundwater may be withdrawn only for reasonable uses, regardless of the needs of other users.

The prior appropriation doctrine has also been applied to groundwater resources to reduce pumping and encourage conservation. Groundwater rights are granted when it is determined that resources are available for use. Senior appropriators are guaranteed resources by requiring that junior users regulate their withdrawals to maintain the water table at a reasonable depth (Ferrey, 1997).

17.3 National Management Agencies

The Environmental Protection Agency (EPA) is a federal agency whose primary charge concerning water resources is to regulate water quality to protect human health and the natural environment. The EPA sets water-quality standards, provides technical assistance to water suppliers, and initiates legal action against suppliers who fail to meet water-quality standards. The EPA manages several programs to assess and protect both surface and underground drinking water sources, including an underground injection-control program to prevent contamination of groundwater from subterranean waste disposal. The EPA also serves to reduce contamination from industrial waste, stormwater, combined sewers, and sanitary sewers through monitoring, reduction, and permitting programs. The EPA also coordinates emergency responses to chemical spills and the cleanup of hazardous materials spills and contaminated sites. In addition to management activities, the EPA maintains scientific research and development programs to improve waste management, prevent pollution, improve ecological protection, and enhance drinking-water treatment processes (EPA, 2000).

The United States Army Corps of Engineers (USACE) is a branch of the Department of Defense (DOD) whose primary goal pertaining to water resources is to manage a network of dams and related facilities for navigation, flood control, and flow augmentation. The USACE operates and maintains approximately 11,000 miles of commercial navigation channels, and maintains 383 major lakes and reservoirs throughout the United States, which contain a total capacity of 329.2 million acre-feet. The Corps also operates 75 hydropower projects, is the fourth largest provider of hydropower in the nation, and has the greatest hydroelectric generation capacity of all major hydropower producers. Finally, USACE maintains several water resource-related research programs focusing on coastal and hydraulic engineering, environmental quality, and infrastructure design, operation, and maintenance (USACE, 2000).

The United States Bureau of Reclamation (USBR) is a branch of the Department of the Interior (DOI) formed in 1902 to promote economic development of the 17 western states (Arizona, California, Colorado, Idaho, Kansas, Montana, Nebraska, Nevada, New Mexico, North Dakota, Oklahoma, Oregon, South Dakota, Texas, Utah, Washington, and Wyoming). The original mission of the USBR was to provide water, power, and land by draining and converting wetlands to arable lands.

The USBR constructed more than 600 dams during the twentieth century; however, its present mission is to manage, develop, and protect water and related resources. The USBR administers 348 reservoirs, with a total storage capacity of 245 million acre-feet, which provide irrigation water for 10 million acres of agricultural land. The USBR operates 58 hydroelectric facilities, and is the second largest producer of hydropower in the western United States (USBR, 2000). It has the second-greatest hydroelectric generation capacity of all major hydropower producers.

The Federal Energy Regulatory Commission (FERC) is a branch of the U.S. Department of Energy (USDOE) that has many duties, including the licensing and inspection of both publicly and privately owned hydroelectric projects, and the oversight of environmental matters related to the generation of hydroelectric power. Licenses for hydroelectric projects are granted by FERC for periods of 30 to 50 years, following a detailed review of the engineering, environmental, and economic aspects of the proposed project. When a hydroelectric project's FERC licenses expire, its operators must go through a similar re-licensing procedure to determine whether the project will continue to maintain a beneficial public use. After re-licensing, dam operators may be required to make changes in order to maximize the public benefit by balancing power needs, environmental protection, and recreation interests (FERC, 2000).

Water resources are also controlled by the National Marine Fisheries Service (NMFS), a branch of the National Oceanographic and Atmospheric Administration (NOAA), which is a branch of the Department of Commerce (DOC), and by the U.S. Fish and Wildlife Service (USFW), a branch of DOI. The NMFS and USFW are guided by the Endangered Species Act (ESA), which grants the power to regulate water quantity and quality for the preservation of aquatic habitat. The primary objective of NMFS is to support domestic and international fisheries management operations, fisheries development, and protected-species and habitat-conservation programs for saltwater and anadromous (sea-going) fish (NMFS, 2000).

The primary functions of the USFW are to protect endangered species, restore nationally significant freshwater fisheries, and conserve and restore wildlife habitat, including wetlands. USFW also manages 90.6 million acres of land as wildlife refuges and 2.6 million acres for waterfowl production (USFW, 1999). A large quantity of fish and wildlife habitat is on non-federal lands; therefore, USFW also maintains a number of partnerships to foster aquatic conservation and habitat development on those lands (USFW, 2000).

The Bureau of Land Management (BLM) is an agency within DOI that seeks to maintain the health, diversity, and productivity of 264 million acres of forest and rangelands in the United States. The agency directly affects water resources through its involvement with the restoration and management of riparian areas, abandoned mined lands, and rangelands (BLM, 2000).

The Forest Service (FS) is a branch of the U.S. Department of Agriculture (USDA) that manages the national forests and grasslands. One of the founding directives of the Forest Service is to manage the natural resources contained in 191 million acres of publicly owned lands for the sustained yield of water and aquatic species. The Forest Service also maintains a research program and provides stewardship assistance to state agencies and private enterprises to improve watershed protection on non-federal lands (USFS, 2000).

The Natural Resources Conservation Service (NRCS) is a branch of the USDA that is concerned mainly with agricultural lands. The mission of NRCS is to conserve, improve, and sustain natural resources and the environment through partnerships with public agencies and private landowners. One of the primary functions of the NRCS with regard to water resources is to provide snow surveys and water-supply forecasts for the western states. The NRCS also maintains a variety of environmental programs to protect watersheds from erosion and flood damage, reclaim water resources impacted by mining, and increase aquatic habitat. The agency also runs a wetland reserve program to protect, enhance, and restore wetlands on privately owned lands (NRCS, 2000).

The Farm Service Agency (FSA) is a branch of the USDA that supports American farmers through a variety of programs. In particular, FSA administers the Conservation Reserve Program (CRP), which protects highly erodible and environmentally sensitive farmlands. The CRP encourages farmers to not grow crops on sensitive areas, thereby directly affecting water and habitat quality of the streams draining farmlands. Up to 36.4 million acres may be enrolled in CRP at any given time. The highest concentrations of CRP acreage are found in the semi-arid and sub-humid regions in the West and Midwest.

The U.S. Geological Survey (USGS) is another branch of DOI, whose primary function regarding water resources is to provide scientific analyses to support effective management. In cooperation with other agencies, USGS maintains a nationwide network of stream gauges to provide data on stream flows and water quantity. The USGS also maintains research programs focused on water quantity, quality, and aquatic ecology to build understanding of the complex functioning and interactions of environmental systems (USGS, 2000*b*).

In addition to the agencies listed above, several other agencies play minor roles in the management of the nation's water resources. The primary function of the Federal Emergency Management Agency (FEMA) is to protect the nation's infrastructure through an emergency management program of mitigation, preparedness, response, and recovery. In addition, FEMA manages development within the nation's floodplains in order to reduce flood damage. Such action can directly influence water resources, since floodplain development can impact water quality, aquatic habitat, and runoff processes (FEMA, 2000). The National Park Service (NPS) is another minor player in water resource management. One of the Park Service's primary goals is to preserve natural resources, including aquatic habitat and water quality, within the 80.7 million acres comprising the National Parks System (NPS, 2000).

17.4 International Management Commissions

The United States shares 18 drainage basins with Canada and Mexico (Figure 17.1). To help prevent and resolve disputes over boundary waters, joint commissions were established with Canada in 1909 and with Mexico in 1944. The International Joint Commission (IJC) helps governments find solutions to problems associated with the lakes and rivers that lie along, or flow across the border between

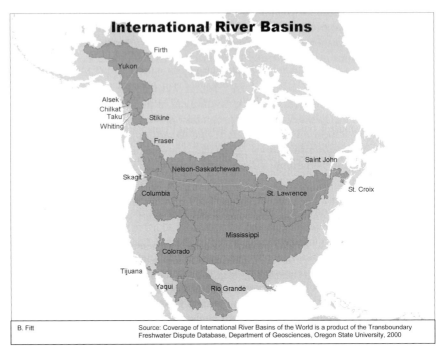

Fig. 17.1. The United States shares 18 drainage basins with Canada and Mexico.

the United States and Canada. The IJC is composed of six members, three appointed by the President of the United States with approval of the Senate, and three appointed by the Governor in Council of Canada on the advice of the Prime Minister.

The IJC rules on proposed projects that may affect transboundary waters, and may regulate the operation of such projects. It also assists the two nations in protecting the transboundary environment, including air quality, and alerts governments to emerging issues that may result in disputes. A major function of the IJC is to help both governments in their goal of improving the water quality and ecological health of the Great Lakes and to prevent any further degradation of the system. Another key function of the IJC is to set conditions for dam operations and allocate water from a number of transboundary waters, including the Columbia, Kootenay, Osoyoos, St. Mark, Milk, Souris, Rainy, and St. Croix rivers (IJC, 2000).

The U.S.-Mexico border area comprises approximately 157,600 square miles, and the hydrologic border extends from 2.8 to 177 miles from the international boundary (Mumme, 1996). The International Boundary and Water Commission

(IBWC) is concerned with water-management issues along the U.S.-Mexico border. The U.S. section of IWBC operates as part of the U.S. State Department, and the Mexican section operates as part of the Mexican Foreign Ministry. The IBWC is designed to facilitate joint action, while protecting the national sovereignty of the two nations. It is responsible for the distribution of water from the Colorado and Rio Grande rivers, joint operation of dams and flood control works on the Rio Grande, solution of border water-quality problems, and stabilization of the river boundaries (IBWC, 2000). Management of transboundary groundwater in the United States is largely a function of state law; as a result, there is limited bi-national cooperation. In isolated cases, however, IBWC has dealt with groundwater issues to settle resource disputes or exchange information concerning domestic projects that may affect water use in the other country (Mumme, 1996). Both Mexico and the United States depend heavily on groundwater in these arid border areas; therefore, bi-national cooperation and effective management strategies are critical in order to sustain the resource and avoid water conflicts in the future.

18. Water Storage and Regulation Infrastructure

Economic development in the United States was greatly assisted by the development of an extensive system of dams that enabled transportation and provided flood protection, irrigation water, and inexpensive power. The number of dams in the United States has been estimated at more than 2.5 million, among them 75,000 dams greater than 6 feet in height (NRC, 1992). That number includes 7,782 major dams 50 feet or more in height, or with a normal storage capacity exceeding 5,000 acre-feet, or with a maximum storage capacity of 25,000 acre-feet (Figure 18.1). Major dams account for 10.4% of the total; 67,405 small and medium dams account for the remaining 89.6%.

Major dams in the United States store more than 6 million cubic meters of water, while the smaller dams store between 60 thousand and 6 million cubic meters of water (USGS, 2000*b*). The surface area of reservoir water is greatest in regions V and IV because of the extensive navigation structures on the Mississippi River system (Figure 18.2). Impounded water surface is relatively high on the large river systems in the western regions (I-III), and is lower in the eastern regions (VI-IX).

18.1 Purpose and Function

Although many major dams may serve multiple functions, the primary function of most major dams is flood control (Figure 18.3). Other primary functions may include public water supply, recreation, irrigation, hydroelectric power, and mine tailings impoundments. More than half (54%) of the dammed surface area of major dams is primarily used for navigation, with 17% used for flood control, and 13% for hydropower. Irrigation and water supply comprise relatively minor amounts of the total surface area, each accounting for approximately 5% of the total (Figure 18.4).

Most major dams are owned by the private sector, state and local governments, or by public utilities; however, the federal government owns the largest and most important dams (Figure 18.5). The era of federal dam building began in 1902 with the establishment of the Bureau of Reclamation, whose primary charge was to provide water and power for the development of the West. Subsequent federal legislation and the establishment of federal power administrations continued to promote dam development in the first half of the twentieth century.

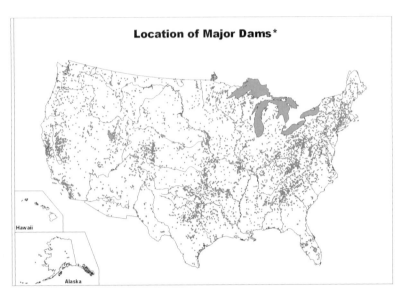

Fig. 18.1. Major dams are defined as dams 50 feet or more in height, or with a normal storage capacity of at least 5,000 acre-feet or a maximum storage capacity of at least 25,000 acre-feet. (Source: U.S. Geological Survey, Major Dams of the United States, 1999.)

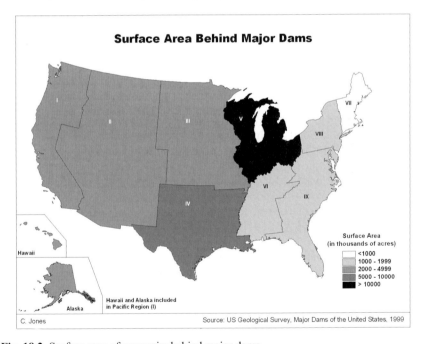

Fig. 18.2. Surface area of reservoirs behind major dams.

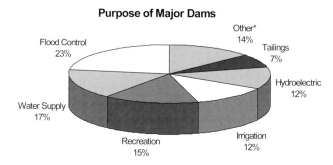

Purpose of Major Dams

Fig. 18.3. Although most major dams serve multiple purposes, the primary purpose of most is flood control.

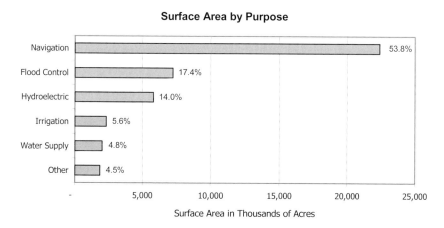

Surface Area by Purpose

Surface Area in Thousands of Acres

Fig. 18.4. Most of the surface area behind major dams is used for navigation. Flood control, the primary purpose of most major dams, constitutes 17.4% of reservoirs' purposes, by surface area. (Source: US Geological Survey, Major Dams of the United States, 1999.)

After the passage of the Flood Control Acts of 1936 and 1944, the Army Corps of Engineers initiated construction of numerous dams. The dam-building era reached a peak in the 1960s, then declined due to lack of public funds for dam projects and concern over environmental issues (Figure 18.6). The age distribution of major dams reflects this history; roughly one-third of all U.S. dams fall into one of three age ranges: 0 to 30, 31 to 50, and 51 to 100 years (Figure 18.6).

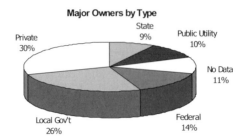

Fig. 18.5. Although most major dams are owned by the private sector and local governments, the largest and most important dams are owned by the federal government. (Source: US Geological survey, Major Dams of the United States, 1999.).

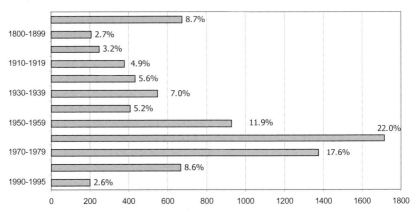

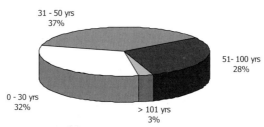

Fig. 18.6. Completion date and age of major dams. (Source: US Geological Survey, Major Dams of the United States, 1999.).

With assistance from NRCS, local communities have constructed more than 10,000 small upstream flood-control dams since 1948. Most of those dams are made of compacted earth, range from 25 and 60 feet in height, and create lakes of a few acres to several hundred acres in size (NRCS, 2000). The dams are located in 46 states, with the majority (43%) located in Texas, Oklahoma, and Arkansas (Buckley et al., 1998). Within the next 10 years, more than 1,300 will reach their original expected lifespan of 50 years. In the near future, many of these small dam projects will require either significant rehabilitation or replacement or removal in order to protect human and natural systems. Currently over 2,200 small dams require rehabilitation, at an estimated cost exceeding $540 million.

18.2 Dam Removal

In the past 40 years, at least 480 dams were removed in the U.S. because of economic, safety, or environmental concerns (American Rivers, 2000). Most of those dams were removed in the 1980s (92 dams) and 1990s (177 dams), and were relatively small, with an average height of 21 feet and length of 224 feet. More than 40 dams exceeding 40 feet have been removed, along with several large dams exceeding 120 feet. The cost of dam removal varies widely, ranging from $1500 for very small dams to $3.2 million for large, complex removals. At least 100 more dams are currently either committed or are under consideration for removal (American Rivers, 2000). The decision to remove a dam is made by varying organizations, depending on the regulatory oversight of the dam. The dam owner typically decides whether to remove a dam or continue operations based on economic considerations. For cases where a dam poses a safety risk, state regulatory agencies may require its removal. FERC may also mandate the removal of a private or publicly owned hydroelectric dam based on environmental and safety concerns raised during the dam re-licensing process. Federal, state, and local governments, dam owners, and environmental mitigation funds have covered costs of dam removals.

Dam removal in the United States is perceived by many to be a reasonable and cost-effective method for river restoration for a variety of reasons. Improved scientific understanding of river systems indicates that in many cases, the negative ecological and economic costs of a dam outweigh the benefits. Negative environmental impacts include reducing river levels, blocking the transfer of nutrients and sediment, altering temperature and oxygen levels, and impeding the movement of fish and wildlife. Society has developed alternatives to some dams, such as improving irrigation systems, restoring wetlands, and maintaining riparian buffers, thereby reducing dependency on dams for irrigation and flood storage.

The recent transition to a competitive electricity market and the elimination of public subsidies for many dams, coupled with mandated environmental controls, will make some dams prohibitively expensive to operate, leading to abandonment.

For many deteriorating, unsafe, or abandoned dams, removal costs are frequently less than the cost of repairs, particularly for cases where the benefits of the dam are limited. Furthermore, public support for dam removal has increased due to widespread media coverage of successful dam removals, which have served to educate the public about the environmental and economic benefits of dam removal.

There are many challenges associated with dam removal, including issues regarding the quantity and quality of stored sediment, as well as social, legal, and economic impacts that must be addressed on a case-by-case basis. The effective management of aging and obsolete structures will therefore require effective scientific analyses coupled with innovative approaches in order to mitigate the economic and social impacts associated with dam decommissioning and removal.

19. Water Availability and Use

Water available for use in the United States consists of renewable sources, such as surface water in lakes and streams, and non-renewable sources, such as the extraction of groundwater exceeding recharge rates. The total amount of renewable water resources in the U.S. is 2,459 cubic kilometers per year, or 8,983 cubic meters per person per year. In addition, 19 cubic kilometers enter the United States each year as river flows from other countries, whereas a negligible quantity flows out to other nations. The average annual groundwater recharge is 1,514 cubic kilometers, or 5,531 cubic meters per person.

Water used or *withdrawn* in the United States each year totals approximately 467 cubic kilometers or 1,839 cubic meters per person, which represents 19% of the total available water resources. Surface water accounts for 81% and groundwater for 19% of the total withdrawals in the United States (Figure 19.1). Annual groundwater withdrawals total 110 cubic kilometers or 433 cubic meters per person, which is 7.3% of the annual *recharge* (World Resources Institute, 1998).

Current and projected water use in the United States can be categorized as either consumptive or non-consumptive. Water use can also be measured by geographic area and by sector, such as industrial or economic. Consumptive use means that water is fully consumed during use. Consumptive use of water is based on the estimated proportion of water that is evaporated, transpired, or incorporated into products or crops.

Total Withdrawals by Source

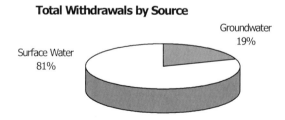

Fig. 19.1 Surface water withdrawals total 321,186 million gallons per day (mgd); groundwater withdrawals total 77,320 mgd. (Source: US Geological Survey National Water Use Data Files, 1999.)

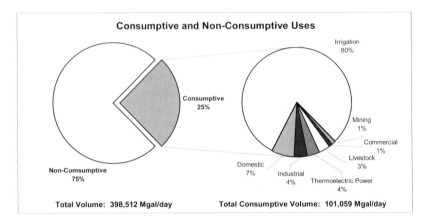

Fig. 19.2. Water use is categorized as consumptive or non-consumptive. The total volume of water used is 398,512 million gallons per day (mgd); total consumptive volume is 101,059 mgd. Consumptive use is further categorized by type of use. Irrigated agriculture constitutes the greatest consumptive use of freshwater resources in the United States. (Source: US Geological Survey, National Water Use Data Files, 1999.)

In 1995, consumptive freshwater use in the United States was 100 billion gallons per day (bgd) and accounted for 25% of the total amount of freshwater withdrawn (Figure 19.2). Irrigation accounts for the largest proportion (80%) of consumptive use, followed by the domestic (7%), industrial (4%), and thermoelectric sectors (4%). Non-consumptive use means that some or all of the water withdrawn is later available for reuse. That water is called the return flow.

19.1 Geographic Region

The United States can be divided into 20 water resource regions (WRRs) for estimations of current and projected water use (Figure 19.3). These WRRs correspond to the nation's major watersheds or to large areas of contiguous coastal watersheds. They are characterized by relatively homogeneous geography, climate, precipitation, and water use characteristics. The 20 WWRs are used to provide geographically relevant delineations for water resource assessments such as the impacts of climate change on available resources. Water use in the United States is also estimated for a reduced number of regions representing aggregated states with similar physical, climate, and water use characteristic (Table 19.1). The water use data files (I-IX region) classification scheme is used to aggregate data from the state water use surveys. In this report, both the WWR and I-IX classification schemes are used to discuss the spatial differences in water sources and uses within the United States.

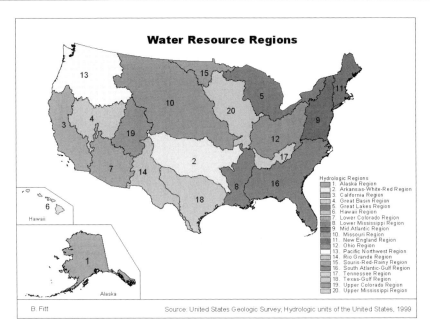

Fig. 19.3. The US Geological Survey divides the country into water resources regions (WWRs). These are large natural drainage basins or hydrologic areas that contain either the drainage area of a major river or the combined drainage areas of two or more rivers.

The arid Pacific and Mountain regions withdraw the most water, 32% of the total national volume withdrawn (Figure 19.4). The East North Central and South Atlantic regions also withdraw large volumes, accounting for 30% of the total. The New England and East South Central regions withdraw the smallest volumes, accounting for just 3.2% and 6.2%, respectively, of the national total. As noted in Section 1, the source of the water withdrawn also varies by region. The more arid western states generally depend more heavily on groundwater, and the more humid eastern states rely more on surface water (Figure 19.5). In the western regions (I-IV), groundwater accounts for 19% to 39% of the total water extracted.

In the eastern regions (V-IX), groundwater accounts for only 6% to 15% of total volume withdrawn. Consumptive water use is highest in the West, accounting for 47% of the total freshwater withdrawals. The high consumption is mainly due to the large amount of irrigated land and high evaporative demand that occurs in the arid western states. Consumptive use in the East is approximately 12% of the freshwater withdrawn and only accounts for 20% of the total U.S. consumptive use (Solley et al., 1998).

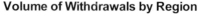

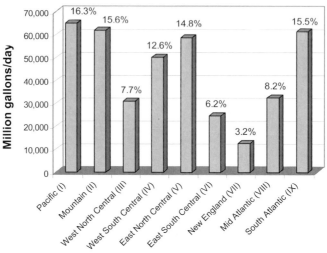

Fig. 19.4. Volume of freshwater withdrawals by water use data files region. (Source: US Geological Survey Water Use Data Files, 1999.)

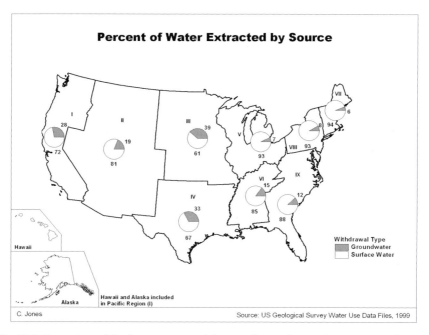

Fig. 19.5. Percentage of freshwater extracted from surface and groundwater sources by water use data files region.

Table 19.1. US Geological Survey water use data files regions.

Region and Location	States
Region I, Pacific	Alaska, California, Hawaii, Oregon, Washington
Region II, Mountain	Arizona, Colorado, Idaho, Montana, Nevada, New Mexico, Utah, Wyoming
Region III, West North Central	Iowa, Kansas, Minnesota, Missouri, Nebraska, North Dakota, South Dakota
Region IV, West South Central	Arkansas, Louisiana, Oklahoma, Texas
Region V, East North Central	Illinois, Indiana, Michigan, Ohio, Wisconsin
Region VI, East South Central	Alabama, Kentucky, Mississippi, Tennessee
Region VII, New England	Connecticut, Maine, Massachusetts, New Hampshire, Rhode Island, Vermont
Region VIII, Mid Atlantic	New Jersey, New York, Pennsylvania
Region IX, South Atlantic	Delaware, Florida, Georgia, Maryland, North Carolina, South Carolina, Virginia, West Virginia

The California WRR was the largest consumer of freshwater, accounting for 25% of the national consumption (Figure 19.8). The large consumptive use in this highly populated agricultural region is for irrigation and public water supplies, coupled with high evaporative demand. Other major consumptive WRRs are the Missouri Basin and Pacific Northwest Regions, which account for 14% and 11% of the total consumptive use, respectively. The high consumptive use in both regions is largely attributable to irrigation.

19.2 By Delivery Sector

The thermoelectric sector extracts the greatest proportion, or 48% of the national total water withdrawn (Figure 19.7). The agriculture (irrigation and livestock) sector extracts the second greatest quantity, accounting for 35% of the national total. The municipal and industrial sectors account for the remaining 17% of the national balance and include withdrawals for commercial (2.4%), domestic (6.6%),

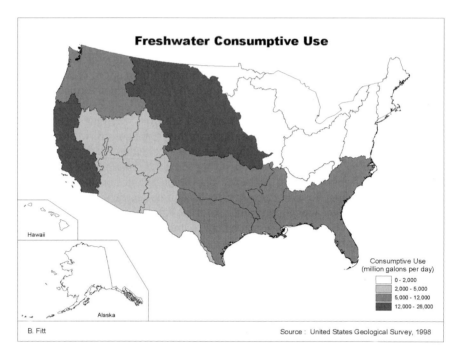

Fig. 19.6. Total consumptive use of freshwater by water resources region.

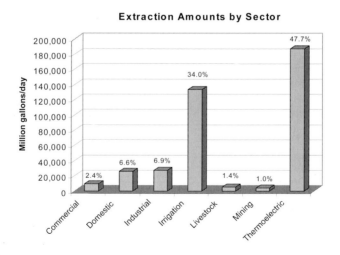

Fig. 19.7. Total freshwater withdrawals by economic sector. (Source: US National Geological survey Water Use Data Files, 1999.)

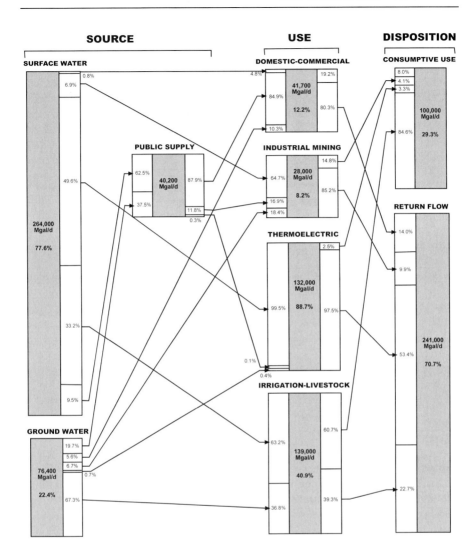

Fig. 19.8. Flow chart summarizing the source, use, and disposition of water resources in the United States. (Source: US Geological Survey.)

industrial (6.9%) and mining (1.0%) uses (USGS, 1999c). The thermoelectric sector accounts for about 50% of total fresh surface water withdrawals, agriculture, including irrigation and livestock, 33%, public supply, 10%, and industry and mining, 7% (Figure 19.8) (Solley, 1998).

Fresh groundwater withdrawals for agriculture account for 67%, public supply for 20%, industry and mining for 6.7%, and domestic and commercial for 5.6% of the total (Figure 19.8) (Solley, 1998). The U.S. produces a negligible amount of desalinated water relative to the total resource.

The largest amounts of thermoelectric power withdrawals occur in the Mid-Atlantic and South Atlantic-Gulf regions, with Illinois accounting for the largest thermoelectric withdrawal. The largest amount of irrigation occurs in the western and northern Great Plains regions, with California accounting for the largest irrigation withdrawal. A summary of water sources, uses, and disposition in the United States is shown in Figure 19.8.

For simplicity, the domestic sector is combined with commercial, industrial with mining, and irrigation with livestock. The relatively large amount of public supply coming from groundwater and going to domestic uses indicates the need to protect and conserve groundwater resources, given the projected future population growth. Although the thermoelectric and irrigation-livestock sectors use a large quantity of water, these sectors also contribute large volumes to the total return flow, which is water available for reuse. Sources of return flows are particularly important for water quality, because the primary impact of thermoelectric uses is increased temperature, whereas agricultural uses may increase chemical constituents such as salts, nutrients and pesticides in the return flow.

19.3 Historical Trends and Future Projections

From 1900 to 1990, population in the United States increased by an annual rate of 1.2%, whereas freshwater withdrawals increased by 2.4%. Per capita withdrawals increased by 284%, from 475 gpd in 1900 to 1,350 gpd in 1990. That increase has been attributed to increases in irrigated land and thermoelectric cooling, which accounted for the greatest use of water during that period (Brown 1999). In 1985, total withdrawals dropped for the first time in the century and remained stable through 1995, despite continued population increases (Figure 19.9). That decrease has been attributed to lower withdrawals by the agricultural, thermoelectric, industrial, and commercial sectors. The drop in 1985 resulted from a combination of above-average precipitation, economic slowdown, increased groundwater pumping costs, and improved efficiency of water use (Solley, 1998). Analyses indicate that the apparent sharp drop in 1985 may also be due to changes in how water use estimates are prepared by the USGS, such that use prior to 1985 may have been overestimated (Brown, 1999).

On a national basis, water withdrawals are not projected to increase substantially within the next 40 years (Figure 19.10). Withdrawals in the year 2040 are expected to be 7% of the 1995 level, despite an anticipated increase in population of 41% and continued economic growth. The changes are expected to be relatively small because of projected increases in efficiency in the thermoelectric, agriculture, industrial, and public sectors (Brown, 2000). Future demands for water resources should also be moderated by a decline and regional shift in irrigated acreage from the West to the East (Table 19.2). The anticipated changes in water withdrawals vary regionally, with the greatest increase (27%) expected in the Southeast. The projected changes for the other regions are relatively small, ranging from a 9% increase in the Northeast, to a slight decrease in the Pacific Coast region.

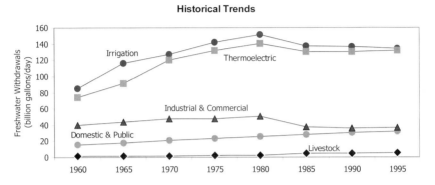

Fig. 19.9. Total freshwater withdrawals by use sector, from 1960 to 1995. (Source: Thomas C. Brown, Projecting U.S. Fresh Water Withdrawals, USDA Forest Service, 2000.)

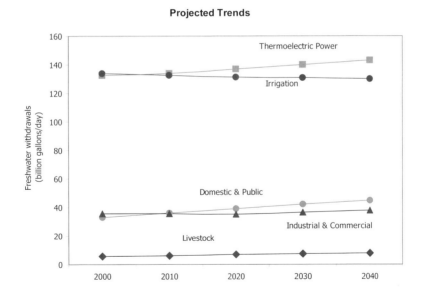

Fig. 19.10. Projections of freshwater withdrawals by use sector from 2000 to 2040. (Source: Thomas C. Brown. Projecting U.S. Fresh Water Withdrawals, USDA Forest Service, 2000.)

Table 19.2. Project percent changes in population, irrigated acreage, and withdrawals from 1995 to 2040.

	North-east	South-east	North Central	Great Plains & Texas Gulf	South-west	Pacific Coast	**National projections (%)**
Population	31	51	30	41	64	52	**41**
Irrigated acres	12	40	40	-6	4	-11	**3**
Withdrawal							
Livestock	31	51	30	41	64	52	**41**
Domestic and public	31	51	30	41	64	52	**41**
Industrial and commercial	-5	14	-4	8	26	12	**6**
Thermoelectric	3	20	2	11	33	120	**9**
Irrigation	-5	41	35	-10	-1	-10	**-3**
Total withdrawal	9	27	4	2	5	0	**7**

Source: Brown, Thomas C., Projecting U.S. Fresh Water Withdrawals, USDA Forest Service, 2000.

Nationally, 19% of surface flows are withdrawn for out-of-stream uses. Despite the apparent water surplus in the United States, water supplies are unevenly distributed and are limited in many areas of the nation (Figure 15-8). Nevertheless, future changes in the water supply infrastructure are assumed to be negligible, due to the fact that most opportunities for water resource development have already occurred, particularly in arid regions, and because of concern over the environmental impacts of new facilities. Furthermore, the supply infrastructure may decrease after dam decommissioning and mandated operational changes resulting from FERC re-licensing activities.

Water levels can vary spatially within an aquifer because land uses above the aquifer often vary. Withdrawals, irrigation techniques, and recharge can all affect groundwater levels. In areas where relatively large volumes of groundwater are used, the excessive withdrawals are limiting available resources because of the increased costs associated with increased pumping lifts. The High Plains Aquifer is an example of a critical groundwater resource that has been heavily exploited for irrigation in recent years (Figure 19.11). The annual area-weighted change in water levels in the aquifer was 0.33 feet per year for 1950 to 1980. Water level changes decreased to 0.19 feet per year from 1980 to 1997, as a result of changing irrigation practices and recharge (USGS, 1999b). The limited options for new surface storage, i.e., reservoirs, and few opportunities to develop new groundwater sources require a continued reduction of demand through improvements in water use efficiency.

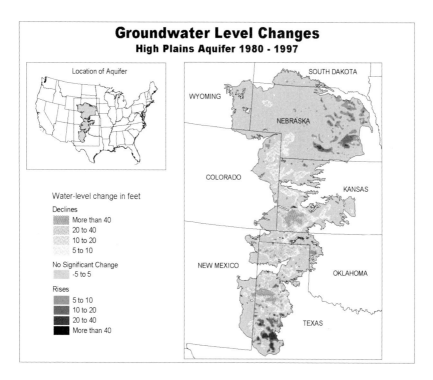

Fig. 19.11. The High Plains Aquifer serves as an example of the types of groundwater level changes that can occur due to changing land-use practices. (Source: US Geological Survey, 1998.)

Many competing users vie for limited water resources throughout the country, particularly in arid regions. Rising concerns over the preservation of aquatic habitat, riparian ecosystems, water quality, recreation, and aesthetics are driving the regulation of minimum stream flows that must be maintained in a given channel. As demands for water resources increase in the future, managers and scientists will continue to be faced with the challenge of how best to balance the competing societal and ecological demands placed on riparian systems and aquifers in the United States.

In addition, water supply projections also assume a constant climate; however, recent analyses indicate that climate change may alter the hydrologic balance of many regions of the United States (IPCC, 1995). The projected effects include changes in the timing and regional patterns of precipitation, evaporation, and transpiration; changes in the frequency and severity of droughts; and an alteration of the winter snow regime in western regions that depend on seasonal snowpack to meet summer water demands. The potential for climate change further emphasizes the need to develop innovative and effective water management policies, in order to address uncertainties in matching future supplies with demands (Frederick, 1997).

20. Electric Power Generation

As of January 1, 1999, the existing generating capacity of the United States totaled 686,692 megawatts (DOE, 1999). In 1996, the U.S. produced 3.629 trillion kWh per year, split between fossil fuel (65.1%), nuclear (18.6%), hydroelectric (9.6%), and other (6.7%) sources (CIA, 2000). During this year, the U.S. imported 46.543 billion kWh, exported 9.02 billion kWh, and consumed a total of 3.666 trillion kWh. The annual per capita energy use has risen from about 4,200 kWh in 1960 to about 11,400 kWh in 1995 (Brown, 1999).

Water used for the generation of electricity is split between off-stream use (water is withdrawn from the waterway, used, then returned) and in-stream use (water is channeled through mechanisms such as turbines, but is not removed from the waterway). Thermoelectric power plants (fossil-fuel, nuclear, and geothermal), which use heat to generate electricity, require off-stream water use. Water used for hydroelectric power generation, including pumped-storage facilities, is classified as an in-stream use. Hydroelectric power plants generate power by channeling moving water through or around turbines (Table 20.1). Thermoelectric power plants extract water for off-stream use, primarily from surface water, whereas hydroelectric power plants primarily run waters through turbines in-stream. Of the total amount of water used for the production of electricity, 94% is used for hydroelectric power, while only 6% is used for thermoelectric power.

Table 20.1. Total amount of water used for thermoelectric* and hydroelectric power by source, in millions of gallons per day (mgd).

Use/Source	Water (mgd)	%
Thermoelectric		
Groundwater off-stream	626	
Surface water off-stream	186,583	
Total	187,209	6
Hydroelectric		
Surface water in-stream	3,071,560	
Surface water off-stream	90,033	
Total	3,161593	94
Total thermoelectric and hydroelectric use	3,348,802	

Note: Thermoelectric power water use includes fossil fuel, geothermal, and nuclear power sources. (Source: US Geological Survey Water Use Data Files, 1999.)

20.1 Thermoelectric Water Use

Most water withdrawn for thermoelectric power generation is used for cooling condensers and reactors. In 1995, thermoelectric withdrawals in the United States were 190 bgd, including 58 bgd of saline water, and 132 bgd of fresh water. Almost all water used for cooling is returned to the natural water system, with only 2% consumed as a result of evaporation during "once-through", cooling tower, or pond cooling. Thermoelectric withdrawals accounted for 47% of the total fresh and saline water withdrawals, and 39% of all off-stream fresh water use. Virtually all water used for thermoelectric power production is self-supplied. Public or municipal water accounts for less than 1% of the total amount of water used in thermoelectric power production. Surface water comprises greater than 99% of all thermoelectric use, and is composed of 69% fresh water, and 31% salt water (Table 20.1).

Thermoelectric power generation is greatest in the eastern regions, with production in the South Atlantic (IX) and East North Central (V) regions exceeding 500 billion kWh per year (Figure 20.1). Thermoelectric power generation consequently represents the largest use of water in the East, where 75% of the total thermoelectric withdrawals occur.

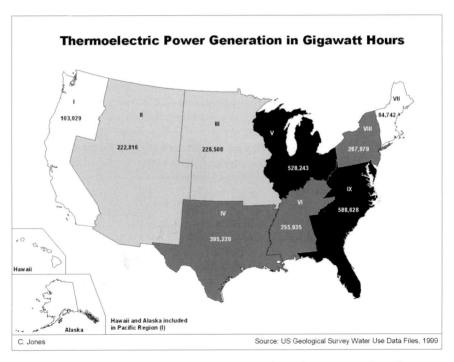

Fig. 20.1. Thermoelectric power generation in gigawatt hours by water use data files region.

Illinois leads the nation in thermoelectric withdrawals, at 17.1 bgd, followed by Texas (9.6 bgd), Michigan (8.4 bgd) and Tennessee (8.3 bgd). Illinois also leads the nation in thermoelectric consumptive use at 407 mgd, followed by Ohio (336 mgd), and Texas (297 mgd).

All states withdraw water for thermoelectric power production from fossil fuels. New York and Texas lead all other states in withdrawals for fossil fuel power, with use exceeding 10 bgd. Consumptive freshwater use for fossil fuel power production is greatest in Ohio at 309 mgd, followed by Texas, Louisiana, and Kentucky, all of which consume more than 200 mgd. Twenty-nine states in the United States withdraw water for nuclear power production. Illinois leads withdrawals with 7.5 bgd, followed by California (4.7 bgd) and North Carolina (4.2 bgd). Illinois also has the greatest amount of consumptive freshwater use for nuclear power production at 263 mgd, followed by Pennsylvania with 119 mgd, and Georgia with 93 mgd (Solley, 1998).

Thermoelectric water withdrawals rose steadily from 1960 to 1980, declined in 1985, and rose slightly through 1995 (Figure 19.9). The declining rate of thermoelectric water use is largely driven by improvements in efficiency. For example, during this period, freshwater withdrawals increased by 78%, whereas freshwater thermoelectric power production increased by 322% (Brown, 1999). By the year 2040, the annual per-capita energy use is projected to reach 13,040 kWhs, rising over 14% from the 1995 level. It is assumed that hydropower production will remain stable through the year 2040; therefore all increases in production are assumed to occur at thermoelectric facilities. The increase in per capita energy use, coupled with the projected increases in the U.S. population, indicates that the annual energy production at freshwater thermoelectric plants must increase from 2.1 trillion kWhs to 3.5 trillion kWhs. Assuming that freshwater use per kWh produced will continue to decrease as efficiency improves, it is estimated that freshwater withdrawals will rise 9%, from 132 bgd in 1995 to 143 bgd in 2040 (Figure 19.10). On a per-capita basis, thermoelectric fresh water withdrawals are projected to decrease by almost 23%, from 504 gpd in 1995 to 389 gpd by 2040.

20.2 Hydroelectric Use

During the twentieth century, hydroelectric generating capacity in the United States increased steadily, from 4,800 megawatts in 1920, to 33,300 megawatts in 1960, to 74,800 megawatts by 1996. During this period, the relative contribution of hydroelectric generating capacity to the national total decreased from approximately 30% in 1920, to 20% by the early 1960s, and to 10% by 1996. However, hydroelectric power currently comprises 96% of the renewable energy production in the U.S.

Hydroelectric water use for 1995 was estimated to be 3.16 trillion gallons per day, corresponding to 2.6 times the average annual runoff of the U.S. Hydroelectric water use exceeds the annual runoff because the water is used several times as it passes through multiple dams on the same river system. Consumptive use during hydroelectric power generation is negligible, consisting only of evaporation from reservoirs. In the U.S., 944 of the 7,781 major dams are mainly used for the generation of hydroelectric power. Most of the hydroelectric dams are located throughout the Appalachian Mountains in the East, and along the Pacific Cordillera in the West (Figure 20.2).

The Pacific and Rocky Mountain regions generate the largest amount of hydroelectric power (Figure 20.3). Most of the hydroelectric power generation in Pacific Region occurs in the Pacific Northwest, which accounts for about 40% of the water used in hydroelectric plants. The greatest use occurs along the Columbia River System in Washington and Oregon, and on the Niagara and St. Lawrence River systems in New York, accounting for 46% of the total use. Other important hydroelectric generating regions are California, which generates the second highest amount of power (47.1 billion kWh) after Washington (82.3 billion kWh), and the Southeast, where approximately 11% of the nation's hydroelectric power is generated by the systems managed by the Tennessee Valley Authority (TVA).

Dams Used for Hydroelectric Power Generation

944 out of 7,781 major dams (12%) are used for hydroelectric power generation

C. Jones

Source: US Geological Survey, Major Dams of the United States, 1999

Fig. 20.2. Twelve percent of the major dams in the United States are primarily for hydroelectric power production. These are located along mountain ranges in the eastern and western parts of the country.

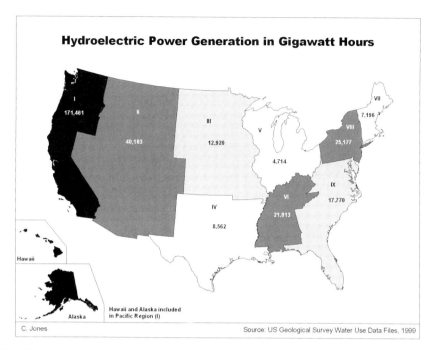

Fig. 20.3. Hydroelectric power production, by gigawatt hours, is greatest in the western United States.

Water use for hydroelectric power generation steadily increased throughout the twentieth century with the construction of dams. Hydroelectric use declined 4% from 1990 to 1995 (Solley, 1998), and is assumed to remain stable through the year 2040 (Brown, 2000). Dam removal or mandated spills (water released over spillways) arising from environmental concerns, such as endangered fish runs, may result in decreased hydroelectric production, which will require a concomitant increase in thermoelectric or other power production.

21. Agricultural Use

Primary agricultural products in the United States are wheat, corn, other grains, fruits, vegetables and cotton; beef, pork, poultry, and dairy products; timber and forest products; and fish (CIA, 2000). The market value of agricultural products is steadily increasing, and is approaching $200 billion per year (Figure 21.1).

21.1 Irrigation

The total amount of cropland in the United States is decreasing slowly and is nearing 430 million acres, down from 450 million acres in 1987 (Figure 21.2). The amount of irrigated land, which includes both cropland and pastureland, is increasing slightly and is nearing 58 million acres (Figure 21.3). Most of the irrigated acreage is in the West, representing 86% of the national total (USGS, 2000b). Similarly, the states with the highest ratio of irrigated land to total farmland are generally located in the West (Figure 21.4). The south central and southeastern states also exhibit a relatively large percentage of irrigated farmlands, ranging from 8% to 15% of total farmlands.

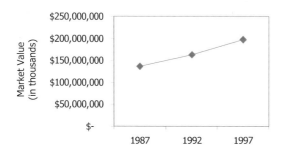

Fig. 21.1. The market value of agricultural products has increased steadily from 1987 to 1997. (Source: US Census of Agriculture, USDA National Agriculture Statistics Service, 1997.)

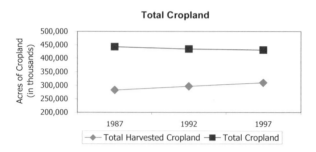

Fig. 21.2. While the market value of agricultural products sold has increased, the total acreage of cropland has decreased slightly over the same period. (Source: US Census of Agriculture, USDA National Agricultural Statistics Service, 1997).

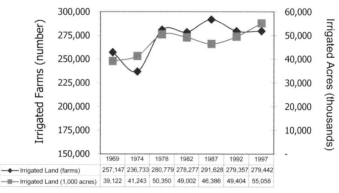

Fig. 21.3. Although both the number of irrigated farms and the amount of irrigated land increased overall from 1969 to 1997, these quantities varied considerably during this period.

The amount of irrigated land in the West is declining slightly, however, and is offset by more rapid increases in the amount of irrigated land in the East (Brown, 2000), as farmers seek to increase yields and lessen susceptibility to drought. On average, irrigated land in the more arid West requires approximately 3 feet of water per acre per year, whereas irrigated land in the more humid East requires approximately only 1.5 feet of water per acre per year.

Although the total amount of irrigated land is increasing, water withdrawals have decreased, due to increased efficiency and the eastward geographical shift in the location of irrigated lands. Improvements in efficiency and technology are driven by rising water costs caused by a combination of a decline in publicly funded infrastructure and increased pumping lifts (Moore et al., 1990). In 1995, irrigation water withdrawals in the United States totaled 150 million acre-feet, or 134 bgd. Irrigation accounts for 39% of the freshwater withdrawals for all categories of off-stream use (Solley, 1998). The Pacific and Mountain regions of the West withdraw the greatest volume of water for irrigation, accounting for 89% of the total amount of water withdrawn nationwide (Figure 21.5).

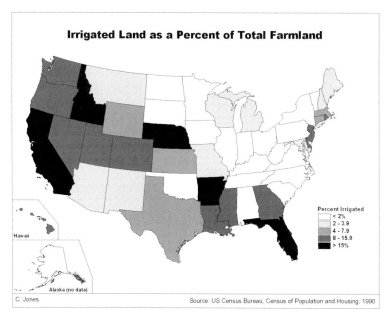

Fig. 21.4. The percentage of irrigated land compared with total farmland is typically highest in the arid western states; however, the percentage of irrigated farmland is expected to increase in the southeaster states.

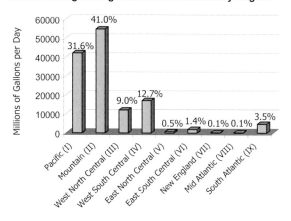

Fig. 21.5. Total volume of water withdrawals for irrigated agriculture by water use data files region. (US Geological Survey Water Use Data Files, 1999.)

Irrigation Water by Source

Fig. 21.6. Most irrigation water is extracted from surface waters. The amount of surface water extracted for irrigation totals 84,627 million gallons per day (gmd); groundwater extractions for irrigation total 49,005 mgd. (Source: US Geological Survey Water Use Data Files, 1999.)

Irrigation with Reclaimed Wastewater

Fig. 21.7. Only a small portion (<1%) of irrigation water is derived from reclaimed wastewater – *wastewater treatment return flow.* Irrigation use of fresh and saline water totals 81,237 million gallons per day (mgd); reclaimed wastewater use totals only 719 mgd. (Source: US Geological Survey Water Use Data Files, 1999.)

21.2 Future Trends in Irrigation

Irrigation withdrawals increased from 1960 to 1980 then declined until the present (Figure 19.9). National irrigation withdrawals are projected to decrease by 4 bgd, or by 3% by the year 2040 (Figure 19.10). On a per capita basis, irrigation withdrawals are expected to decrease from 514 gal/day, to 354 gal/day by the year 2040 (Brown, 1999). Projected changes in irrigation water withdrawals vary across the United States, reflecting changes in irrigated acreage. Irrigation withdrawals in the western regions are generally expected to decrease more than the national average, with the largest decreases of 19% occurring in the Pacific Northwest and Rio Grande regions, and decreases of 14% occurring in the Arkansas-White-Red and Texas-Gulf regions.

Irrigation water use in the eastern regions is expected to increase more than the national average, with increases exceeding 100% in the Ohio and Tennessee regions. Many factors control the anticipated decreases in irrigation withdrawals, including a geographical shift in irrigated acreage from arid western regions with high evapotranspiration demands to humid eastern regions with lower evapotranspiration demands, transfer of agricultural lands to higher valued uses, and increased irrigation efficiency (Brown, 1999). Furthermore, high energy and operating costs, environmental and land-use laws, pollution from salt and chemicals, and competition for urban uses will continue to limit development of new irrigated acreage, particularly in the West.

21.3 Sources of Irrigation Water

Most irrigation water (63%) comes from surface water, with the remaining 37% from groundwater, and a small portion (<1%) from reclaimed wastewater (Figures 21.6 and 21.7). Most of the irrigation water withdrawn (61%) is consumptively used, 19% is lost during conveyance, and 20% makes up the return flow (Solley, 1998). Conveyance losses can be either consumptive losses (e.g., evaporation) or can contribute to the return flow (e.g., seepage from unlined irrigation ditches). In the West, a large proportion of the total amount of irrigation water comes from surface water stored in reservoirs created when dams were built. Surface water resources are replenished each year when the snowpack melts in the mountains. The high cost of extracting deep groundwater also serves to encourage the use of surface water for irrigation in arid regions.

Irrigation water comes mainly from surface water in the East as well, due to the presence of dams and reservoirs in mountainous areas that capture and store precipitation. Groundwater from several large aquifer systems is the major source for irrigation water in the central and midwestern regions of the United States (see Figure 1-6). In some areas, the extraction of groundwater, mainly for irrigation has resulted in large declines in the water table in excess of 100 feet (USGS, 1999b). Over-pumping of groundwater reserves results in many problems, including increased pumping lifts and associated costs, land subsidence, and loss of aquifer storativity.

21.4 Irrigation Techniques

Irrigated surface acreage is classified as surface (flood, furrow, and ditch), sprinkler (center pivot and traveling gun), and micro (trickle and drip). On a nationwide basis, surface irrigation accounts for 55% of the irrigated acres, sprinkler systems account for 42%, and micro-irrigation accounts for 3%. In the eastern and Great Lakes regions, farmers mainly use sprinkler irrigation (use ranges from 85% in New England to 99% in the Upper Mississippi region). In the South-Atlantic Gulf region, sprinkler irrigation accounts for 52% of the irrigated acreage. In the central and western regions, surface applications comprise a much greater proportion of the irrigated land, from 33% in the Pacific Northwest region, to 86% in the Upper Colorado region (Solley, 1998).

In terms of irrigation efficiency, micro-irrigation systems are the most efficient, followed by sprinkler and surface applications (Schwab et al., 1981). Clearly irrigation withdrawals may be reduced by the conversion to more efficient irrigation techniques, particularly in the West, where water resources are more limited. Higher efficiency systems typically are more expensive than lower efficiency systems; therefore the conversion will most likely be driven by economic factors, as water becomes a more expensive commodity.

21.5 Soil Salinity

Soil salinity, which can affect crop germination and growth, is a problem on agricultural lands. Nine percent (48,500 million acres) of U.S. crop and pastureland is affected by salinization or has sodic (containing sodium) soils, including 12.8 million irrigated acres (NRC, 1993). Although saline soils are mainly a problem in arid and irrigated areas, saline seeps and seawater intrusion into coastal lands can affect non-irrigated and humid areas. Salinization problems are greatest in the Mountain and Northern Plains states, where a respective 30% and 20% of crop and pastureland is affected (Table 21.1). Saline soils are currently a minor problem in the eastern states, where less than 1% of crop and pastureland is affected.

Table 21.1. Pecent of land affected by salinization, by region.

Region	States in Region	% Affected
Northeast	Maine, New Hampshire, Vermont, New York, Connecticut, Rhode Island, New Jersey, Pennsylvania, Maryland, Massachusetts, Delaware	<1
Appalachia	West Virginia, Virginia, Kentucky, Tennessee, North Carolina	<1
Southeast	South Carolina, Florida, Georgia, Alabama	<1
Lake States	Wisconsin, Michigan, Minnesota	5
Corn Belt	Ohio, Indiana, Illinois, Iowa, Missouri	<1
Delta	Mississippi, Louisiana, Arkansas	2
Northern Plains	North Dakota, South Dakota, Nebraska, Kansas	20
Southern Plains	Texas, Oklahoma	7
Mountain	Utah, Nevada, Idaho, Montana, Wyoming, Colorado, New Mexico, Arizona	30
Pacific	Oregon, California, Washington, Alaska	16
Other	Hawaii, Caribbean	<1
Total Land Affected		9

Source: National Academy of Sciences, Soil and Water Quality, 1993.

Livestock Water Use by Source

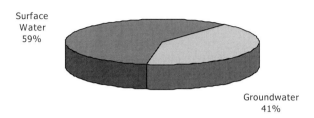

Surface
Water
59%

Groundwater
41%

Fig. 21.8. Sources of water used for livestock production. Surface water use totals 3,229 million gallons per day (mgd); groundwater use totals 2,268 mgd. (Source: US Geological Survey Water Use Data Files, 1999.)

21.6 Livestock Use

Livestock water withdrawals represent a similar source distribution, with 59% obtained from surface waters, and 41% from groundwater (Figure 21.8). Of the water withdrawn for livestock, 58% is consumptively used and 42% comprises the return flow. Livestock accounts for 29% of freshwater withdrawals for all categories of off-stream use (Solley, 1998). Withdrawals of water for livestock totaled 6.1 million acre-feet or 5.5 bgd.

Water used for the livestock sector remained relatively constant over time, but then doubled in 1985. The doubling was not due to increased use, however, but to the inclusion of animal specialties (mainly aquaculture) in the livestock category. Withdrawals of water for livestock have climbed steadily from 1960 to 1995, in response to increasing numbers of animals (Figure 19.9). The largest withdrawals for livestock are in the Pacific Northwest and Lower Mississippi regions.

Water use is projected to increase by 2 bgd, or 41%, by 2040, which is equivalent to the projected population growth (Figure 19.10). This increase assumes that the population's preferences for beef, pork, lamb, chicken, eggs, and farm-raised fish will not decrease or change over the next 40 years.

21.6 Aquaculture Use

Aquacultural production, the farming of fish, shellfish, and algae, has grown considerably in the last 25 years in the United States. In 1974, the total value of all products sold was $45 million. In 1998 the total value of all aquaculture products sold in the United States was approximately $978 million from 4,028 farms.

The largest component of aquacultural products is food fish, accounting for 71% of the total sales revenues, followed by mollusk, accounting for 9%, and ornamental fish, accounting for 7% (Figure 21.9). The category shown as Other Animal Aquaculture includes algae and sea vegetables, alligators, frogs, and other products. Catfish is the major type of food fish produced by aquaculture farms, accounting for 65% of total sales, followed by other types of food fish, including carp, perch, and salmon (25%), and trout (10%).

Most aquaculture facilities (68%) are located in the southern states, which produce 65% of the total value of products sold. In 1998, the top aquacultural states were Mississippi and Arkansas, with combined sales of catfish and baitfish totaling $290 and $84 million, respectively. In Florida, sales of ornamental fish, clams, oysters, and alligators totaled $77 million. Maine was fourth, with sales totaling $67 million, mainly of Atlantic salmon, followed by Alabama, with sales of $59 million almost exclusively of catfish (USDA, 1999).

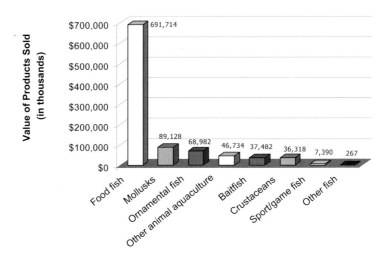

Fig. 21.9. Monetary value of aquaculture products sold, by type. (Source: US Census of Aquaculture, Department of Agriculture-National Agricultural Statistics Service, 1998.)

Approximately half of the aquaculture farms are small operations, selling between $1,000 and $25,000 worth of products annually; however these small farms account for only 2% of the total national sales. About 10% of the farms are large operations, selling over $500,000 per year, and accounting for 77% of the total sales in the United States.

Over 63% of the aquaculture farms in the country use ponds as the primary method of production. Additional off-stream methods include flow-thru raceways or tanks and closed recirculation tanks or cages. Additional in-stream methods include prepared bottoms and net pens. Current aquaculture production uses 385,000 acres of surface water, including approximately 321,000 acres of freshwater (83%) and 64,000 acres (17%) of saltwater. Aquaculture in Mississippi accounts for almost 34% of the national freshwater aquacultural acreage, associated mainly with the production of catfish for human consumption. Connecticut, Washington, and Virginia maintain the most saltwater acreage, mainly for the production of mollusks and crustaceans, which account for more than 81% of the national total.

Most (77%) aquaculture operations in the United States use groundwater and surface water on the farm itself. About 20% use saltwater, and the remainder use water from external sources (Figure 21.10). In the Northeast, water for aquaculture is evenly split between groundwater, on-farm surface water, and saltwater. In the north central states, about half of the farms use groundwater and half use surface water. Approximately 50% of the aquaculture farms in the southern region obtain water from groundwater sources, 33% from on-farm surface sources, and 17% from saltwater sources. In the western region, 10% of the farms obtain water from external sources, 25% from saltwater, and 65% are evenly split between local groundwater and surface water (USDA, 1999)

Sources of Water Used for Aquaculture

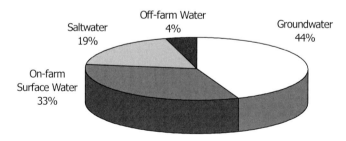

Fig. 21.10. Sources of water used for aquaculture products production (percent of the number of farms). (Source: US Census of Aquaculture, Department of Agriculture-National Agriculture Statistics Service, 1998.)

The total volumes of water withdrawn and consumed for aquaculture production are not quantified or estimated for the future independently, but are included within the livestock sector estimates as part of the animal specialties sector. If aquaculture production continues to grow at the rate observed over the past 25 years, this sector may have an impact on water resource management that warrants special consideration in future water use estimates and management. Sales of U.S. aquaculture products depend mainly on the strength of the economies of the United States, Canada, and Asia; therefore, it is difficult to predict with certainty the amount of water that will be required by the aquaculture sector in the future (USDA, 2000).

22. Municipal Use

Municipal use of water in the United States refers to public water. Public water represents both an intermediate *source* of water and a *use* of water.

22.1 Public Water Supply

As a *source,* the municipal or public supply refers to water withdrawn from groundwater or surface resources for systems that supply a minimum of 25 people or have a minimum of 15 connections. In 1995, 40.2 bgd of water were withdrawn to meet public water supplies, accounting for 12% of the total fresh water withdrawn in the United States. Public suppliers serve about 84% of the U.S. population, with the remaining 16% supplying their own water from wells or other sources (Solley, 1998).

The largest withdrawals were made in the Pacific, South Atlantic, and East North Central regions, accounting for 50% of the total public-supply withdrawals (Figure 22.1). The greatest withdrawals were made by the four most populous states, California, Texas, New York, and Florida. These states account for 31% of the total population, and 35% of the total public water withdrawals. Per capita withdrawals of public water supply are greatest in the relatively arid and sparsely populated states in the west (Figure 22.2). Withdrawals are lowest in the northeastern, north central plains, and Appalachian states.

Surface water makes up 62% of the public supply and groundwater comprises 38% (Figure 22.3). Extraction of surface waters for the public supply is larger in the East, where surface waters account for 59% to 79% of the public supply budget (Figure 22.4). In the West, surface waters account for a smaller fraction of the public supply, ranging from 46% to 66%. Surface water withdrawals are greatest in New York, California, and Texas, and groundwater withdrawals are greatest in California, exceeding 2 bgd in all cases.

22.2 Public Water Use

Use of the public water supply is divided among the domestic (56%), commercial (17%), public (15%), and mining/industrial (12%) sectors (Figure 22.5). The public water supply is mainly used by the domestic, commercial, and public sectors, which together comprise the municipal sector. Water use by mining and industrial sectors is described in Section 23. Domestic water use includes all water for normal household purposes, including lawn and garden irrigation, washing clothes and dishes, flushing toilets, bathing, cooking and drinking. Commercial water use includes water for all civilian, military, and commercial facilities, such as hotels, restaurants, and office buildings. Public water use includes water used for fire-fighting, street cleaning, and parks, as well as conveyance losses.

22.2.1 Domestic Use

Domestic water use in the United States is 26.1 bgd, accounting for about 8% of the total off-stream freshwater use. Water originating from the public supply is the main source of domestic water, accounting for 87% of the total. The remaining 13% of domestic water is self-supplied, almost entirely from groundwater wells. Self-supplied withdrawals are highest in the east, and are lowest in the arid southwest. Total domestic use is closely tied to population, with California, Texas, New York, Florida, and Illinois accounting for the largest withdrawals. Of the domestic water withdrawn, 26% (or 6.7 bgd) is consumptively used; with the remaining 74% making up the return flow (Solley, 1998).

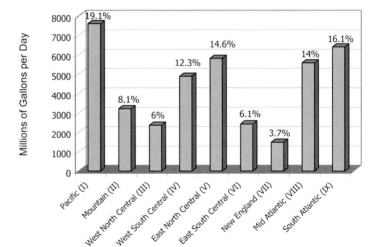

Fig. 22.1. Total volume of withdrawals for public water supply by water use data files region. (Source: US Geological Survey Water Use Data Files, 1999.)

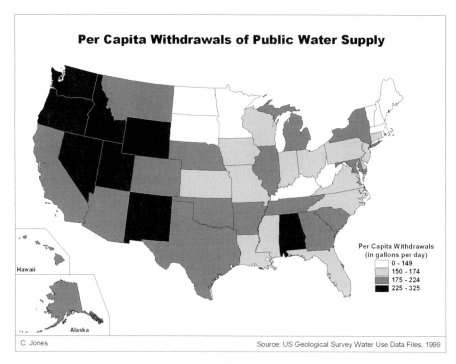

Fig. 22.2. Daily per capita withdrawals for public water supply by state. (Source: US Geological Survey Water Use Data Files, 1999.)

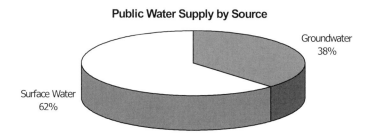

Fig. 22.3. The public water supply is derived primarily from surface waters. Surface water withdrawals total 28,808 million gallons per day (mgd); groundwater withdrawals total 14,976 mgd.

Average per capita domestic water use from public supplies is 101 gal/day, and from self-supplied water is 80 gal/day. In many areas, the largest household uses of water is for lawn and garden irrigation (AWWARF, 1999), therefore the warmer, arid western and southern regions have the highest per capita domestic use. Similarly, the cooler, humid regions in the East and North have the lowest per capita domestic use. Per capita domestic use from public supplies ranges from 62 gal/day in Pennsylvania to 213 gal/day in Nevada. Per capita domestic use from self-supplied sources ranges from 39 gal/day in Alaska to 168 gal/day in Idaho (Solley, 1998).

Domestic and public water use rose steadily from 1960 to 1995, in response to population growth (Figure 19.9). During this period, per capita domestic water use also increased from 89 gal/day in 1960 to 122 gal/day in 1990, largely in response to a decrease in household size. A certain amount of water is necessary to support the infrastructure of a household, regardless of the number of residents in that household. Other factors resulting in the per capita rise include the conversion of households to complete plumbing and increase in the use of appliances such as washing machines, dishwashers, swimming pools, and lawn sprinkler systems.

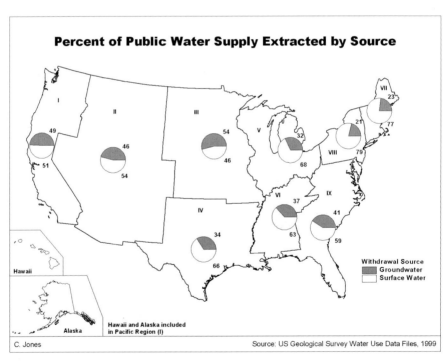

Fig. 22.4. Percentage of public water supply extracted from ground- and surface waters by water use data files region.

Public Water Supply Delivery

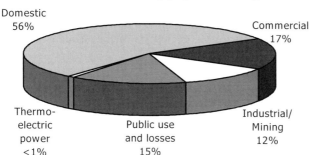

Domestic
56%

Commercial
17%

Thermo-
electric
power
<1%

Public use
and losses
15%

Industrial/
Mining
12%

Fig. 22.5. Public water use by sector. (Source: US Geological Survey Water Use National Data Files, 1995.)

Recent trends indicate that per capita domestic use is stable in many areas, and is decreasing in some regions as a result of conservation programs, water metering, installation of efficient plumbing fixtures, and stabilization in the number of people per household (Brown, 1999). Although per capita domestic water use is expected to remain constant through 2040, total domestic and public water use is expected to increase by 13 bgd (or 42%), as the population increases (Figure 19.10).

22.2.2 Commercial Use

Commercial water use in the United States is 9.6 bgd, accounting for 3% of the total off-stream freshwater use. Most of the commercial water used (70%) comes from the public supply; the remaining 30% self-supplied from surface water (20%) and groundwater (10%) sources. Only 14% of the commercial water used is consumed; the remaining 86% makes up the return flow.

The Pacific Northwest, California, and South Atlantic-Gulf regions have the greatest total withdrawals of commercial water, exceeding 1.2 bgd in each region. The largest consumptive use of commercial water occurs in the California, South Atlantic-Gulf, Mid-Atlantic, and Lower Colorado regions, all of which exceed 100 mgd. Consumptive commercial water use is less than 80 mgd in all states except California, which consumes 259 mgd (Solley, 1998). All resource regions in the United States have self-supplied commercial withdrawals of less than 400 mgd, except for the Pacific Northwest, which totals 1.1 bgd in withdrawals. The large commercial withdrawals are generally associated with off-stream fish hatcheries that are included in the commercial category.

Future changes in commercial water use in the United States are linked to both population and economic trends, whereas domestic withdrawals are assumed to be only a function of population changes. Future changes in industrial water use are also assumed to be functions of population and economic trends; therefore commercial water-use projections for the United States are subsumed under industrial water-use projections in the following section.

23. Industrial Use

Industrial water use includes water for processing, washing, and cooling in manufacturing facilities. Major water-using industries in the United States include steel, chemical, and paper product manufacturing and petroleum refining. The total industrial water use in 1995 was 27.1 bgd, including 25.5 bgd of fresh water and 1.6 bgd of salt water. Industrial water use represents about 7% of the total off-stream freshwater withdrawals.

Consumptive use totals only 15% of withdrawn industrial water, with the remaining 85% making up the return flow. The greatest industrial withdrawals in the country are in the West-South-Central and East-North-Central regions, which account for 48% of the total industrial withdrawals (Figure 23.1). Louisiana, Indiana, and Michigan are the largest users of freshwater for industrial purposes, with withdrawals exceeding 2.2 bgd in each state. Other major industrial users of fresh water are West Virginia, Texas, and Pennsylvania, where withdrawals exceed 1 bgd each. Texas, Louisiana, California, and West Virginia consume the most fresh water for industry, exceeding 200 mgd.

Surface water comprises 82% of the nation's industrial water withdrawals and groundwater accounts for 18% (Figure 23.2). Surface water sources represent a larger fraction of industrial water used in the East and South, ranging from 65% to 88% of the industrial supply (Figure 23.3). In the West, the fraction of industrial water originating from surface water is slightly less, and ranges from 42% to 57% of the industrial supply.

Eighteen percent of industrial water used in the United States is supplied by public systems, and 82% is self supplied. Surface water accounts for 82% of the self-supplied industrial water, groundwater accounts for 18%, and reclaimed wastewater accounts for less than 1% of the self-supplied water used. Nine percent of the self-supplied surface water is saline, and less than 1% of the self-supplied groundwater is saline.

The mining industry uses water for the extraction of minerals or petroleum, as well as for quarrying, milling, washing, dust control, and other mine-related activities. Dewatering is not considered a mining use unless the water is used in other mine operations. All mining water is self-supplied. Twenty-seven percent of water used in mining operations in consumed, and 73% comprises the return flow.

Volume of Industrial Withdrawals by Region

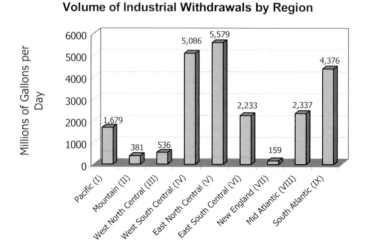

Fig. 23.1. Total volume of water withdrawals for industrial use by water use data files Region. (Source: US Geological Survey National Water Use Data Files, 1999.)

Industrial Water by Source

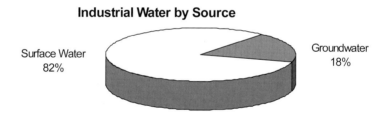

Fig. 23.2. Percentage of water used in industry that is extracted from surface water and groundwater. Surface water withdrawals total 18,277 million gallons per day (mgd); groundwater extractions total 4,092 mgd. (Source: US Geological Survey National Water Use Data Files, 1999.)

Water used in mining accounts for less than 1% of total off-stream freshwater uses. Mining operations withdraw 3.8 bgd, including 1.2 bgd of saline water.

Fifty-six percent of the water used in mining is groundwater, approximately half of which is saline, 44% is surface water, 11% of which is saline (Figure 23.4). The Texas Gulf and Great Lakes regions withdraw the greatest amount of water for mining, accounting for 24% of the national total. Minnesota, Florida, Texas,

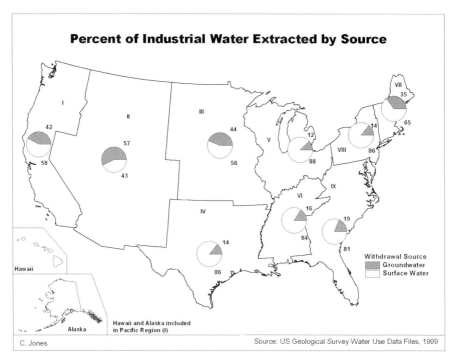

Fig. 23.3. Percentage of industrial water extracted from ground- and surface waters by water use data files region.

Mining Water Use by Source

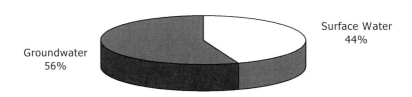

Fig. 23.4. Percentage of water used in mining derived from surface water and groundwater. Surface water withdrawals total 1,692 million gallons per day (mgd). Groundwater withdrawals total 2,076 mgd. (Source: US Geological Survey National Water Use Data Files, 1999.)

and Pennsylvania have the greatest freshwater withdrawals for mining, accounting for 32% of the national total. Texas and Arizona consume the largest amount of freshwater for mining, totaling over 200 mgd and 100 mgd, respectively.

Industrial and commercial withdrawals rose gradually from 1960 to 1980, and then decreased sharply after 1980 (Figure 19.9). Since 1960, trends in industrial and commercial withdrawals per dollar of personal income have steadily dropped, due to a shift away from water-intensive heavy industry and an increase in the efficiency of water use. Much of the improvements in efficiency are due to environmental pollution legislation that regulated discharges and encouraged industry to pioneer recycling techniques, thereby reducing withdrawals (Brown, 1999).

Total industrial and commercial withdrawals are expected to remain relatively stable in the upcoming years, rising 5%, from 37 bgd to 39 bgd, by 2040 (Figure 19.10). Although projected water withdrawals for industry and commerce are positively correlated with population, efficiency and recycling technology are also expected to improve. Such improvements should reduce projected increases in industrial and commercial water use caused by population and economic growth.

24. Water Quality

Water quality in the United States is assessed on a national level by the Index of Watershed Indicators (IWI), compiled by the EPA, and by the National Water Quality Assessment Program (NAWQA), administered by the USGS. The IWI uses data from many sources in order to map 15 indicators of water quality on a watershed basis across the entire United States (EPA 2000). The IWI is used to assess the level of watershed health and to determine whether activities on the surrounding lands are making the waters more vulnerable to pollution. The NAWQA program is designed to assess historical, current, and future water-quality conditions in 60 representative river basins and aquifers nationwide. The primary focus of NAWQA is to investigate the relationships between natural processes, human activities, and water quality, to provide a basis for improved decision making by water resource managers, planners, and policy makers. This chapter describes water quality conditions in the United States, investment aspects of water quality, and challenges facing water resource managers.

24.1 Assessments

Water quality conditions in the United States were compiled for 2,111 watersheds in the contiguous 48 states and are cataloged in the IWI (EPA, 2000). Watershed condition was assessed on the basis of the overall health of the watershed and its vulnerability to degradation by land-use activities occurring in the basin. The most recent IWI indicates that 16% of the watersheds have good water quality, 36% have moderate water quality problems, 21% have serious water quality problems, and 27% lack sufficient data to fully characterize the watershed.

Because of the large number of water bodies in the United States, subsets were surveyed for the water quality assessment. The IWI surveys covered 693,905 miles of rivers and streams, accounting for 53% of the perennial river miles or 19% of all river miles. Over 16.8 million acres, or 40%, of the nation's lakes, ponds, and reservoirs, and 28,819 square miles of estuaries, or 73%, were surveyed. In addition, 3,651, or 6%, of ocean shoreline miles, and 5,186 miles, or 94%, of Great Lakes shorelines were surveyed. Water quality in rivers, lakes, and estuaries was evaluated on the basis of how well the waters support designated beneficial uses. The individual beneficial uses were defined as follows:

- Aquatic life support. The water body provides suitable habitat for the protection and propagation of desirable fish, shellfish, and other aquatic organisms.
- Fish consumption. The water body supports fish free from contamination that could pose a human health risk to consumers.
- Shellfish harvesting. The water body supports a population of shellfish free from toxicants and pathogens that could pose a human health risk to consumers.
- Drinking water supply. The water body can supply safe drinking water with conventional treatment.
- Primary contact recreation - swimming. People can swim in the water body without risk of adverse human health effects (such as catching waterborne diseases from raw sewage contamination).
- Secondary contact recreation. People can perform activities on the water (such as boating) without risk of adverse human health effects from ingestion or contact with the water.
- Agriculture. The quality of the water is suitable for irrigating fields or watering livestock.

States, Tribes, and other jurisdictions may also define their own individual uses to address special concerns. For example, many Tribes and States designate their waters for the following beneficial uses:

- Ground water recharge. The surface water body plays a significant role in replenishing ground water, and surface water supply and quality are adequate to protect existing or potential uses of groundwater.
- Wildlife habitat. Water quality supports the water body's role in providing habitat and resources for land-based wildlife as well as aquatic life.

Tribes may designate their waters for special cultural and ceremonial uses:

- Culture: Water quality supports the water body's role in Tribal culture and preserves the water body's religious, ceremonial, or subsistence significance. (EPA 2000)

The individual U.S. states, American Indian Tribes, and other jurisdictions assessed the ability of their water bodies to support the designated beneficial uses described above, and compiled the information into a summary use support determination (Table 24.1).

Table 24.1. Levels of summary use support help states, tribes, and other regulatory agencies characterize the degree to which their waterways meet their designated beneficial uses.

Use support level	Water quality condition	Definition
Fully supports all uses	Good	Water quality meets designated use criteria
Threatened for one or more uses	Good	Water quality supports beneficial uses now, but may not in the future unless action is taken
Impaired for one or more uses	Impaired	Water quality fails to meet designated use criteria at times
Not attainable	—	The State, Tribe, or other jurisdiction has performed a use-attainability analysis and demonstrated that use support is not attainable due to one of six biological, chemical, physical, or economic/social conditions specified in the Code of Federal Regulations

Source: US Environmental Protection Agency, 2000.

Table 24.2. Five leading pollutants and processes impairing water quality in surveyed rivers, lakes, and estuaries.

Rank	Rivers	Lakes	Estuaries
1	Siltation	Nutrients	Nutrients
2	Nutrients	Metals	Pathogens
3	Pathogens	Oxygen-depleting substances	Priority toxic organic chemicals
4	Oxygen-depleting substances	Siltation	Oxygen-depleting substances
5	Pesticides	Noxious aquatic plants	Oil and grease

Source: US Environmental Protection Agency, Office of Water, National Water Quality Inventory: 1996 Report to Congress.

Table 24.3. Five leading sources of water quality impairment due to human activities in rivers, lakes, and estuaries.

Rank	Rivers	Lakes	Estuaries
1	Agriculture	Agriculture	Industrial point sources
2	Municipal point sources	Unspecified non-point sources	Urban runoff/storm sewers
3	Hydrologic modification	Atmospheric deposition	Municipal point sources
4	Habitat modification	Urban run-off/storm sewers	Upstream sources
5	Natural resource extraction	Municipal point sources	Agriculture

Source: US Environmental Protection Agency, Office of Water, National Water Quality Inventory: 1996 Report to Congress.

24.2 Primary Contaminants

Where possible, the states, Tribes and other jurisdictions identified the pollutants and processes leading to water quality impairment. The five leading pollutants and processes that impair water quality in rivers, lakes, and estuaries are given in Table 24.2. The leading sources of water quality impairment in rivers, lakes, and estuaries related to human activities are given in Table 24.3. Water body surveys did not use statistical or probabilistic methods to characterize water quality, therefore results are specific to the water bodies surveyed and cannot be extrapolated to the entire set of water bodies in the United States.

Excess nutrients were identified as a major source of impairment in all three classifications of water bodies. Excess nutrients such as nitrogen and phosphorus compounds over stimulate the growth of aquatic weeds and algae. Noxious weeds and algae can clog navigable waters, interfere with swimming and boating, outcompete native vegetation, and lead to oxygen depletion. The common sources of the nutrients are crop and lawn fertilizers, sewage, manure, and detergents. High levels of nitrogen generally occur in streams and groundwater in agricultural areas, whereas high levels of phosphorus occur mainly in agricultural and urban areas (USGS, 1999a).

Oxygen-depleting substance can impair all types of water bodies. When levels of dissolved oxygen are too low, adult fish, eggs, and larvae can suffocate. Low oxygen levels also can suffocate the insects developing fish need for food, thereby starving developing fish. Severe oxygen depletion in water bodies typically results from the introduction of biodegradable organic materials into surface waters.

Sedimentation and siltation caused by eroding land surfaces impair both river and lake water quality. Siltation is particularly detrimental to fish populations, and can suffocate eggs and larvae, fill critical spawning sediments, and clog and abrade fish gills. Sediment sources include plowed fields, construction and logging sites, urban areas, and mined lands. Sedimentation is exacerbated by the removal of streamside vegetation, resulting in increased streambank erosion.

The presence of pathogens is important in both rivers and estuaries. Pathogens can cause a range of human illnesses when the water is ingested or otherwise contacted. Sources of pathogens include inadequately treated sewage and storm water, septic systems, runoff from livestock pens, and sewage dumped overboard from recreational boats.

Toxic organic chemicals, pesticides, metals, oil and grease can impair all types of water bodies to varying degrees. Many of these compounds are known to cause cancer in humans and birth defects in many species. Health risks associated with combinations of these substances and their degradation products are poorly understood and may impact aquatic life more severely than human health. In addition, many of these compounds do not break down in natural ecosystems, and tend to persist and accumulate in the environment, particularly in sediments and animal tissues.

In 1998 alone, more than 3.6 million tons of toxic materials were released to the environment, according to the Toxic Release Inventory (TRI) (EPA, 1998). Although a relatively minor portion of toxic releases occurred directly to underground or surface water (Figure 24.1), water quality is highly susceptible to contamination by atmospheric deposition and land disposal of wastes (EPA, 2000).

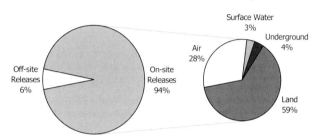

Fig. 24.1. Percentage of off- and on-site toxic releases and a breakdown of on-site toxic releases. Total off-site releases: 444.2 million pounds; total on-site releases: 6.863.1 million pounds. From the EPA's Toxic Release Inventory, 1998. (Source: US Environmental Protection Agency, Office of Pollution, 1998.)

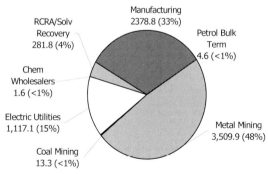

Fig. 24.2. Total and percentage of toxic releases by type of industry. From EPA's TRI, 1998. (Source: US Environmental Protection Agency, Office of Pollution, 1998.)

The metal mining industry is responsible for almost half of the toxic releases in the United States (48%), and manufacturing is responsible for roughly one-third of the releases (Figure 24.2). Other entities responsible for toxic releases are electric utilities, solvent recovery operations, coal mining, petroleum terminals, and chemical wholesalers. Toxic release trends for manufacturing and Federal facilities exhibit a 45% decrease, corresponding to a reduction of 1.5 billion pounds since 1988, as a result of environmental regulations. Releases to air comprise the greatest proportion of the reduction, making up 1.3 billion pounds of the total decrease (EPA, 1998).

24.2.1 Rivers and Streams

In the U.S., 19% of all river miles were surveyed for the IWI. Of these, 56% were categorized as exhibiting good water quality, fully supporting all uses, 8% were listed as threatened for one or more uses, 36% were listed as impaired for one or more uses, and less than 1% were listed as not attainable (Figure 24.3). Siltation was the most widespread contaminant, affecting aquatic habitat, treatment processes for drinking water, and recreational uses rivers and streams. Siltation impacted 18% of the surveyed river miles, followed by nutrients (14%), bacteria (12%), and oxygen-depleting substances (10%). Other leading factors were pesticides (7%), habitat alterations (7%), suspended solids (7%), and metals (6%) (EPA, 2000).

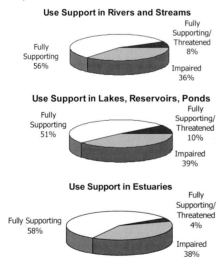

Fig. 24.3. The rivers, lakes, and estuaries surveyed by the National Water Quality Inventory are characterized by their levels of use support. The survey assessed 19% of river and stream miles, 40% of lake acres, and 72% of estuary square miles in the United States. Most of the surveyed waters of each type fully support their designated beneficial use. (Source: US Environmental Protection Agency, Office of Water, National Water Quality Inventory: 1996 Report to Congress.)

Agricultural activities impaired the largest percentage (20%-25%) of surveyed river miles (Table 24.3; Figure 24.4). Non-irrigated crop production impaired the most river miles, followed by irrigated crop production, rangeland, pastureland, feedlots, animal operations, and animal holding areas (Figure 24.5). Major sources of agricultural pollution in rivers and streams are associated with non-irrigated crop production, animal operations, and irrigated crop production. Pesticides can also persist in stream sediments and aquatic biota for many years after cessation or reduction in use. Nationally, levels of most organochlorine pesticides in whole freshwater fish have declined since the 1960s, but are still present. For example, mean DDT concentrations in whole fish decreased from approximately 1,800 g/kg in 1969, to 900 g/kg in 1972 when use was discontinued, to 250 g/kg in 1986 (USGS, 2000a).

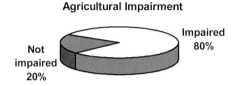

Fig. 24.4. Agricultural impairment of surveyed rivers and streams in the United States.

Table 24.4. Total miles of surveyed rivers and streams impaired by agricultural sources.

Source	Miles
Non irrigated crop production	61,950
Irrigated crop production	37,562
Rangeland	20,316
Pastureland	19,765
Feedlots	13,994
Animal operations	12,835
Animal holding areas	8,170
Total impaired miles*	173,629
Total miles surveyed	693,905
Estimated total river miles in U.S.	3,634.152

* Note: In Table 24.4, river miles impacted by individual agricultural activities do not add up to the total river miles impaired by agriculture in general for the following reasons : 1) Less than half of the 49 States, Tribes, and Territories that reported impacts from agriculture, in general, identified specific agricultural activities contributing to water quality impacts; 2) The 22 States that did provide more detailed information could not identify specific agricultural activities causing impacts in all waterbodies impacted by agriculture, in general; and 3) The river miles impacted by individual agricultural activities are not additive because more than one agricultural activity may impact a single river or stream segment and EPA tabulates the miles impacted by each activity separately. (Source: US Environmental Protection Agency, Office of Water, *National Water Quality Inventory: 1996 Report to Congress.*)

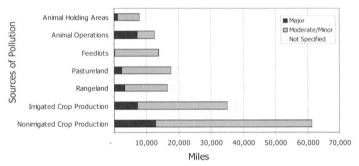

Fig. 24.5. Miles of surveyed rivers and streams that are impaired from different types of agricultural activities. Each source of agricultural pollution is also specified as either major or moderate/minor impairment. (Source: US Environmental Protection Agency, Office of Water, *National Water Quality Inventory: 1996 Report to Congress.*)

Municipal *point sources* were the second largest source of water quality impairment in rivers and streams (5%), despite a large number of wastewater treatment facilities (Figure 24.6), recent improvements, and permit controls on discharges (EPA 2000). Municipal sources remained significant as a result of population increases that place a burden on existing facilities. Total nitrogen levels below wastewater treatment plants have remained stable since the early 1970s, however treatment has altered the form of nitrogen from ammonia to nitrate, which is less toxic to fish, but still promotes excessive growth of noxious weeds (USGS, 1999*a*).

Hydrologic and habitat modifications are found to impair 5% of the surveyed river miles each. Such modifications include activities that alter the volume of flow in river, such as channelization, damming, and dewatering of channels. Habitat modifications include the alterations or removal of streamside vegetation that protects the stream from high temperatures, and reduces stream bank erosion. Resource extraction, urban runoff, and industrial point sources are found to impair a significant number of river miles. In addition to contamination resulting from human activities, surveys also noted that natural processes such as low flows and soils with naturally occurring metal deposits affect a significant number of river miles, preventing waters from supporting designated beneficial uses.

Hydrology and land use are the most important factors controlling the input of nutrients and pesticides to surface waters. The primary factors controlling the vulnerability of surface waters to contamination are slope and soil type, which interact to control the amount and timing of runoff (USGS, 1999*a*). Streams in areas characterized by high rainfall, steep slopes, and a large proportion of poorly drained or impervious areas have the highest potential to receive contaminated surface runoff.

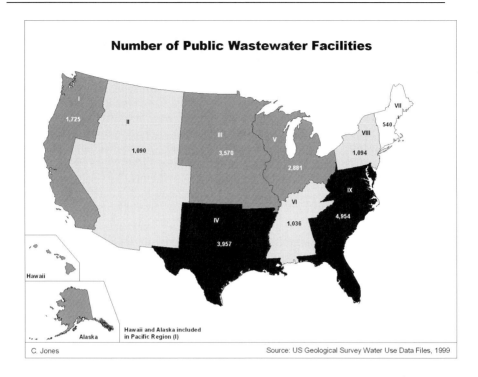

Fig. 24.6. Total number of public wastewater treatment facilities by water use data files region.

24.2.2 Lakes, Ponds, and Reservoirs

National surveys covered 40% of all lake areas. Of the surveyed areas, 51% were categorized as fully supporting all beneficial uses, 10% were listed as threatened for one or more uses, 39% were listed as impaired for one or more uses, and less than 1% were listed as not attainable (Figure 24.3). Excess nutrients and metals accounted for 51% of the total impaired lake area and 40% of the surveyed areas. In addition, siltation was found to impair 10% of the surveyed lake area, and enrichment by organic wastes that deplete oxygen was found to affect 8% of the surveyed area. The remaining areas were impacted by noxious aquatic weeds (6%), suspended solids (5%), and toxic compounds (5%). Agricultural activities were the largest source of contamination lakes, impacting 19% of the surveyed lake area (EPA 2000). Other major sources included unspecified *non-point sources*, which affected 9% of the areas, atmospheric deposition (8%), urban runoff and storm sewers (8%), municipal point discharges (7%), and hydrologic modifications (5%). Minor sources of contamination were construction activities and land disposal of waste (i.e., landfills), each affecting 4% of the surveyed area.

The Great Lakes (Superior, Michigan, Huron, Erie, and Ontario) were surveyed separately from the other lakes in the United States. Water quality in the Great Lakes is of great concern because the lakes contain one fifth of the world's fresh surface water and are a relatively closed system. Ninety-four percent of the Great Lakes shoreline was surveyed. Of the surveyed shoreline, only 2% was categorized as fully supporting all beneficial uses, 1% was listed as threatened for one or more uses, 97% was listed as impaired for one or more uses, and less than 1% was listed as not attainable. Great Lakes shorelines have been heavily impacted by chemicals, including toxic organic chemicals, mainly polychlorinated biphenyls (PCBs), which have affected 31% of the surveyed shoreline; pesticides (20%); and other organic chemicals (20%). Other constituents were nutrients (7%), metals (6%), and oxygen-depleting substances (6%). The major sources of contamination in the Great Lakes were atmospheric deposition (20%), discontinued point discharges (20%), and contaminated sediments (15%). Other important sources were land disposal of wastes (9%), unspecified non-point sources (6%), other point sources (6%), and urban runoff and storm sewers (4%).

24.2.3 Estuaries, Wetlands, and Ocean Shorelines

The IWI surveys covered 72% of all estuaries in the United States. Of those, 58% were listed as fully supporting all beneficial uses, 4% were listed as threatened for one or more uses, 38% were listed as impaired for one or more uses, and less than 1% were listed as not attainable (Figure 24.3). Nutrients were the major constituent of concern in estuaries, affecting 22% of the surveyed waters, and 57% of the impaired waters. Bacteria contaminated 22% of the surveyed estuary areas and 16% of the impaired estuarine waters, indicating that the waters were contaminated with sewage that may also contain viruses that cause diseases in humans. Other pollutants impairing estuaries were priority organic toxic chemicals (15%), oxygen depletion from organic wastes (12%), oil and grease (8%), salinity (7%), and habitat alterations (6%).

The most widespread source of pollution in estuaries was industrial discharge, which affected 21% of surveyed waters. Urban runoff and storm sewers impaired 18% of the surveyed estuary surface, municipal discharges impaired 17%, and upstream sources impaired 11% of the surveyed areas. Other sources were agriculture (10%), combined sewer overflows (8%), and land disposal of municipal and hazardous wastes (7%). Urban, municipal, and combined sewer overflow sources are a more significant source of pollution in estuaries than in rivers and lakes because large urban centers are located on many of the nation's major estuaries.

Wetlands are recognized as some of the most important lands in the nation for their function in providing food and habitat for a wide range of species, and their ability to intercept and remove contaminants from surface waters and buffer floodwaters. It is estimated that more than half of the nation's wetlands have been lost, mainly due to the conversion to agricultural and urban uses. Limited data indicates that sedimentation/siltation, nutrients, filling and draining, pesticides, flow alterations, habitat alterations, elevated metals, and salinity are the major causes of wetland degradation. Agriculture, hydrologic modification, urban runoff, filling and draining, construction, natural processes, dredging, resource extraction, and livestock grazing are the main sources of wetland degradation.

Six percent of the nation's ocean shorelines were surveyed, totaling 3,651 miles. Of the shoreline surveyed, 79% had good water quality, fully supporting all beneficial uses, 9% was considered as threatened for one or more uses, 13% was listed as impaired for one or more uses, none was listed as not attainable. The primary constituents impairing ocean water quality are bacteria, turbidity, nutrients, oxygen-depleting substances, suspended solids, acidity, oil and grease, and metals. The primary sources are urban runoff and storm sewers, land disposal of wastes, septic systems, municipal sewer discharges, industrial discharges, recreational marinas, spills, and illegal dumping of wastes.

24.3 Groundwater

Groundwater quality impairment is strongly controlled by hydrology and land use, and is most common in developed areas, agricultural areas, and industrial complexes. Groundwater contamination frequently occurs as well-defined plumes emanating from specific sites, such as spills, landfills, and waste lagoons. Non-point sources of pollution such as agricultural fertilizer and pesticide applications, septic systems, urban runoff, animal operations, and mining activities can impair groundwater over large areas.

Groundwater quality assessments indicate that nitrates, metals, volatile organic compounds, and semi-volatile organic compounds are the constituents that most commonly degrade water as a result of human activities. Groundwater degradation mainly affects drinking water sources due to high quality standards for water used for human consumption. Poor groundwater quality also affects irrigation, commercial, livestock and industrial water sources, and can be a non-point source of surface water contamination.

The most important source of groundwater contamination in the United States is leaking underground storage tanks (USTs). Leakage from USTs is commonly due to improper installation or subsequent corrosion of tanks and distribution lines. Over 300,000 releases from USTs have been confirmed nationwide, 60% of which impact groundwater supplies. The second largest source of groundwater impairment is from municipal and industrial landfills. Landfills contain metals, halogenated solvents, petroleum compounds, and pesticides, which are all constituents of concern. Septic systems are another primary source of groundwater impairment. Releases from septic systems occur because of faulty designs and installations or when such systems are improperly located and maintained. Contaminants associated with septic systems include bacteria, nitrates, viruses, phosphates, and other chemicals originating from household products.

Groundwater is the most susceptible to nitrate contamination in areas of high rainfall or in well-drained soils, areas where crop-management practices slow runoff, and where organic matter is low and dissolved oxygen levels are high (USGS, 1999b). The largest risk of groundwater contamination by nitrate occurs throughout the agricultural regions in the Midwest, central valley of California, Columbia Plateau, and isolated in areas in the East (Figure 24.7). The lowest risk of nitrate contamination occurs mainly in mountainous regions and rangelands, where nitrogen inputs are low (Nolan et al., 1997). Increased knowledge of the risks of groundwater contamination can help water-resource managers and private landowners identify the need to protect water supplies. Furthermore, it represents a significant step forward in linking scientific analyses with policy decisions to reduce non-point source pollution of groundwater.

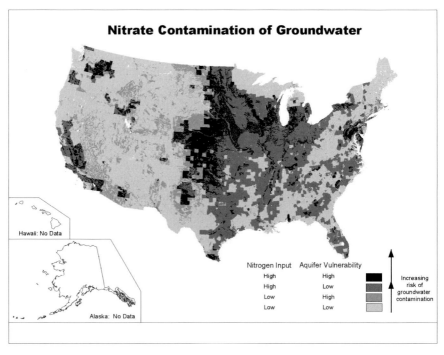

Fig. 24.7. Risk of nitrate contamination of groundwater as a function of nitrogen inputs and aquifer vulnerability

24.4 Investment Aspects of Water Quality

Since the early 1970s, private and public sectors have spent more than $500 billion on quality control measures, most of which have been directed toward municipal and industrial point sources (Knopman and Smith, 1993). In 1994, the estimated current and planned spending for water quality improvement under the Clean Water Act (CWA) was $63.7 to $65.1 billion per year (EPA, 1994). The expenditures were for all aspects of pollution control, including development of water quality standards, water quality assessment, characterization of pollution sources, development of wasteload allocations, implementation of source controls, and evaluation of control effectiveness.

Private entities are responsible for roughly half of CWA-related expenditures, and municipalities are responsible for approximately one-third of the total CWA expenditures (Figure 24.8). Federal agencies comprise about 15% of the total CWA expenditures to administer water quality programs. Expenditures by agriculture and by state water programs currently represent minor portions of the total CWA expenditures.

The greatest portion of CWA-related expenditures ($52.6 billion per year) are to comply with pre-1987 base programs, including Water Quality Standards (WQS), Total Maximum Daily Load (TMDL) monitoring, and the National Pollutant Discharge Elimination System (NPDES). Expenditures for the base programs are mainly made by private sources (48%), municipalities (33%), and federal agencies (18%), with minor expenditures by agriculture and state water programs (<1%). Expenditures of $9.1 to $10 billion per year are estimated for the construction of combined sewer overflow (CSO) systems and for the treatment of storm water by private entities and municipalities.

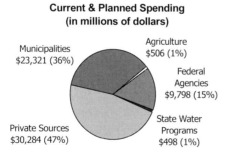

Current & Planned Spending
(in millions of dollars)

Municipalities
$23,321 (36%)

Agriculture
$506 (1%)

Federal
Agencies
$9,798 (15%)

State Water
Programs
$498 (1%)

Private Sources
$30,284 (47%)

Fig. 24.8. Current and planned spending by private and public sectors under the existing (1994) Clean Water Act. (Source: US Environmental Protection Agency, 1994.)

Expenditures of $1.0 to $1.3 billion are made to comply with post-1987 CWA programs including non-point source controls and watershed protection measures. Additional expenditures of $1.0 to $1.2 billion are made to cover other miscellaneous costs.

Although reductions in non-point source releases are noted to be a high priority, expenditures comprise approximately 2% of the CWA total. Non-point source expenditures are equally split between municipalities, agriculture, and federal agencies. Federal agency expenditures for non-point sources are dominated by the Department of Agriculture, with minor expenditures by the Department of the Interior and EPA (Figure 24.9).

Obligations Addressing Non-point Source Pollution by Agency

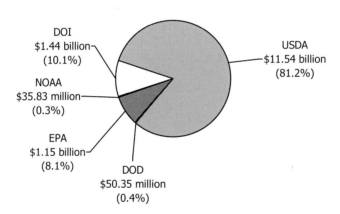

Fig. 24.9. Expenditures by the Department of Agriculture (USDA), Department of Defense (DOD), Environmental Protection Agency (EPA), National Oceanic and Atmospheric Administration (NOAA), and the Department of the Interior (DOI) to address non-point source pollution. The U.S. Government Accounting Office (GAO) estimated total, based on agencies data, is $14.2 billion (total and individual amounts may not sum due to rounding). Many programs do not have specific non-point source pollution objectives, but address non-point pollution through other program objectives. (Source: GAO, RCED, *Report on the Federal Role in Addressing Nonpoint Source Pollution*, 1999.)

24.5 Future Challenges

Water quality in the U.S. greatly improved after the passage of the CWA in 1972, largely due to improvements in municipal wastewater treatment and technology-based standards for industrial point sources (Boyd, 2000). Yet despite the improvements in water quality since 1972, 36% of surveyed rivers and streams, 39% of surveyed lakes, and 38% of surveyed estuaries are categorized as fully or partially impaired by pollution (EPA, 2000). Most point sources of water pollution are currently regulated; therefore non-point sources of now of paramount concern in the United States.

To address non-point sources, a total maximum daily load (TMDL) approach is being implemented. This method specifies the maximum amount of given pollutants that may be present in a water body, and allocates the allowable amount of pollution among sources. Instead of focusing on individual sources of contaminants, the TMDL approach focuses on monitoring and controlling water quality in individual bodies of water. Although this approach may have a significant effect on the regulation of point sources, it is even more important for the regulation of non-point sources of pollution. The regulation and reduction of non-point sources are particularly difficult from legal, scientific, and technical perspectives, however. Federal authority to mandate the control of non-point sources is weak, and there are frequently geographical differences between the pollution sources and legal jurisdictions. Furthermore, implementation of the analytical procedures required by the TMDL process are costly and difficult, and scientific understanding of the processes affecting and controlling the distribution of non-point materials are lacking (Boyd, 2000).

The politics, economics and implementation of water quality regulation in the United States will change significantly as federal and state agencies seek to control non-point sources through the TMDL program. For example, federal authority to implement non-point source controls will have to be enhanced, which may be accomplished by reclassifying certain non-point sources as point sources. Non-point source enforcement tools at the state level have failed to yield acceptable water quality in many water bodies; therefore, the implementation of non-point controls at the state level is also important for the improvement of water quality. Non-point source control is frequently implemented through the use of best management practices (BMPs), which are general standards that guide the management of land-use activities such as forestry, agriculture, and construction for the benefit of water quality. Many BMPs are used as general guidelines, but are not directly enforceable at present; however, recent legal cases indicate that BMPs may increasingly be used to determine compliance with statutory land-management requirements.

From the technical perspective, the scientific analysis of non-point source loadings in support of the TMDL approach requires a sophisticated understanding of the interactions between complex environmental systems. An increased scientific understanding of the relationships between control practices and quantities of contaminants is necessary in order to increase the certainty that control measures will be effective (Boyd, 2000). Analysis of preliminary data from the NAWQA program has begun to enhance the understanding of process and pattern interactions that affect water quality. Continued NAWQA monitoring using nationally consistent techniques is necessary to track long-term water quality trends to evaluate the effectiveness of various control measures.

In many situations, water quantity decisions have a direct impact on water quality. Legal issues concerning the interrelations between water quality and quantity will present additional challenges for the implementation of the TMDL program. Interjurisdictional conflicts will have to be resolved in areas where water quality problems in upstream regions are translated to downstream jurisdictions, and in areas where airborne contaminants represent significant problems. Another consequence of the TMDL program is that it may prohibit or raise the cost of developing new pollution sources in watersheds that exhibit impaired water quality. The unintended incentive to locate new point and non-point sources in relatively pristine areas will also need to be addressed (Boyd, 2000).

Water quality regulation in the future may also encompass new classes of contaminants that are not currently being monitored in the United States. The USGS has implemented a national survey to provide baseline information on the occurrence of emerging contaminants in surface and groundwater. These potential contaminants include human and veterinary pharmaceuticals (including antibiotics, prescription and non-prescription drugs), industrial and household wastewater products (including antioxidants, detergents, and plasticizers), and sex and steroidal hormones. Sampling locations are expected to have a high susceptibility to contamination, to provide an indication of the potential for these compounds to enter the environment. Analytical methods are currently being developed and refined to measure these compounds at very low concentrations, less than 1 μg/L. This program will be the first nationwide assessment of the occurrence of these emerging contaminants, and will provide a basis for the design of future monitoring programs, and will promote the development of new analytical techniques for measuring these compounds in environmental samples (USGS, 2000).

Finally, the role of invasive species will garner more attention as exotic plants and animals continue to impact water quality and ecosystem integrity, increasing the financial burden on society. For example, zebra mussels, an exotic species, have aggressively populated much of the Mississippi River basin in just over a decade and are now a major concern (Figure 24.10). The Eurasian zebra mussel was unintentionally introduced, first appearing in 1988 in Lake St. Clair, between Lake Huron and Lake Erie. The species probably arrived in the ballast water of a ship traveling from Europe to the Great Lakes region. Zebra mussels can negatively affect the viability of native clam and mussel populations as well as threaten

Invasive Species:
Zebra Mussel Distribution

● Locations of confirmed zebra mussel sightings in the US from 1988 to 1998

C. Jones   Source: US Geological Survey, Florida Caribbean Science Center, 2000

Fig. 24.10. The rapid spread of zebra mussels in the eastern United States over a 10-year period is an example of the often-costly consequences of invasive species on water resources.

navigation and recreation. Colonies of zebra mussels can be so dense as to restrict the water flow of intake pipes to industries and municipalities that withdraw surface water. Billions of dollars have been spent to remove mussels from pipes and remedy other associated damages.

Recognition of the widespread presence and influence of invasive species throughout the United States has recently mounted. In response, many federal agencies and private organizations have begun to organize campaigns to study, control, and, where feasible, eradicate particular invasive species.

25. Public Health

Waterborne disease outbreaks (WBDOs) associated with drinking and recreational water are tracked with a passive surveillance system maintained by the Centers for Disease Control and Prevention (CDC), EPA, and the Council of State and Territorial Epidemiologists. The objectives of the surveillance system are to characterize the epidemiology of WBDOs, identify the etiologic agents associated with WBDOs, and train health personnel in how to detect and investigate WBDOs, as well as in how to collaborate with other government entities on initiatives to prevent waterborne diseases. The information collected by the system is also useful for evaluating the effectiveness of current technologies used to provide safe drinking water and for establishing research priorities and needs for improved water-quality regulations (Levy et al., 1998).

One of the greatest problems with tracking WBDOs, however, is that many cases are probably unreported. Each year, there are an estimated 7.1 million mild-to-moderate infections, 560,000 moderate-to-severe cases, and 1,200 deaths attributable to waterborne infectious disease in the United States (Morris and Levin, 1995). The large number of suspected unreported cases indicates the need for improved surveillance systems, improved water treatment, pathogen-specific monitoring, improved risk assessment methodologies, and an improved understanding of the role of susceptible populations in the transmission of waterborne disease (Ford, 1999).

Although cases of waterborne disease decreased dramatically through the early 1900s, over the past 30 to 40 years, the number of cases per outbreak and total number of cases has increased (Ford, 1999). The collaborative surveillance system has been used to monitor WBDOs in the U.S. since 1971. The highest numbers of outbreaks were reported between 1970 and 1983 (Levy et al., 1998). Gastrointestinal illnesses (AGI) of unknown etiology made up the largest proportion, accounting for nearly half of all outbreaks. In the United States, WBDOs are typically limited to isolated cases and are mainly linked to biological agents, and to lesser degree, chemical agents. All waterborne disease outbreaks are classified as either recreational-exposure related or drinking water-related.

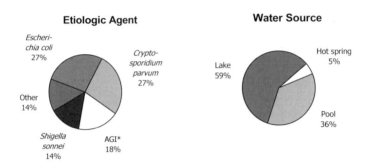

Fig. 25.1. Waterborne disease gastroenteritis outbreaks associated with recreational water, by etiologic agent and by water source. AGI = acute gastrointestinal illness of unknown etiology. (Source: U.S. Center for Disease Control and Environmental Protection Agency, Surveillance for Waterborne-Disease Outbreaks, 1998.)

25.1 Water Recreation

During 1995 and 1996, 37 waterborne disease outbreaks in 17 states were linked to recreational exposure, resulting in 9,129 illnesses and 6 deaths. The fatal cases were all individual occurrences of primary amoebic meningoencephalitis attributed to *Naegleria fowleri,* which accounted for 16% of the total number of illnesses. All fatalities were children who contracted the disease while swimming in natural water bodies.

Cases of gastroenteritis accounted for 60% of the total number of illnesses; cases of dermatitis made up the remaining 24%. Of all the recorded illnesses, 8,449 cases were associated with two large outbreaks of *Cryptosporidosis parvum* originating at public bathing facilities. *Cryptosporidium parvum* and *Escherichia coli* O157:H7 were the most commonly identified etiologic agents in gastroenteritis outbreaks (Figure 25.1). Gastroenteritis outbreaks were most commonly associated with lakes and to a lesser degree, swimming pools, with one outbreak occurring due to exposure at a hot spring. Recent trends of waterborne disease outbreaks indicate that illnesses caused by *Escherichia coli,* associated with recreational activities in lake water have increased. This increase indicates a need for improved water quality monitoring and source identification. Although outbreaks caused by parasites such as *Cryptosporidium* and *Giardia* have decreased, these agents are still responsible for a large number of individual illnesses and are associated primarily with swimming pools. Such outbreaks indicate a need for improved filtration techniques and education regarding the hazards of fecal accidents in public bathing facilities (Levy et al., 1998).

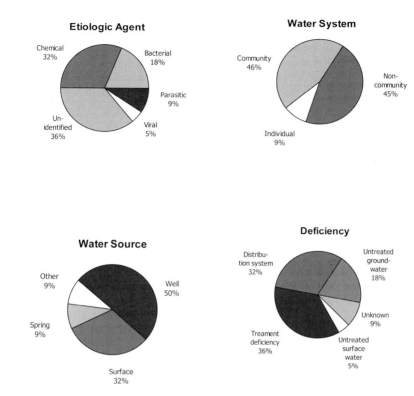

Fig. 25.2. Waterborne disease outbreaks associated with drinking water by type of etiologic agent, water source, water system, and deficiency. (Source: U.S. Center for Disease Control and the Environmental Protection Agency, Surveillance for Waterborne-Disease Outbreaks, 1998.)

25.2 Drinking Water

The most recent survey indicates that between 1995 and 1996, 22 outbreaks related to drinking water occurred in 13 states, resulting in 2,567 illnesses and no deaths. Approximately one-third of the outbreaks were related to biological agents, mainly *Giardia lamblia* and *Shigella sonnei* (Figure 25.2). Chemical contamination, specifically copper, nitrate, sodium hydroxide, chlorine, and concentrated soap, caused approximately one-third of the outbreaks, affecting 90 individuals. The agents responsible for the remainder of the outbreaks were unidentified, however the symptoms were noted to be consistent with viral agents. Of the outbreaks related to drinking water, roughly half were linked to private wells, and half were linked to community water systems (Figure 25.2).

Where disease outbreaks were linked to community water systems, 70% of the cases were identified as resulting from water distribution systems and plumbing at individual homes and businesses, whereas 30% occurred at the water treatment plant. Problems at individual facilities were typically associated with defective or improperly installed equipment. Recent trends indicate that outbreaks originating at water treatment plants have steadily declined since the late 1980s. However, the number of problems still occurring with distribution systems indicates the need to install and monitor appropriate equipment, such as backflow prevention devices.

25.3 Drinking Water Infrastructure Needs

The EPA sponsored the first national survey of capital improvement plans and engineering reports for 4,000 community water systems to assess drinking water infrastructure needs in the United States (EPA, 1997). Based on results of the survey, it is conservatively estimated that $138 billion will be required to upgrade and maintain community water systems through the year 2014. Of this total, $76.8 billion is for current needs to protect public health; the remaining $61.6 billion is for future needs, to provide safe drinking water through the year 2014. The largest portion of the need for improvements is at large systems, with medium, small, and American Indian and Alaska Native systems requiring smaller portions (Figure 25.3).

Total 20-Year Need by System Size
(in billions of dollars)

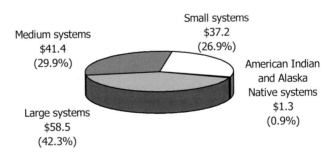

Fig. 25.3. Total 20-year investment need by system size, according to the EPA's first national survey of drinking water infrastructure needs. Large systems are those serving more than 50,000 people; medium systems serve 3,301 to 50,000; and small systems serve less than 3,300. (Source: U.S. Environmental Protection Agency, Office of Water, 1997.)

Although small supply systems have the smallest total need of the three system sizes, the per-household need is the largest of all systems (Figure 25.4). American Indian and Alaska Native systems have the highest per household need because the systems typically service small communities located in arid or permafrost regions where water sources are difficult to obtain and where systems are very expensive to construct. Water from many small systems poses a public health risk because components were improperly designed, constructed, and installed, and because many systems were built without adequate standards and plan review. It is estimated that 81% of small systems require distribution system upgrades, 67% require source improvement, and 10% of groundwater systems require the installation or replacement of treatment facilities.

The geographical distribution of total drinking water infrastructure needs is roughly correlated with state populations (Figure 25.5). The most populous states have the greatest infrastructure needs, with California, Texas, and New York exceeding $10 billion in required improvements. The sparsely populated western states have the smallest financial needs, requiring less than $1 billion.

The greatest requirement is for the improvement of transmission and distribution systems, accounting for 56% of the total funds needed (Figure 25.6). Leaking distribution pipes can lead to a loss of pressure in the system, resulting in contamination due to back-siphonage of contaminated water into the system. Leaks also waste water and energy due to transmission losses of treated water. Deteriorated transmission and distribution systems are common throughout the nation, particularly in older systems.

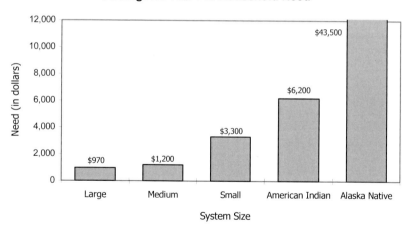

Fig. 25.4. Average 20-year investment needs per household by system sizes. Although large systems comprise 42.3% of system needs (Figure 25.3), the cost per household is lower because there are many more people served by large systems. (Source: US Environmental Protection Agency, Office of Water, 1997.)

Fig. 25.5. Total drinking water infrastructure needs, by state, in millions of dollars.

The second largest requirement is for improved water treatment facilities, accounting for 26% of the total need. Fifty-five percent ($20.0 billion) of the treatment requirements are for the treatment of microbiological contaminants that can cause acute health effects. Thirty-five percent of community water systems using surface water supplies need to install, upgrade, or replace filtration plants to reduce the risk of microbiological contamination. The amount needed for the treatment of contaminants that can cause chronic health problems, including cancer and birth defects, is $10.7 billion. The primary constituents of concern are byproducts of disinfection and lead. An additional $5.3 billion is needed for the treatment of secondary contaminants that affect the taste, odor, and color of water. Treatment of contamination from nitrate requires $0.2 billion. Nitrate can acutely affect health and can interfere with the ability of an infant's body to carry oxygen.

The construction of new storage facilities and the rehabilitation of existing facilities account for $12.1 billion, or 9% of the total need. About two-thirds of water systems need improvements in storage facilities. These improvements are mainly for conventional storage tanks and for covering (i.e., enclosing) finished-water reservoirs. Water storage is particularly important because it maintains positive pressure within the supply system, preventing the entrance of contaminants. Storage facilities also provide water for periods when demand exceeds the capacity of source or treatment systems.

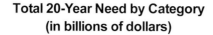

**Total 20-Year Need by Category
(in billions of dollars)**

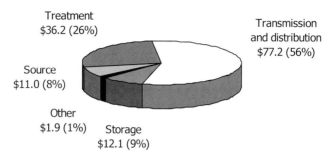

Treatment
$36.2 (26%)

Transmission
and distribution
$77.2 (56%)

Source
$11.0 (8%)

Other
$1.9 (1%)   Storage
$12.1 (9%)

Fig. 25.6. Transmission and distribution comprises the largest category of drinking water infrastructure need. (Source: US Environmental Protection Agency, Office of Water, 1997.)

The need for source development or rehabilitation accounts for $11.0 billion, or 8% of the total need. Although needed expenditures for source enhancement are relatively small, these expenditures are particularly important, because poor source quality can pose a public health risk or require expensive treatment. Additional source capacity is also needed to maintain adequate supplies during dry conditions, and maintain system pressure during periods of high water demand. Source development needs cover a wide range of scales, from the creation of large surface reservoirs to the installation of wells for small systems.

Approximately $12.1 billion of the current need is for systems to come into compliance with the Safe Drinking Water Act (SDWA). The largest component of the need, or 84%, is for treatment of microbiological contaminants, 14% is for treatment of contaminants that pose a chronic health risk, and 2% is to meet the concentration standards for nitrate. An additional $4.2 billion will be needed over the next 20 years to meet existing SDWA regulations, mainly for treatment of microbiological contaminants and for the replacement of lead distribution lines. An estimated $14.0 billion will be needed to comply with new and proposed drinking water standards for radionuclides, arsenic, radon, and sulfate (EPA, 1997). In order to meet the increasing demand for safe water supplies, water mangers will be required to protect existing infrastructure and to develop new water supply facilities.

Managers will be faced with the challenges of updating aging systems, balancing increased demand with the decreasing availability of new, high quality water sources, and meeting quality demands for microbiological and chemical contaminants, in an economically feasible manner. The development of multiple, innovative approaches is therefore critical for providing clean water for the growing population of the United States.

26. Conclusions

Many challenges are currently confronting water resource managers in the United States, and the number and complexity of these challenges will increase as demands for water supplies and improved water quality increase. The primary challenges are to improve efficiency and balance water withdrawals, while maintaining ecosystem health, protecting and improving water quality (particularly in regard to non-point source pollution), and maintaining an aging infrastructure of dams and supply facilities.

The population of the United States is projected to increase 41% by the year 2040, whereas freshwater withdrawals are projected to increase by 7%. Withdrawal projections assume that water-use efficiency in many sectors will continue to increase during this period. Continued efficiency improvements therefore are critical to meet future demands, particularly in the irrigated-agriculture and domestic-commercial sectors that consume the largest amounts of water. Improved water-use efficiency may have the added benefit of enhancing water quality by reducing the amount of contaminated return flows to the hydrologic system. Despite continued improvements in efficiency, it is highly likely that new water sources and novel storage and management techniques will need to be developed and employed to meet the demands of the growing population. Management strategies such as water marketing, banking, transfers, off-stream storage, and aquifer recharge are particularly important in arid regions, where all water supplies may currently be appropriated. Where new sources are available for development, the primary challenge will be to maintain a sustainable supply, while balancing the requirements for ecosystem health.

The improvement and protection of water quality and watershed health is a major challenge confronting U.S. water managers. Recent watershed surveys indicate that 57% of the watersheds surveyed have moderate to serious water quality problems. Water quality can improve as a result of increased point source regulation; however non-point source reduction is imperative if continued improvements are to be realized. Non-point source reductions present many legal, technical, and scientific challenges that mangers must overcome. Management strategies are needed to account for geographical patterns in land use, contaminant releases, chemical applications, and natural factors, on local to regional scales. Development of new or revised environmental policies must consider the entire hydrologic system and its many complexities, including surface and groundwater interactions, atmospheric contributions of contaminants, and new challenges such as invasive species.

Water quality standards and monitoring programs should reflect environmental conditions, including seasonal variations and contaminant mixtures. Finally, reliable and cost-effective predictive models are required to estimate water quality conditions that cannot be directly measured for a wide range of possible scenarios (USGS, 1999a).

A large portion of the water management and supply infrastructure in the United States requires repair, rehabilitation, or improvement, with a high associated cost to society. Water managers will increasingly be confronted with the need to provide safe water supplies while demands increase and as existing facilities deteriorate. Managers of dams and associated storage facilities will also be charged with both infrastructure and environmental protection, as facilities age and as societal demands change for water supplies, power, flood protection, ecosystem health and recreation. Technological and scientific advances are needed to assist managers in the assessment and improvement of existing infrastructure.

Future challenges within certain regions and watersheds are likely to be very complex where water resources are limited, demands are high, water quality is poor or particularly vulnerable, or where infrastructure needs are great. Management challenges from year to year also will vary significantly as a result of climate variability. Policy makers and managers will be required to develop novel strategies to control water use and quality for different contingencies and for areas with many competing uses. Development of innovative management practices combined with an improved understanding of how complex natural systems function is therefore imperative for the effective management of water resources in the United States.

Appendix: Quick Guide to Equivalents and Conversion Factors

EQUIVALENTS

U.S. Customary	Metric Equivalent	Water Relations in Inch-pounds Units (Approximate Equivalents)	
1 inch	2.54 centimeters	1 million gallons	3.07 acre-feet
1 foot	30.48 centimeters		
1 yard	0.9144 meters	1 cubic foot	62.4 pounds
1 mile	1.609 kilometers		7.48 gallons
1 nautical mile	1.853 kilometers		
1 pound	0.4536 kilograms	1 acre-foot	325,851 gallons
1 gallon	3.79 liters		43,560 cubic feet
1 acre	0.4047 hectares		
1 acre	4050 meters^2	1 inch of rain	17.4 million gallons per square mile
1 square mile	259.008 hectares		27,200 gallons per acre
1 gallon	8.34 pounds		100 tons per acre

CONVERSIONS

Temperature
	Conversion Equation
Fahrenheit to Celsius	$C^\circ = 5/9(F^\circ - 32)$
Celsius to Fahrenheit	$F^\circ = (9/5C^\circ) + 32$

Area

Multiply	By	To obtain
Acres	43,560	square feet (ft^2)
	4,047	square meters (m^2)
	0.001562	square miles (mi^2)

Flow

Multiply	By	To obtain
Gallons per day (gal/d)	3.785	liters per day
Million gallons per day (Mgal/d)	1.121	thousand acre-feet per year
	0.001547	thousand cubic feet per second
	0.6944	thousand gallons per minute
	0.003785	million cubic meters per day
	1.3815	million cubic meters per year
Thousand acre-feet per year	0.8921	million gallons per day
	0.001380	thousand cubic feet per second
	0.6195	thousand gallons per minute
	0.003377	million cubic meters per day

Glossary of Technical Terms

The following definitions are derived primarily from the United States Geological Survey (USGS) and the Environmental Protection Agency (EPA). Most terms associated with water use, quantity, and distribution are from USGS, *Estimated Use of Water in the United States in 1990 Glossary of Water-use Terminology,* available on-line at http://water.er.usgs.gov/public/watuse/wuglossary.html. Most terms associated with water quality are from EPA, *Terms of Environment,* available on-line at http://www.epa.gov/OCEPAterms/.

acid deposition (atmospheric deposition, acid rain) — a complex chemical and atmospheric phenomenon that occurs when emissions of sulfur and nitrogen compounds and other substances are transformed by chemical processes in the atmosphere, often far from the original sources, and then deposited on earth in either wet or dry form. The wet forms, popularly called "acid rain," can fall to earth as rain, snow, or fog. The dry forms are acidic gases or particulates.

acre-foot (acre-ft) — the volume of water required to cover 1 acre of land (43,560 square feet) to a depth of 1 foot.

animal specialties water use — water use associated with the production of fish in captivity except fish hatcheries, fur-bearing animals in captivity, horses, rabbits, and pets. *See also* livestock water use.

aquaculture — farming of organisms that live in water, such as fish, shellfish, and algae. *See also* animal specialties water use.

aquifer — a geologic formation, group of formations, or part of a formation that contains sufficient saturated permeable material to yield significant quantities of water to wells and springs.

aquifer storativity — the volume of water released per unit area of aquifer and per unit drop in head

commercial water use — water for motels, hotels, restaurants, office buildings, other commercial facilities, and institutions. The water may be obtained from a public supply or may be self supplied. *See also* public supply and self- supplied water.

consumptive use — that part of water withdrawn that is evaporated, transpired, incorporated into products or crops, consumed by humans or livestock, or otherwise removed from the immediate water environment. Also referred to as water consumed.

conveyance loss — water that is lost in transit from a pipe, canal, conduit, or ditch by leakage or evaporation. Generally, the water is not available for further use; however, leakage from an irrigation ditch, for example, may percolate to a ground-water source and be available for further use.

cooling water — water used for cooling purposes, such as of condensers and nuclear reactors.

dewater — to remove water from a waterbody, aquifer, conveyance system, or storage system

discharge — flow of surface water in a stream or canal or the outflow of ground water from a flowing artesian well, ditch, or spring. Can also apply to discharge of liquid effluent from a facility or to chemical emissions into the air through designated venting mechanisms.

domestic water use — water for household purposes, such as drinking, food preparation, bathing, washing clothes and dishes, flushing toilets, and watering lawns and gardens. Also called residential water use. The water may be obtained from a public supply or may be self supplied. *See also* public supply and self-supplied water.

drought — a period of abnormally dry weather that persists long enough to produce a serious hydrologic imbalance (for example, crop damage, water supply shortage, etc.). The severity of the drought depends upon the degree of moisture deficiency, the duration, and the size of the affected area.

estuary — region of interaction between rivers and near-shore ocean waters, where tidal action and river flow mix fresh and salt water. Such areas include bays, mouths of rivers, salt marshes, and lagoons. These brackish water ecosystems shelter and feed marine life, birds, and wildlife.

evaporation — process by which water is changed from a liquid into a vapor.

evapotranspiration — a collective term that includes water discharged to the atmosphere as a result of evaporation from the soil and surface-water bodies and as a result of plant transpiration. *See also* evaporation and transpiration.

freshwater — water that contains less than 1,000 milligrams per liter (mg/L) of dissolved solids; generally, more than 500 mg/L of dissolved solids is undesirable for drinking and many industrial uses.

groundwater — generally all subsurface water as distinct from surface water; specifically, that part of the subsurface water in the saturated zone (a zone in which all voids are filled with water) where the water is under pressure greater than atmospheric.

hydroelectric power water use — the use of water in the generation of electricity at plants where the turbine generators are driven by falling water.

hydrology — the science dealing with the properties, distribution, and circulation of water.

hydrologic modifications — structures or activities that alter the physical properties of hydrologic systems affecting the distribution and/or transport of water. Examples include dams, stream channelization, and land use changes such as the creation of impervious surfaces (e.g., parking lots).

impervious — resistant to infiltration. The property of a material or soil that does not allow, or allows only with great difficulty, the movement or passage of water.

in-channel use — see instream use.

industrial water use — water used for industrial purposes such as fabrication, processing, washing, and cooling, and includes such industries as steel, chemical and allied products, paper and allied products, mining, and petroleum refining. The water may be obtained from a public supply or may be self supplied. *See also* public supply and self- supplied water.

infiltration — the penetration of water through the ground surface into subsurface soil or the penetration of water from the soil into sewer or other pipes through defective joints, connections, or manhole walls.

instream use — water that is used, but not withdrawn, from a ground- or surface-water source for such purposes as hydroelectric power generation, navigation, water-quality improvement, fish propagation, and recreation. Sometimes called non-withdrawal use or in-channel use.

irrigation water use — artificial application of water on lands to assist in the growing of crops and pastures or to maintain vegetative growth in recreational lands such as parks and golf courses.

landfills — 1. Sanitary landfills are disposal sites for non-hazardous solid wastes spread in layers, compacted to the smallest practical volume, and covered by material applied at the end of each operating day. 2. Secure chemical landfills are disposal sites for hazardous waste, selected and designed to minimize the chance of release of hazardous substances into the environment.

land subsidence — subsidence of the land surface caused by pumping of ground-water and subsequent compaction of the underlying aquifer, collapse of underground caverns or decomposition of peat soils.

livestock water use — water for livestock watering, feed lots, dairy operations, fish farming, and other on-farm needs. Livestock as used here includes cattle, sheep, goats, hogs, and poultry. Also included are animal specialties. *See also* animal specialties water use.

mining water use — water use for the extraction of minerals occurring naturally including solids, such as coal and ores; liquids, such as crude petroleum; and gases, such as natural gas. Also includes uses associated with quarrying, well operations (dewatering), milling (crushing, screening, washing, floatation, and so forth), and other preparations customarily done at the mine site or as part of a mining activity. Does not include water used in processing, such as smelting, refining petroleum, or slurry pipeline operations. These uses are included in industrial water use.

non-consumptive use — a type of water use that does not significantly diminish the quantity of water originally withdrawn. Water is available for reuse.

non-point sources — diffuse pollution sources (i.e., without a single point of origin or not introduced into a receiving stream from a specific outlet). The pollutants are generally carried off the land by storm water. Common non-point sources are agriculture, forestry, urban, mining, construction, dams, channels, land disposal, saltwater intrusion, and city streets.

offstream use — water withdrawn or diverted from a ground- or surface-water source for public-water supply, industry, irrigation, livestock, thermoelectric power generation, and other uses. Sometimes called off-channel use or withdrawal.

per capita use — the average amount of water used per person during a standard time period, generally per day.

point source — a stationary location or fixed facility from which pollutants are discharged; any single identifiable source of pollution; e.g., a pipe, ditch, ship, ore pit, factory smokestack.

pollutant — generally, any substance introduced into the environment that adversely affects the usefulness of a resource or the health of humans, animals, or ecosystems..

pollution — generally, the presence of a substance in the environment that because of its chemical composition or quantity prevents the functioning of natural processes and produces undesirable environmental and health effects. Under the

Clean Water Act, for example, the term has been defined as the man-made or man-induced alteration of the physical, biological, chemical, and radiological integrity of water and other media.

public supply — water withdrawn by public and private water suppliers and delivered to users. Public suppliers provide water for a variety of uses, such as domestic, commercial, thermoelectric power, industrial, and public water use. *See also* commercial water use, domestic water use, thermoelectric power water use, industrial water use, and public water use.

public water use — water supplied from a public-water supply and used for such purposes as firefighting, street washing, and municipal parks and swimming pools. *See also* public supply.

pumping lift — the vertical distance that water must be pumped from an aquifer or surface water body to reach the surface or conveyance system. Declining groundwater levels cause pumping lifts to increase, thereby requiring greater energy costs and/or equipment upgrades to pump the same quanity of water.

recharge — the process by which water is added to a zone of saturation, usually by percolation from the soil surface; e.g., the recharge of an aquifer.

reclaimed wastewater — wastewater treatment plant effluent that has been diverted for beneficial use before it reaches a natural waterway or aquifer.

reservoir — any natural or artificial holding area used to store, regulate, or control water.

return flow — the water that reaches a ground- or surface-water source after release from the point of use and thus becomes available for further use.

runoff — 1. The portion of precipitation that does not infiltrate the soil and ultimately drains to a surface water body (surface runoff). 2. The amount of water that leaves a drainage basin via a surface watercourse (basin runoff).

saline water — water that contains more than 1,000 milligrams per liter of dissolved solids.

self-supplied water — water withdrawn from a surface- or ground-water source by a user rather than being obtained from a public supply.

sewage — the waste and wastewater produced by residential and commercial sources and discharged into sewers.

surface water — an open body of water, such as a stream or a lake.

thermoelectric power water use — water used in the process of the generation of thermoelectric power. The water may be obtained from a public supply or may be self supplied. *See also* public supply and self-supplied water.

total maximum daily loads (TMDLs) — Total Maximum Daily Loads are a tool for implementing state water quality standards and are based on the relationship between pollutants and in-stream water quality conditions.

transpiration — process by which water that is absorbed by plants, usually through the roots, is evaporated into the atmosphere from the plant surface. *See also* evaporation and evapotranspiration.

wasteload allocations (WLAs) — The maximum load of pollutants each discharger of waste is allowed to release into a particular waterway. Discharge limits are usually required for each specific water quality criterion that is or is expected to be violated. The portion of a stream's total assimilative capacity assigned to an individual discharge. Pollutant loads are allotted to existing and future point sources such as discharges from industry and sewage facilities.

wastewater — the spent or used water from a home, community, farm, business, or industry that contains dissolved or suspended matter.

wastewater treatment — the processing of wastewater for the removal or reduction of contained solids or other undesirable constituents.

wastewater-treatment return flow — water returned to the hydrologic system by wastewater-treatment facilities.

watershed — the land area that drains into a stream; the watershed for a major river may encompass a number of smaller watersheds that ultimately combine at a common point.

water-resources region — designated natural drainage basin or hydrologic area that contains either the drainage area of a major river or the combined drainage areas of two or more rivers; of 21 regions, 18 are in the conterminous United States, and one each are in Alaska, Hawaii, and the Caribbean.

water use — 1) in a restrictive sense, the term refers to water that is actually used for a specific purpose, such as for domestic use, irrigation, or industrial processing. In this report, the quantity of water use for a specific category is the combination of self-supplied withdrawals and public-supply deliveries. 2) More broadly, water use pertains to human's interaction with and influence on the hydrologic cycle, and includes elements such as water withdrawal, delivery, consumptive use, wastewater release, reclaimed wastewater, return flow, and instream use. *See also* offstream use and instream use.

wetlands — area saturated or nearly saturated by surface or groundwater for most of the year with vegetation adapted for life under those soil conditions; examples include swamps, bogs, fens, marshes, and estuaries.

withdrawal — water removed from the ground or diverted from a surface-water source for use. *See also* offstream use and self-supplied water.

References

American Rivers, Friends of the Earth, and Trout Unlimited. 1999. Dam removal success stories. (web document).

American Water Works Association Research Foundation. 1999. Residential end uses of water [Project #241]. http://www.awwarf.com/exsums/90781.htm

Boyd, James. 2000. The new face of the Clean Water Act: A critical review of the EPA's proposed TMDL rules. Resources for the Future Discussion Paper 00-12, 35 pp.

Brown, Thomas C. 2000. Projecting U.S. fresh water withdrawals. *Water Resources Research,* Vol. 36, No. 3, p. 769–780.

Brown, Thomas C. 1999. *Past and Future Freshwater Use in the United States: A Technical Document Supporting the 2000 USDA Forest Service RPA Assessment.* Gen. Tech. Rep. RMRS-GTR-39. Fort Collins, CO: U.S. Department of Agriculture, Forest Service, Rocky Mountain Research Station. 47 pp.

Buckley, Robert G., J. Randy Young, and Mike Thralls. 1998. Aging watershed projects: A growing national concern. *Conservation Voices*, pp. 22–24.

Bureau of Economic Analysis. 1999. 1998 State per capita personal income and state personal income. http://www.bea.doc.gov/bea/newsrel/spi0499.pdf

Bureau of the Census. 1996. *Population Projections of the United States by Age, Sex, Race, and Hispanic origin: 1995 - 2050,* Report P25-1130. U.S. Department of Commerce, Washington, D.C.http://www.census.gov/prod/1/pop/p25-1130/p251130.pdf

CIA. 2000. *The World Factbook 1999-United States.* (http://www.odci.gov/cia/publications/factbook/us.html)

DOE. 1999. Inventory of Electric Utility Power Plants in the United States 1999, With Data as of January 1, 1999. http://www.eia.doe.gov/cneaf/electricity/ipp/ipp-sum.html

EPA. 1994. *President Clinton's Clean Water Act Initiative: Analysis of Costs and Benefits.* EPA 800-S-94-001. Office of Water, Washington D.C.

EPA. 1997. *Drinking water infrastructure needs survey, First report to Congress.* U.S. Environmental Protection Agency, Office of Water, 42 pp.

EPA. 1998. 1998 Toxics Release Inventory Data Release: Questions and Answers.

EPA. 2000. National Summary of Water Quality Conditions. http://www.epa.gov/OW/resources/9698/sec-one.pdf

Ferrey, Steven. 1997. *Environmental law, Examples and explanations.* Aspen Law and Business, New York. 525 pp.

Ford, Timothy E. 1999. Microbiological safety of drinking water: United States and global perspectives. *Environmental Health Perspectives* 107, 191–206.

Frederick, Kenneth. 1997. Water resources and climate change. Resources for the Future. http://www.rff.org/issue-briefs/PDF-files/ccbrf3.pdf

IBWC. 2000. http://www.ibwc.state.gov

IJC. 2000. http://www.ijc.org

Intergovernmental Panel on Climate Change (IPCC). 1996. *Climate Change 1995: Impacts, Adaptation and Mitigation of Climate Change: Scientific-Technical Analyses: Contribution of Working Group II to the Second Assessment Report of the Intergovernmental Panel on Climate Change,* Cambridge University Press.

Knopman, D. S. and R. A. Smith. 1993. Twenty years of the Clean Water Act, *Environment,* Vol. 35, No. 1, 17–34.

Levy, Deborah A., M. S. Bens, G. F. Craun, R. L. Calderon, and B. L. Herwaldt. 1998. Surveillance for Waterborne-Disease Outbreaks, United States, 1995-1996, MMWR 1998/47(SS-5); 1–34. Center for Disease Control.

Moore, M.R., W.M. Crosswhite, and J.E. Hostetler. 1990. Agricultural water use in the United States, 1950-85. In: *National Water Summary 1987 - Hydrologic Events and Water Supply and Use.* Carr, J.E., E.B. Chase, R.W. Paulson and D.W. Moody, compilers. U.S. Geological Survey Water Supply Paper 2350.

Moreland, Joe. 1993. Drought. USGS Open-file report 93–642. http://water.usgs.gov/pubs/FS/OFR93-642/index.html)

Morris, R. D. and R. Levin. 1995. Estimating the incidence of waterborne infectious disease related to drinking water in the United States. In: *Assessing and Managing Health Risks from Drinking Water Contamination: Approaches and Applications* (Reichard E. G. and Zapponi, G. A., eds.). IAHS Publ no. 233. Wallingford, UK: International Association of Hydrological Sciences, 75–88.

Mumme, Stephen P. 1996. *Groundwater Management on the Mexico-United States Border.* Report submitted to the Commission on Environmental Cooperation, Montreal (Quebec), Canada under Contract, Project 95.23.01, The Mexico-United States Canada Transboundary Inland Water Project.

National Center for Health Statistics, 1998. *Deaths: Final Data for 1996. National Vital Statistics Reports,* Vol. 47, No. 9, p. 22. National Center for Health Statistics, Hyattsville, Maryland.

National Research Council (NRC). 1999. *Restoration of Aquatic Ecosystems: Science, Technology, and Public Policy.* Committee on Restoration of Aquatic Ecosystems: Science, Technology, and Public Policy. National Academy Press: Washington, D.C., p. 26.

National Research Council (NRC), Committee on Long-Range Soil and Water Conservation Policy. 1993. *Soil and Water Quality: An Agenda for Agriculture.* National Academy Press. http://www.nap.edu/catalog/2132.html

Nolan, Bernard T., Barbara C. Ruddy, Kerie J. Hitt, and Dennis R. Helsel. 1997. Risk of nitrate in groundwaters of the United States - A national perspective. *Environmental Science and Technology,* 31, 2229-2236.

NRCS. 2000. Upstream flood control dams: A growing national concern. http://www.ftw.nrcs.usda.gov/p1566/agingwater/

Schwab, Glenn O., Richard K. Frevert, Talcott W. Edminster, and Kenneth K. Barnes. 1981. *Soil and Water Conservation Engineering,* 3rd Ed. John Wiley and Sons, New York, 525 pp.

Solley, Wayne B., Robert R. Pierce, and Howard A. Perlman. 1998. Estimated Use of Water in the United States in 1995. USGS Circular 1200, 71 pp.

Tarlock, A. Dan, James N. Corbridge, Jr., and David H. Getches. 1993. *Water Resource Management: a Casebook in Law and Public Policy,* 4th Ed. The Foundation Press, Inc. Westbury, NY, 930 pp.

US Census Bureau. 1998. Poverty in the United States. (http://www.census.gov/prod/99 pubs/p60-207.pdf)

USDA. 2000. Aquaculture outlook.http://usda.mannlib.cornell.edu/reports/erssor/livestock/ ldp-aqs/2000/aqs1100.pdf

USDA. 1999. 1998 Census of aquaculture. http://www.nass.usda.gov/census/census97/ aquaculture/aquaculture.htm

US Fish and Wildlife Service. 1999. Report of lands under control of the U.S. Fish and Wildlife Service as of September 30, 1999. http://realty.fws.gov/093099.pdf

USGS. 1999a. The quality of our nation's waters: Nutrients and pesticides. Circular 1225. http://water.usgs.gov/pubs/circ/circ1225/pdf/index.html

USGS. 1999b. Water-level changes, 1980 to 1997, and saturated thickness, 1996–97, in the High Plains aquifer. USGS Fact Sheet 124–99.

USGS. 1999c. Water Use Data Files.

USGS. 2000a. Pesticides in stream sediment and aquatic biota: Current understanding of distribution and major influences. USGS Fact Sheet 092–00.

USGS. 2000b. Status and trends of the nation's biological resources. http://biology. usgs.gov/ s+t/SNT/index.htm

USGS. 2000. http://toxics.usgs.gov/regional/emc.html

World Resources Institute. 1998. World Resources 1998–99 Database.

Agency Web Pages:

BLM. 2000: http://www.blm.gov/nhp/index.htm

EPA. 2000: http://www.epa.gov/

FEMA. 2000: http://www.fema.gov/

FERC. 2000: http://www.ferc.fed.us/default3.htm

NMFS. 2000: http://www.nmfs.gov/

NPS. 2000: http://www.nps.gov/

NRCS. 2000: http://www.nrcs.usda.gov/

US Census Bureau: http://www.census.gov

USACE. 2000: http://www.usace.army.mil/

USBR. 2000: http://www.usbr.gov/main/index.html

USDA FS. 2000: http://www.fs.fed.us/

USFW. 2000: http://www.fws.gov/

USGS. 2000: http://www.usgs.gov/

27. General Aspects

27.1 Physical Context

The United Mexican States, Mexico's official name, lies in the northern portion of the American continent between the coordinates: 32° 43' 06'' N (monument number 206 on the US-Mexico border), 14° 32' 27 S (mouth of the Suchiate river), 118° 27' 24'' W (Guadalupe island) and 86° 42' 36'' E (Isla Mujeres). It borders the United States to the north and Guatemala and Belize to the south-east. The Pacific Ocean sets the western limit and the Gulf of Mexico and the Caribbean the east. It covers a surface area of 1,953,162 km². The 11,122.5 km of coastline is washed by the Pacific, Gulf of California, Gulf of Mexico and the Caribbean. The Federal Law of the Sea sets a limit to territorial waters of 12 nautical miles, while the exclusive economic zone covers 3,149,920 km². The Tropic of Cancer runs through the centre of the country at the confluence of two bio-geographic regions, the neo-Arctic and the semi-tropical. As a result, the climate is dry in almost half of the country, while warm and temperate climates account respectively for 28 percent and 23 percent of national territory (Figure 27.1; INEGI 1998a; INEGI 2000a).

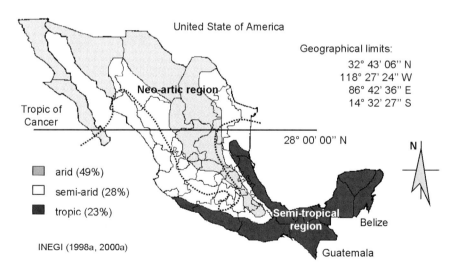

Fig. 27.1. Mexico geographical localization and climates

Mexico's diversity of forms of physical relief make it one of the world's most diverse countries in terms of topographical features and variety. Some 64 percent of the territory is hilly and only 36 percent has slopes of less than 10 percent; altitudes vary from sea level to 5,000 metres. Because of its topography and latitude, Mexico has an enormous variety of zones in terms of vegetation. At altitude, vegetation is scarce, but in zones such as the Lacandona jungle in the state of Chiapas, flora and fauna abound. Between these two extremes, there lies a tremendous variety of plant life that includes extensive areas of scrub, grassland, conifer and hardwood forests in almost all the mountainous zones, palm groves and jungle with a high degree of biodiversity, well-developed mangroves along the southern coasts and communities of pioneer plants amid the coastal sand dunes, to name but a few. Mexico is one of a handful of countries that has mega-diversity (Figure 27.2; INEGI 1998a; OECD 1998; INEGI 2000a).

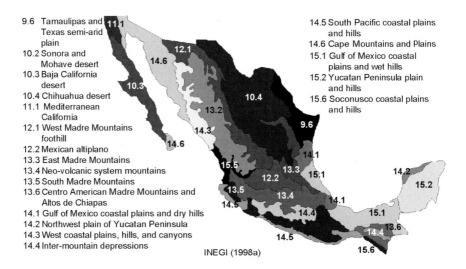

9.6 Tamaulipas and Texas semi-arid plain
10.2 Sonora and Mohave desert
10.3 Baja California desert
10.4 Chihuahua desert
11.1 Mediterranean California
12.1 West Madre Mountains foothill
12.2 Mexican altiplano
13.3 East Madre Mountains
13.4 Neo-volcanic system mountains
13.5 South Madre Mountains
13.6 Centro American Madre Mountains and Altos de Chiapas
14.1 Gulf of Mexico coastal plains and dry hills
14.2 Northwest plain of Yucatan Peninsula
14.3 West coastal plains, hills, and canyons
14.4 Inter-mountain depressions

14.5 South Pacific coastal plains and hills
14.6 Cape Mountains and Plains
15.1 Gulf of Mexico coastal plains and wet hills
15.2 Yucatan Peninsula plain and hills
15.6 Soconusco coastal plains and hills

INEGI (1998a)

Fig. 27. 2. Ecological Regions

Grassland, which occupies about 39 percent of the nation's territory, represents the most important land use, followed by the 30 percent of woods and forests. Cultivable land regularly sown to crops represents 13 percent, and 18 percent of land is devoted to other uses. Changes in land use have had a considerable impact on Mexico's eco-systems and natural resources. Between 1970 and 1990, the extent of arable land grew by 40 percent and pasture by 15 percent while forests declined by at least 15 percent. Mexico still accounts for 1.3 percent of the world's woodland resources but deforestation, despite slowing in recent years, continues at one of the highest rates in the Americas. Estimates vary greatly, but 1-2 percent, some 500,000 hectares, of forest, most of it tropical, are lost each year. In addition, as a result of fragmentation, the area classified as degraded forest increased from 17.8 million of hectares to 22.2 million of hectares between 1985 and 1994 (OECD 1998).

27.2 The Socio-economic Context

Current demography is the result of rapid population growth up to the 1970s. Mexico has some 100 million of inhabitants, and its population rates as relatively young, though the average age rose from 15.7 to 20 years between 1970 and 1992. In 1996, the nation had 20.5 million of households with an average 4.5 persons in each of them. (INEGI 1996a; INEGI 1996b; INEGI 1998a; OECD 1998). Following rapid growth for much of the last century, the rate declined from 3.2 percent in 1950 to 1.8 percent in 1995. However, there are marked differences between urban and rural areas in the way the birth-rate has declined. While the average number of births in Mexico City was 2.2 per woman, it was four in states such as Oaxaca and Chiapas (INEGI 1998a). Life expectancy rose from 50 years in 1955 to 75 in 1998, as a result of declines in the rates of both births and deaths. The mortality rate fell from 9.5 to 4.5 per thousand between 1970 and 1995. Factors that have helped the rate to decline include a growth in education and health services (INEGI 1998a; OPS 1994; SEGOB 1998). Distribution of the population has historically been irregular, with heavy concentrations in the cities and wide dispersion in rural villages (Figure 27.3).

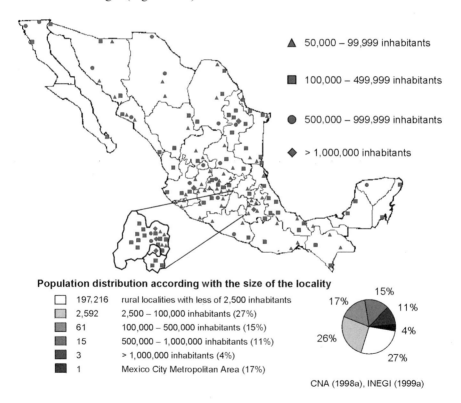

Fig. 27. 3. Geographical distribution of population

Each state's growth is accompanied by a complex process of migration within and between regions. These movements are the result of the growth in the socio-economic power that some urban and regional centres exercise over their periphery as a result of the assignation of economic resources and support by federal and state development policies. In recent decades, inter-state migration has generated major flows of population: in 1960, 5.2 million of people lived in a state different from the one of their birth but by 1995 that had grown to almost 18 million (19 percent of the population) (INEGI 1998a). At present, the centre and north, border municipalities in particular, act as powerful magnets because of their economic development and the lack of development programmes and financial incentives in other regions.

The population has gone from largely rural (70 percent in 1900) to predominantly urban (75 percent in 1995) (INEGI 1999a). There are states in which the population density is less than 30 inhabitants per square kilometre, indicative of low population in terms of territory but with a high concentration in the cities. Of the 31 states and the Federal District, the urban population accounts for 60 percent in 25. This points to a clear trend towards concentration in urban zones, a phenomenon that sets major challenges in terms of the provision of adequate infrastructure and health care. (Figure 27.4). Education has fallen badly behind. In 1992, the average educational level was the fifth year of primary school. (CONAPO 1992). By 1996, 63 percent of the schoolage population had yet to complete or receive primary education. Some 20 percent had completed primary school, 12 percent had completed the first three years of secondary education, 4 percent had reached university or college-entrance level, and 1 percent had completed higher education (BANAMEX-ACCIVAL 1999a). Some 40 percent of children who enter primary school drop out later for social or economic reasons (Urquidi 1996).

Education has become concentrated in some cities and certain social groups. Education is billed as the best way to prepare the human resources that the country needs for its development, but the educational system fails to implant values that promote the work ethic. The main task of Mexican schools is supposed to be the formation of citizens prepared to take part in democracy, but in practice they are authoritarian. Education has long been the hope for a better future, but its present leaves much to desire.

In 1997, gross domestic product (GDP) was 2,922,035 million pesos (1,273,383 million pesos at 1993 prices). while GDP per capita was 30,995 pesos (13,507 pesos at 1993 levels). From 1980 to 1990, real annual GDP growth was minus 0.3 percent, and plus 1 percent from 1991 to 1998, a rate inferior to that of the growth in population. GDP per capita grew by an average of 0.3 percent in recent years, while prices rose by an annual average of 37.4 percent in the last 25 years (Gutiérrez 1999a, INEGI 1999a; Solís 1999). Nearly 70 percent of GDP came from services, 24 percent from industry and 5.7 percent from agriculture, forestry and fisheries. (Figure 27.5). Four economic activities contributed 77 percent: manufacturing with 21 percent, retail restaurants and hotels 20 percent; community services 18 percent, and financial services 18 percent. In value terms, 84 percent of Mexican exports go to the United States and 5 percent to other Latin American countries (INEGI 1999a; OECD 1998; SEMARNAP 1997).

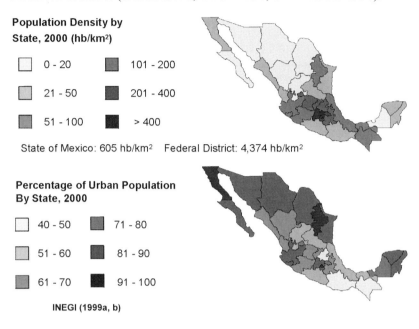

Population Density by State, 2000 (hb/km²)

☐ 0 - 20   ■ 101 - 200

☐ 21 - 50   ■ 201 - 400

■ 51 - 100   ■ > 400

State of Mexico: 605 hb/km²   Federal District: 4,374 hb/km²

Percentage of Urban Population By State, 2000

☐ 40 - 50   ■ 71 - 80

■ 51 - 60   ■ 81 - 90

■ 61 - 70   ■ 91 - 100

INEGI (1999a, b)

Fig. 27. 4. Population density and percentage of urban population by state

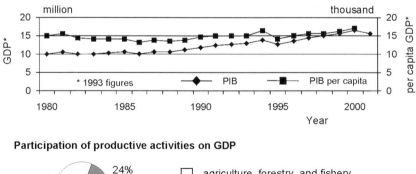

Participation of productive activities on GDP

24%   ☐ agriculture, forestry, and fishery

70%

6%   ■ Industry

■ Services

OCDE (1998), INEGI (1999a)

Fig. 27.5. GDP, per capita GDP, and economic participation of productive sectors

Growth in the Mexican economy in recent years has been based on external markets, with exports making the main contribution, while the domestic market has lagged behind. Domestic consumption has lost considerable momentum: in 1998 it grew by 7.3 percent while in 1999 it will do so by 1.6 percent. (Gutiérrez 1999b). Beginning in 1980, the economy changed from one in which the state had a broad and deep-going role in productive processes to one in which the principal role in generating growth and development was taken by the private sector, both national and foreign. Decision-making, in part, has moved abroad or into the domestic private sector. Agriculture, finance and transport and communications have been deregulated and rely now on private investment to ensure their expansion. However, problems persist in all three sectors of the Mexican economy:

1. Agriculture remains sunk in a deep crisis as a result of 30 years of capital loss. Of the 27 million hectares that can be cultivated, 75 percent is rainfed land worked with only a handful of technical resources at low levels of productivity and profitability and with little or no financial support. Its contribution to GDP in recent years has scarcely been 6 percent (OECD 1998; Urquidi 1996).
2. Some branches of the industrial sector, autos, glass, electronics, machinery, information technology, have developed to international levels of competitivity. On the other hand, low technology and outdated management techniques prevail at the overwhelming majority of micro, small and medium industrial establishments that face a serious crisis as a result of the depressed domestic market, a lack of credit on favourable terms and other financial and commercial limitations.
3. The services sector combines large modern companies, such as those in tourism and retail, with thousands of small and micro-firms without any basis of financial support and no hope of growth. The finance and insurance sectors face serious risks, seeking to survive through the introduction of foreign capital and costly restructurings (Urquidi 1996).

Some 99 percent of taxes are collected by the federal government, whose outlays form four-fifths of total state spending. The rest of the income is distributed among states and municipalities in accordance with a formula based mainly on size of population and the state's ability to generate income directly (OECD 1998). Income distribution is among the most unequal in Latin America: 63.2 percent of the population earns less than twice the minimum wage, while 64.1 percent of the nation's total income is concentrated in 30 percent. Average income per inhabitant was 7.6 percent lower in the 1990s than in the previous decade and scarely 4.6 percent higher than in the 1970s; purchasing power has fallen by 62 percent. As a result, 38 percent of the population lives in extreme poverty and 15 percent in relative poverty (Figure 27.6; BANAMEX-ACCIVAL 1999a; BANAMEX-ACCIVAL 1999b; Fabre 1999; Gutiérrez 1999b; INEGI 1998b).

The social consequences of the economic strategy that has been followed have been unimaginable: income inequalities have been exacerbated as have poverty and extreme poverty; the informal economy has grown and social stability has come under threat. Despite some positive aspects, the economic growth strategy, it cannot properly be called a development strategy, has failed to generate conditions that could significantly reduce social inequality. If sustainable and equitable development is taken to be the long-term aim of society, it is clear that the economic strategy is far from having taken that route.

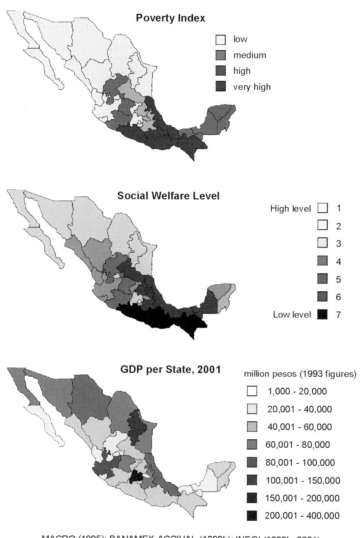

MACRO (1995); BANAMEX-ACCIVAL (1999b); INEGI (1999b, 2001)

Fig. 27.6. Poverty index, social welfare level, and GDP per state

28. Water Resources

As mentioned in the previous chapter, the result of the geographical spread and its physiographic features is that Mexico's water resources form a complex and varied mosaic. Authorities have defined 314 watersheds grouped into 37 water regions and 13 administrative entities defined in strictly hydrological terms. (Figure 28.1; SEMARNAP/CNA 1996).

VI. Bravo River
VII. Central Basins of North
VIII. Lerma – Santiago - Pacific
IX. North Gulf
X. Central Golf
XI. South Border
XII. Yucatan Peninsula
XIII. Valley of Mexico

I. Baja California Peninsula
II. Northwest
III. North Pacific
IV. Balsas
V. South Pacific

CNA (2001)

——— Hydrologic Regions   ——— Administrative Regions

Fig. 28.1. Hydrologic and administrative regions

28.1 Rainfall

Average annual rainfall is 780 mm, equivalent to 1,522 km³. Its distribution is far from homogenous: rainfall is as low as 50 mm in some northern areas, but abundant in the south-east, where some areas on the Gulf of Mexico and on the Pacific to the south of the Tropic of Cancer receive as much as 3,000 mm. In 42 percent of the nation's territory, rainfall is less than 500 mm. Likewise, rainfall is unequally distributed throughout the year: most comes between June and the end of September, except for an area in the north-east that has rainy seasons in both summer and winter. Besides the monthly variations, there are differences from year to year, including extraordinary periods of drought that appear on average once every 10 years then last for two or three years. Extreme meteorological phenomena include tropical hurricanes, and irregular sleet and snow showers (Figure 28.2; SEMARNAP/CNA 1996; CNA 1999a).

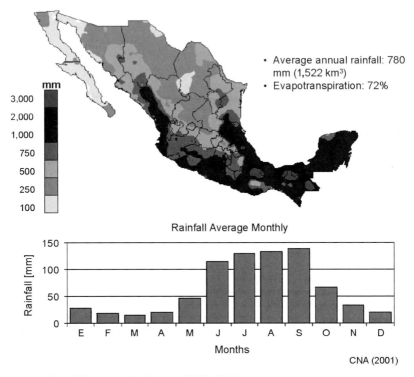

Fig. 28.2. Rainfall spatial distribution (1941-1997)

28.2 Runoff

Some 72 percent of the rain that falls returns to the atmosphere through evaporation, while the rest provides an annual primary runoff of 410 km³. The distribution of the runoff follows a similar pattern to that of the rainfall, ranging from zero in the Altar desert of Sonora to more than 3,000 mm a year in some areas of the Rio Papaloapan's coastal plains. As a result, 50 percent of the volume is generated in only 20 percent of the country in the south-east, while 4 percent is generated in 30 percent in the north (SEMARNAP/CNA 1996). As far as altitude goes, 4 percent of filtration is generated at above 2,000 m, compared with 50 percent at less than 500 m.

Because of its geographical location, our country is subject to extreme hydro-meteorological phenomena, especially hurricanes and intense rains. Each year an average of 24 hurricanes form off Mexico's shores and about two or three of them hit landfall, causing severe damage. The high volume of filtration during the rainy season and the inefficiencies of the drainage system cause severe flooding, in the Gulf of Mexico region in particular. Damage is made worse by the presence of human settlements in areas prone to flooding where water gathers more rapidly because of the loss of plant cover through deforestation. The outcome is loss of human life, and severe damage to drinking-water infrastructure, roads, agriculture, etc. At the opposite extreme, long-lasting droughts affect water supplies to the population and damage agriculture, stock-rearing and other economic activities.

Historical data shows that these phenomena are likely to occur with greater intensity every 10 years and their duration varies. In Mexico, the worst-hit area is the north because of its location in the Northern Hemisphere's desert belt. Studies indicate that the main disasters that occurred during the 1980-1998 period cost the country an annual average of 4.5 billion pesos. Apart from the 1985 earthquake, these losses were incurred as a result of extreme hydro-meteorological phenomena (CNA 2001). Table 28.1 shows the annual average runoff for each administrative region as well as its equivalent in water-bearing stratum as an indirect measure of the annual availability of surface water per region.

Table 28.1. Average runoff per administrative region

Administrative Regions	Runoff [km^3]	Water-bearing[mm]
Baja California Peninsula	2.600	18
Northwest	5.210	25
North Pacific	21.000	141
Balsas	24.900	212
South Pacific	36.812	462
Bravo River	6.738	18
North Central Basins	2.067	10
Lerma - Santiago - Pacific	28.169	147
North Gulf	22.860	180
Central Gulf	98.063	931
South Border	155.948	1,533
Yucatan Peninsula	3.250	23
Valley of Mexico	2.293	140

CNA (1999a, 2001)

It has to be pointed out that, because of the variability of runoffs in time and geographical spread, it is impossible to take full advantage of the primary runoff. Some northern zones have almost no surface runoff, and all the water in a rainy season can fall in one shower. In other cases, the fain falls in areas which lack the infrastructure to store it and runoff takes place with no way of turning it to use (Figure 28.3).

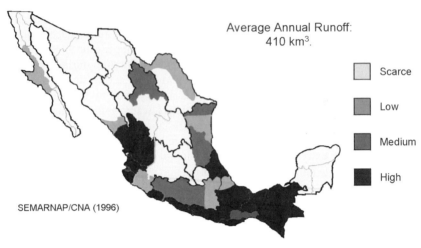

Fig. 28.3. Surface water availability

28.3 Groundwater

Some 459 aquifers have been identified in 340 geo-hydrologic regions, in which the amount of water stored is reckoned at 170-250 billion m³. But not all this water can be used because the depth at which it is found and its quality characteristics render it uneconomical. That reduces the economically available resource to about 27.8 billion m³. Most of the aquifers studied are in the north because the north is the main source of supply. Natural replenishment is 48 km³, plus 15 km³ brought in for irrigation, giving a total of 63 km³. As with runoff, the distribution of filtration follows a similar pattern to that of rainfall: 79 percent of filtration is in the south-east (Figure 28.4; Table 28.2; SEMARNAP/CNA 1996; CNA 2001).

28.4 Availability of Water

Taking all the aforementioned data into account, we can establish the following balance sheet. Each year, 1,522 km³ of rain falls, of which 1,045 km³ returns to the atmosphere through evaporation, runoff accounts for 410 km³ and filtration for 67 km³, of which 15 km³ is due to replenishment induced by irrigation. Of the runoff volume, 223.3 km³ flows into the sea, and the difference of 186.7 km³ is the volume extracted and put to use in various economic activities. In general terms, water supply is greater than demand. However, this takes no account of the regional shortages that occur in some zones and the over-abundance in others because rainfall is far from homogenous in either time or geographical spread. This can be seen in Tables 28.1 and 28.2. Water is scarce in the north and centre of the country and abundant in the states of the south-east (Figure 28.5).

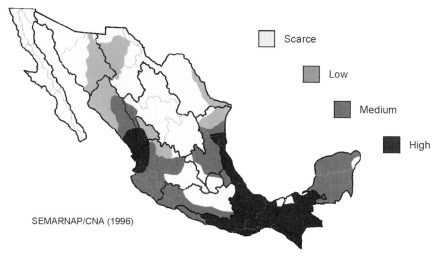

SEMARNAP/CNA (1996)

Fig. 28.4. Groundwater availability

Table 28.2. Annual groundwater recharge per administrative region

Administrative Region	Groundwater Recharge [km³]	Water-bearing[mm]
Baja California Peninsula	1.394	10
Northwest	2.767	13
North Pacific	1.392	9
Balsas	3.351	28
South Pacific	1.726	22
Bravo River	5.321	14
North Central Basins	1.662	8
Lerma - Santiago - Pacific	7.113	37
North Gulf	1.207	10
Central Gulf	2.349	22
South Border	6.220	61
Yucatan Peninsula	31.053	223
Valley of Mexico	2.070	126

CNA (1999a, 2001)

In addition, regional balance sheets must take into account the volumes extracted for different uses and the geographical distribution of economic activities and population centres. These factors determine the development process of the regions because when water use exceeds the volume replenished by natural means, over-exploitation of supply sources endangers medium and long-term development capacity, above all where groundwater supplies are the main source, as they are in northern Mexico.

Water availability per administrative region

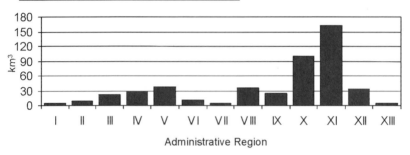

Water availability according with source

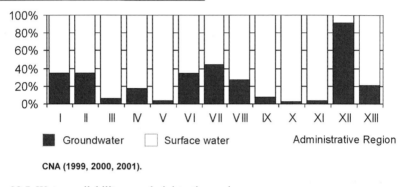

CNA (1999, 2000, 2001).

Fig. 28.5. Water availability per administrative region

29. Uses of Water

The uses to which water is put are classified in two groups: consumptive and non-consumptive (SEMARNAP/CNA 1996). Consumptive uses are those in which, put simply, the amount of water that goes out is less than what comes in. Non-consumptive uses are those in which there are no losses, the amount of water that comes in is roughly the same as what the process consumes (Table 29.1).

Table 29. 1. Water uses consumptive

Consumptive	Non-consumptive
Irrigation	Hydropower
Domestic	Aquaculture activities
Industrial	Recreational activities
Pecuario	Navigation
Thermo power	Transport

According to the CNA (2001), abstraction for the main uses totalled 198.4 km^3, equal to 42 percent of the water supplies that are renewed each year. Of the volume extracted, some 40 percent a year corresponded to consumptive uses and the remaining 60 percent to non-consumptive (Figure 29.1).

In absolute terms, the balance sheet shows demand running at less than water availability. However, unequal distribution of water leads to less availability in some regions, mainly in the states of the centre and north. In regions with less availability but with high concentrations of population and of water-consuming economic activities, per capita availability varies from 227 to 994 m^3/year. On the other hand, per capita availability in areas with more water and less population ranges from 10,056 to 28,453 m^3/year (Table 29.2; CNA 1999a).

Abstraction volumes are indirect indicators of the intensity of economic activity and the size of human settlements, as well as of the geographical distribution of both. Table 29.3 shows the volume abstracted per administrative region and its relative weighting at national level. Sixty-five percent of water abstracted nationally is of surface source and the remaining 35 percent groundwater.

It can be seen that in the north (regions II, III and VI) and the centre (regions IV, VIII and XIII), volumes of abstraction run highest, at 34 and 36 percent respectively, indicating zones with a high concentration of productive activity and of human settlements.

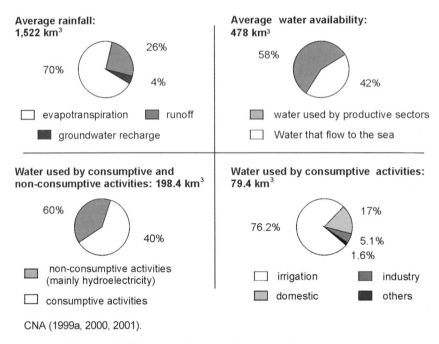

Average rainfall:
1,522 km³

26%
70%
4%

☐ evapotranspiration   ■ runoff
■ groundwater recharge

Average water availability:
478 km³

58%
42%

◻ water used by productive sectors
☐ Water that flow to the sea

Water used by consumptive and
non-consumptive activities: 198.4 km³

60%
40%

◻ non-consumptive activities
    (mainly hydroelectricity)
☐ consumptive activities

Water used by consumptive activities:
79.4 km³

17%
76.2%
5.1%
1.6%

☐ irrigation   ■ industry
◻ domestic   ■ others

CNA (1999a, 2000, 2001).

Fig. 29. 1. Water availability and volume used by economic sector

Table 29. 2. Water availability in Mexico

Region	Population [million]	Water availability [km³]		Per capita water availability [m³/year]
		Surface water	Groundwater	
Baja California Peninsula	2.87	2.600	1.394	1,438
Norwest	2.37	5.210	2.767	3,436
North Pacific	3.87	21.000	1.392	5,840
Balsas	9.93	24.900	3.351	2,903
South Pacific	3.88	36.812	1.726	10,056
Bravo River	9.25	6.738	5.321	1,327
Central Basins of the North	3.79	2.067	1.662	994
Lerma-Santiago-Pacific	18.83	28.169	7.113	1,902
North Gulf	4.83	22.860	1.207	5,062
Central Gulf	9.18	98.063	2.349	11,077
South Border	5.80	155.948	6.220	28,453
Yucatan Peninsula	3.24	3.250	31.053	10,872
Valley of Mexico	19.44	2.293	2.070	227

CNA (1999a, 2000, 2001)

Except for region XIII, the regional balance sheets show availability as greater than demand (Figure 29.3). However, on analysing regional abstraction in relation to its source, we can see that sources are over-exploited in some regions (Table

29.4). The Table shows that in regions I, II, VII and XIII, groundwater is being exploited beyond the possibility of natural replenishment, while in regions III, VIII and IX, abstraction comes close to availability. The availability of surface water leaves little margin between supply and demand in regions I, II, VI, VII and XIII. However region XIII merits special attention: the regional surface-water balance sheet is positive, but that is because water is imported from the Lerma and Balsas basins in order to meet demand.

Table 29. 3. Annual abstraction by administrative regions (consumptive uses)

Region	Water abstraction [km^3]	%	Source [km^3]	
			Ground-water	Surface water
Baja California Peninsula	4.139	5	2.291	1.848
Norwest	7.044	9	2.988	4.056
North Pacific	9.557	12	0.791	8.766
Balsas	8.366	11	2.551	5.815
South Pacific	1.674	2	0.392	1.282
Bravo River	10.142	13	3.737	6.405
Central Basins of the North	4.084	5	2.265	1.819
Lerma–Santiago–Pacific	15.220	19	6.276	8.944
North Gulf	6.634	8	0.949	5.685
Central Gulf	4.474	6	1.312	3.162
South Border	1.978	2	0.754	1.224
Peninsula of Yucatan	1.308	2	1.182	0.126
Valley of Mexico	4.737	6	2.633	2.104

CNA (1999a, 2000, 2001)

Regions IV, VI and VIII present no apparent problems, but this is because of the very large areas that they comprise. In all three cases, economic activities and human settlements are concentrated in very specific zones (the upper and middle Lerma basin, the upper Balsas and the border municipalities of the north) where semi-arid conditions prevail, leading to severe shortages and competition for water. By contrast, in the middle and lower Balsas and in the Santiago basin, the physiographic conditions favour higher rainfall, leading apparently to greater availability of water in relation to its extraction. In the Lerma basin, annual recharge of groundwater is 3.98 km^3/year and abstraction of the order of 4.53 km^3. In the case of surface water, annual filtration runs at 6.022 km^3, while demand is 7.030 km^3 (CNA 1999a), leading to a deficit in surface and groundwater by contrast with the region's overall positive balance.

The major consumer is agriculture, with an average of 76 percent of the total, while domestic use accounts for 17 percent and industrial for 5 percent, with others running at 2 percent (Table 29.5). Extraction for agricultural activity is concentrated in regions I, II, III, IV, VI, VII, VIII and IX in the northern and central states. Extraction for industrial use is very pronounced in regions V, X, XII and XIII. Extraction for public use is concentrated in regions IV, V, VI, VIII, X, XI, XII and XIII.

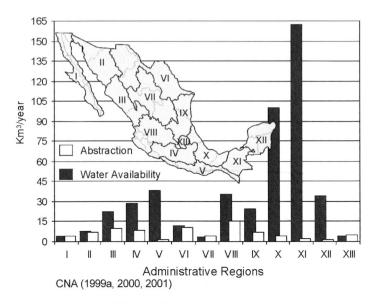

Fig. 29. 2. Water availability by administrative region

Table 29. 4. Water availability and abstraction according with the source

Region	Water Availability [km³]		Abstraction [km³]	
	Surface water	Groundwater	Surface water	Groundwater
Baja California Peninsula	2.600	1.394	1.848	2.291
Norwest	5.210	2.767	4.056	2.988
North Pacific	21.000	1.392	8.766	0.791
Balsas	24.900	3.351	5.815	2.551
South Pacific	36.812	1.726	1.282	0.392
Bravo River	6.738	5.321	6.405	3.737
Central Basins of the North	2.067	1.662	1.819	2.265
Lerma–Santiago–Pacific	28.169	7.113	8.944	6.276
North Gulf	22.860	1.207	5.685	0.949
Central Gulf	98.063	2.349	3.162	1.312
South Border	155.948	6.220	1.224	0.754
Peninsula of Yucatan	3.250	31.053	0.126	1.182
Valley of Mexico	2.293	2.070	2.104	2.633

CNA (1999a, 2000, 2001), INEGI (2000a), SEMARNAT (2002)

Table 29. 5. Abstraction by consumptive uses [km^3]

Region	Agriculture	Domestic	Industrial	Aquaculture	Thermo power
Baja California Peninsula	3.705	0.382	0.048	0.003	0.001
Norwest	6.667	0.339	0.036	0.002	0.000
North Pacific	8.831	0.560	0.106	0.060	0.000
Balsas	6.407	1.284	0.335	0.334	0.006
South Pacific	1.133	0.440	0.101	0.000	0.000
Bravo River	8.127	1.538	0.403	0.008	0.066
Central Basins of the North	3.540	0.446	0.076	0.001	0.021
Lerma–Santiago–Pacific	11.103	2.709	0.846	0.541	0.021
North Gulf	5.570	0.636	0.344	0.071	0.013
Central Gulf	2.175	1.259	1.015	0.022	0.003
South Border	0.947	0.819	0.200	0.012	0.000
Yucatan Peninsula	0.665	0.565	0.069	0.002	0.007
Valley of Mexico	1.617	2.517	0.504	0.044	0.055

CNA (1999a, 2000, 2001), INEGI (2000a), SEMARNAT (2002)

29.1 Water Infrastructure

In order to achieve an equilibrium of shortage, excess and demand in the different water regions, great efforts have been needed in order to develop infrastructure that allows for the capture, regulation, transport, distribution, collection and treatment of water for different uses, as well as public works that protect people and productive zones against flooding. This capacity enables the regulation of seasonal and annual variations in hydraulic resources and the availability of the resource in the volumes that are needed.

In terms of agriculture, 6.3 million hectares have irrigation distributed in 80 districts and more than 30,000 units. The irrigation districts have 138 storage dams, 393 distribution dams, 2,970 wells, 456 pumping stations and 46,230 km of canals. The irrigation units have 1,515 storage facilities, 2,509 for distribution, 24 755 wells and 3,292 pumping stations. Nationally, drinking water is available to 82.4 percent of the population and sewage systems to 72.1 percent. More than 2,500 km of aqueducts have an annual capacity to transport 1.6 billion km^3. In addition there are 356 plants to raise water to drinking standard and 981 for the treatment of wastewater. The total installed capacity of electricity generating plants is 35,255 MW, of which hydro plants account for 9,700 MW (CNA 1998a; Paz 1999).

Of this entire infrastructure, the large dams stand out as the axis of hydraulic activity in Mexico. These are the elements that enable a balance to be kept between supply and demand in time and space by storing and regulating the volumes over water that are later used in different productive activities and for domestic consumption.

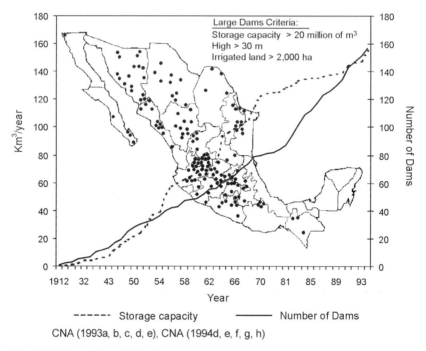

CNA (1993a, b, c, d, e), CNA (1994d, e, f, g, h)

Fig. 29. 3. Large dams in Mexico

To date, the country has more than 4,000 dams of which 850 are classified as large dams. Together they store and regulate 155 km³ in addition to the 14 km³ naturally present in lakes and lagoons (Figure 29.3). Of present capacity, 42 percent is for agriculture, 39 percent for electricity generation, 9 percent for supplies of drinking water and the 10 percent remaining is for unblocking capacity; in addition 20 percent of super-storage is available for avenue control (CNA 1994a; Guerrero 1999).

The construction of the major dams can be broken down into three historic moments:

- **1910 - 1945**. The construction of dams began to take off in Mexico from 1910, first to resolve the supply problems of the leading cities, later for the development of agriculture. Most of the effort went into the region along the US border with the aim of strengthening its social and economic development and integrating its economy with that of the rest of the nation. By the end of this period, storage capacity stood at 12 km³, irrigation covered 775,000 hectares in 44 major districts and in more than 42,000 through small-scale irrigation projects (CNA 1994b).

- **1946 - 1976**. More water had to be made available to meet growing development needs. Executive commissions were set up to promote the hydraulic development of the major watersheds. By 1976, storage capacity had risen to 125 km³ and the area under irrigation had grown to 4.85 million of hectares. The construction of multi-purpose dams helped to provide more electricity and more protection against flooding in the villages and productive regions most affected by the problem (CNA 1994b).

- **1977 – at present**. The financial crisis, changes in development policies and the deregulation of agriculture drew capital resources away from the water sector. As a result, government investment dried up and strict limits were placed on new infrastructure. During this period only 1.35 million of hectares was added to the area under irrigation, with investment mainly channelled towards the maintenance and rehabilitation of existing infrastructure.

Investment in dam construction has fallen considerably in the last 10 years, mainly because of the economic crisis. Since 1983, the majority of dams constructed have been of medium height and moderate storage capacity, most of them built for small and medium-sized irrigation projects. By 1995, 105 large dams had been completed with a total storage capacity of 27.7 billion m³, of which only seven accounted for 21.9 billion m³, and only eight were more than 80 m in height. Now, Mexico's water policy is to reduce to the minimum the construction of large dams, with investment centred on operations, conservation and rehabilitation, as well as raising height, in order to maintain the existing infrastructure in the best possible condition. (Oliva 1999; Ramos 1999a; Vega 1999).

Mexico's large dams hold 58 percent of water storage, while medium-sized and minor dams make up the rest. The age of 50 percent of the large dams ranges from zero to 30 years, which rates as relatively young. However, 67 percent of the storage is in dams more than 30 years old (Figure 29.4).

The National Water Commission (CNA) says it has a portfolio of 500 projects relating to dams, and the Federal Electricity Commission (CFE) reckons that the nation's hydroelectric potential has only been 30 percent exploited, leaving 550 sites with unexploited potential. The nation's development would be inconceivable without the large dams. The multitude of services which the country currently enjoys include the following:

- Seventy percent of the water used for irrigation comes from surface deposits that have to be regulated by a storage basin. As a result, benefits are derived by 529,000 and 574,000 users in Irrigation Districts and Units respectively, and more than 6 million jobs are directly provided (INEGI 1998b; Paz 1999).

- Drinking-water supplies in the main cities are totally or partially based on storage basins. To this end, more than 2,500 km of aqueducts have been built bringing 1.6 billion km³ a year to more than 14 million people (Paz 1999).

- Dams with control capacity in the principal watersheds provide protection against recurring floods to productive zones and their infrastructure, enabling riverbank populations to add more than 500,000 hectares to production.
- Multi-purpose hydroelectric dams generate 20 percent of the nation's electricity needs and most of the peak-hour requirement.
- Better use is made of aquifers by allowing greater volumes to be recharged.

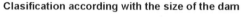

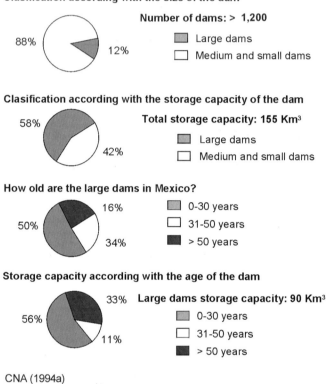

CNA (1994a)

Fig. 29.4. Status of large dams in Mexico

As demand grows, so will the cost of provision of water supplies for both new projects and those aimed at repairs and maintenance. If the over-exploitation of some aquifers is to be reduced, additional measures must be taken. These include the development of sufficient storage and regulation capacity for surface water. However the construction and maintenance of dams is ever more complex as a result of a combination of factors:

- In irrigation projects, the area that benefits reflects diminishing returns in proportion to the water that flows in;

- The impact on productive activity and human settlements grows ever greater in the areas that are flooded. Environmental impact studies basically take no account of the need to relocate population centres that are obliged to move because of the construction of dams. Aspects such as the place to which they ought to be moved, the economic compensation they should receive, and who bears the responsibility for the relocation and its costs are never taken into account (Tortajada 1999a);
- Now that all the most economically and technically favourable sites have been taken up, problems that have to do with the geology of basins and outlets have become increasingly serious;
- Not enough soil suitable for irrigation is located near the dams, so the water has to be transported over ever-greater distances;
- Severe environmental damage can be caused;
- Some of the existing dams are close to the end of their useful life or have already surpassed it, and the build-up of sediment has totally or partially reduced their capacity. It should be noted that the deforestation of river basins has shortened the useful life of many dams and built up sediment to an extent that is much more than was forecast (Vega 1999);
- In some cases, the original aim of dams has been changed. Some that were planned for the management of river water are currently handling residuals which attack the concrete and steel reinforcements, so weakening the structure;
- Materials have aged or been degraded through time, so reducing coefficients both of labour and security.

29.2 Analysis

By the year 2030, Mexico is expected to have a population of more than 130 million, most of them concentrated in the northern and central states where the availability of water is low. At this rate, per capita availability of water in these zones will be less than 1,000 m^3 a year, similar to that of people who live in countries with severe water shortages. In the future this means that, if recent demographic trends continue, pressure on the water resources of the arid and semi-arid zones of the north will become ever-more severe. It will become increasingly difficult to satisfy increased demand, and competition among uses and users will become fierce, obliging the authorities to take strict regulatory measures in order to ensure continued development of these regions.

As a result, action will have to be taken on three fronts: 1) Development of hydraulic infrastructure to meet the new needs, 2) Adequate maintenance of existing infrastructure in order to continue to meet current demand and ensure optimum use, and 3) The establishment of policies for the efficient management of the infrastructure. The last point is just as important as the first two, though scant heed is usually paid to it. Generally, it is assumed that maintenance of the physical infrastructure in optimum conditions is all that is needed in order to make efficient use of water. However, in the absence of adequate management practices, it can lead, among other things, to wastage, irrational distribution, damage to the infrastructure as a result of mis-use, and under-utilisation of plant.

As mentioned in the previous chapter, because of geographical and seasonal variations in the runoffs of water, much goes to waste. Those who talk about the volumes available usually refer to the gross runoff, as though all could be used. But the true availability of water (according to use) is limited by four main factors: 1) The cost-benefit relationship determines the extent to which use can be made within a financial scheme that ensures profitability and the social and economic benefits that derive from it; 2) The geographical and seasonal variations determine where and when runoffs can be put to use; 3) Water quality determines the uses to which the water can be put and/or the treatment that is needed; and 4) The environment uses water, minimum levels of quality and quantity have to be set for water bodies so as to ensure the conservation of the related eco-systems. In view of all this, the true availability of water is likely to be well below the level mentioned in official figures and that severe restrictions on water availability already exist in the north and centre of the country.

The large dams, like all other great infrastructure projects, incur both costs and benefits in social, economic and environmental terms. In a country like Mexico, where water availability and the development of productive activities are inversely related, there is no alternative but to build whatever major dams may be necessary. Otherwise, economic development and improvements in living standards will be impossible. Nor will the country be able to reap the other benefits that the major dams offer: dependable supplies for irrigation and drinking water; protection against flooding and recurring damage to production and the infrastructure of river and lakeshore communities; the generation of electricity by renewable means; greater recharge rates for aquifers; and the possibility of adding to a region's economic activities through the addition of fishing and tourism. In Mexico's case, the debate is not over whether or not major dams should be built, but over the nature of the planning process and management in order to maximise benefits and minimise costs.

30. Use in Agriculture

The country has some 44.5 million hectares of arable land, of which 24.9 percent has high productive potential, 14 percent moderate and the remaining 61.1 percent is of low potential. In other words, land of low productive potential predominates, particularly in the north (Garcés et al. 1997). About 21 million hectares are used for agriculture but only 30 percent of that land has infrastructure for irrigation. Even so, that 30 percent accounts for 50 percent of agricultural output, 70 percent of farm exports, 80 percent of the sector's employment, and consumes 76 percent of all water extracted (Novelo 1998; SEMARNAP/CNA 1996).

For administrative purposes, 60 percent of the irrigated area is divided into 81 Irrigation Districts, which are jointly managed by the government and associations of users. The remaining 40 percent is divided into 30,000 Irrigation Units, exlu- sively operated and managed by the users' associations (Takeda 1998). Irrigation Districts account for about 75 percent in terms of area. Most are in the north of the country and they vary from 10,000 ha to more than 200,000 ha, though most are in the 10,000–50,000 ha range (Figure 30.1).

Region	Districts	Area [000 ha]	%	Region	Districts	Area [000 ha]	%
I	2	244.6	7.2	VIII	13	437.6	12.9
II	7	504.4	14.9	IX	10	265.5	7.8
III	8	802·7	23·7	X	2	36·0	1·1
IV	8	174.7	5.2	XI	4	36·4	1·1
V	5	74.7	2.2	XII	3	46.4	1.4
VI	12	540.7	16.0	XIII	5	86.9	2.6
VII	2	134.4	4.0				

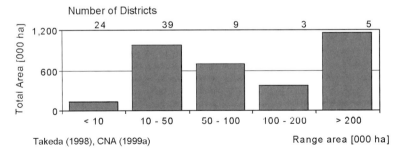

Takeda (1998), CNA (1999a)

Fig. 30.1. Classification of irrigation districts by size

Mexico has the seventh largest area under irrigation in the world. As mentioned in the previous chapter, the Irrigation Districts and Units are served by 1,650 storage dams, 2,902 distribution dams, close on 30,000 wells and 3,292 pumping stations (Garcés et al. 1997; Paz 1999). In the central region, where average annual rainfall ranges from 200 to 500 mm, agriculture is impossible without irrigation. Indeed, it is reckoned that agriculture is possible without irrigation and without risk of water shortages in only one percent of the nation's territory. Except for region XII (where no surface water is used), agriculture is highly dependent on surface water (Figure 30.2).

Water abstraction per administrative region

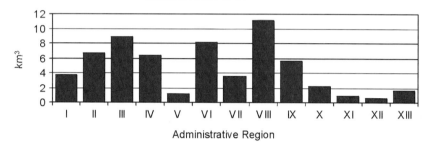

Water abstraction according with source

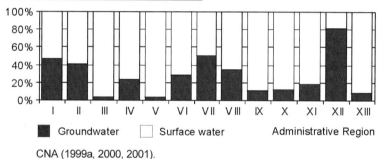

CNA (1999a, 2000, 2001).

Fig. 30.2. Water abstracted for agricultural purposes

Some 60.487 km³ of water a year is extracted for agricultural use, 73 percent from surface sources and the remainder from groundwater. Extraction is greatest in regions I, II, III, IV, VI, VII, VIII, IX and X in the centre and north of the country, where the areas under irrigation are largest and agriculture that depends only on rainfall is impossible.

Despite the development that Mexico's agricultural sector has undergone, in recent years its role in the national economy has diminished. A whole series of problems have been revealed by the lack of support from the government and other institutions, the successive economic crises, and the lack of policies in keeping with globalisation and the opening of markets. From the 1970s on, capital has been steadily drained from the countryside because of the low profitability of the crops most commonly grown, the lack of financial support for the sector's development and the state's inadequate management policies. Until 1970 agriculture, forestry and fisheries together accounted for 11.2 percent of GDP and 35.6 percent of all the jobs in the country, making it the number one sector in employment. By 1980, the contribution to GDP and employment had fallen to 8.2 and 29 percent respectively. By 1990 those figures had been reduced to 6.6 and 24 percent, and by 1996 to 5 and 22.3 percent (INEGI 1998a).

In the past, it was the state's job to build, operate and manage the facilities of the Irrigation Districts. But a transfer of responsibility for operations and management is now under way as part of the policy of deregulation in agriculture and as a result of a lack of finance. This process involves making the users' associations responsible for the management, operation and maintenance of the systems of secondary irrigation and drainage, while the National Water Commission (CNA) takes charge of storage systems and the main canals. Thus far, 76 percent by area has been transferred in this way, mostly in the north (Gorriz 1996). In the south-east, where there is no shortage of water, but the necessary technical training and organisation are in very short supply, a large number of Irrigation Units have been abandoned, while most of the Irrigation Districts have maintenance problems because of the users' lack of interest. As a result, only 12 percent of the irrigated area in the south-east has been transferred from state hands (Gorriz 1996; SEMARNAP/CNA1996).

A great number of projects have failed to achieve the aims and objectives that were set for them at the planning stage, partly because the targets were over-optimistic in the first place. It has also taken longer than expected for these projects to reach maturity. In some cases, finance has been slow to come on stream. In others, there has been a lack of communication and continuity between the phases of planning, construction and operation. In addition, the economic and financial feasibility of the projects has been undermined by a failure to take account of socio-political and technological factors (Figure 30.3) (Jaimes et al. 1998).

The low profitability of the principal crops has discouraged investment. Of 7.8 million hectares studied, 2.2 million are unprofitable. The prices of the products grown in 5.8 million hectares are uncompetitive. In particular, of 5.8 million hectares sown to maize, beans, wheat, rice, soy and sorghum, 1.2 million hectares of rainfed land are unprofitable as are 1.8 million irrigated hectares (SEMARNAP/CNA, 1996). The private banks, because of the size of their agricultural bad-debt portfolios (18-20 percent) rate the sector as one of high risk. As a result, the banks concentrate on recovering debt and restructuring their portfolios. They have become much more selective in their choice of projects and borrowers, including more rigid conditions and increased guarantee requirements. The average term of credits has been reduced, as have the number of production lines and regions for which loans are available, and the number of agricultural technicians employed by the banks has been reduced. The result of all this is that credits only benefit farmers who already are economically and technically more developed.

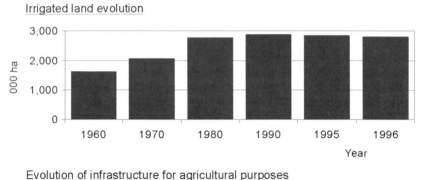

Irrigated land evolution

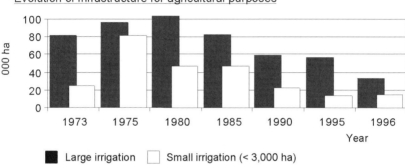

Evolution of infrastructure for agricultural purposes

■ Large irrigation   ☐ Small irrigation (< 3,000 ha)

INEGI (1999a, 1999b)

Fig. 30.3. Evolution of irrigated land

As for the producers, many of their clients in suspension of payments, frequently cases where a bad debt has been restructured only to fall into delinquency once again. The producers believe that their relationship with the banks has worsened and perceive interest rates as being too high. Where they have to choose, they meet their production costs rather than their obligations to the banks (Novelo 1998). The lack of financial resources has not only limited growth, the sector has contracted. Low profitability has meant a gradual but sustained reduction in the area sown to crops. The main cause is the increase in production costs, the lowering of prices, few facilities for the marketing of crops and the lack of credit. A lack of financial resources has delayed the development an estimated nearly 4 million hectares of new irrigation zones, mainly in huge tracts of the north-west, in medium-sized projects throughout the country, and in the humid tropics. The humid tropics in particular contain large areas that cannot be exploited because of a lack of technical and financial infrastructure. The cost of providing irrigation has been estimated at more than 90,000 pesos a hectare (Figure 30.4; CNA 1994a; González et al. 1998).

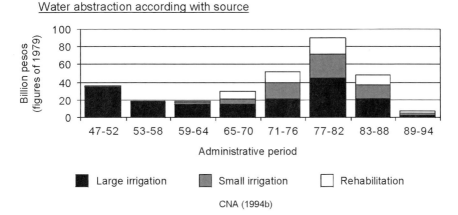

Fig. 30.4. Evolution of irrigation project investment

Efficiency rates in the transport systems are reckoned at 72 percent, and at 52 percent for irrigation itself. Pumping systems have average efficiency of 45 percent, while some old installations manage only 30 percent, leading to considerable waste of both power and water. Overall efficiency runs at 38-45 percent. The deterioration of the infrastructure in the Irrigation Districts is basically due to a shortage of funds for proper maintenance; a lack of organisation, user participation and bad practice in efforts to cut operating, maintenance and management costs; shortcomings in the control of soil humidity, and problems of erosion, salination, alkalination and drainage (Fragoza 1998). Some 25 percent of irrigation infrastructure is not used because of a lack of complementary installations, or as a result of institutional, social or legal barriers. Delivery of water to users is uneven as a result of a lack of measuring and volumetric systems (CNA 1994a; Novelo 1998; Solís and Arenas 1998).

Some 92 percent of irrigated land uses gravity applications, while the rest is technified: aspersion, micro-aspersion and drip-feed. As a result, crop repetition patterns run at 0.8 to 1.2. With adequate development and management of the agricultural infrastructure, it has been estimated that the area under irrigation could be increased to 10-11 million hectares, with a resulting pay-off in increased production. The use of gravity and surface applications, in many cases, uncontrolled flooding, leads to substantial water losses and an increase in subsoil deposits whose consequences are progressive salination of the soil and a drop in the output of crops. Some 20 percent of irrigated land is estimated to have problems of salination. However, the problem may be greater: some Irrigation Districts are affected by salination, to different degrees, in up to 40 percent of the land they encompass (González et al. 1988).

Even allowing for the increase in the land under cultivation, the low profitability of the countryside has generated migration within and between regions, adding to the concentration in urban areas in the north and centre of the country, with all the social problems that such a phenomenon entails. In 1980, the rural population was estimated at 22.5 million and the rural at 44 million. By 1995, the rural population was 24 million and the urban 67 million (INEGI 1999a). Despite the flow of migration, the population in the countryside was maintained by very high rural birth-rates. At the same time, living standards in the countryside have fallen in recent years. In states with predominantly rural populations (Chiapas, Oaxaca, Guerrero, Zacatecas, Michoacan, Veracruz), poverty indices are among the highest in the country. On the other hand, unregulated growth of the area under cultivation in some parts of the country leads to competition and conflicts over resources, generating social problems and water over-exploitation (Figure 30.5).

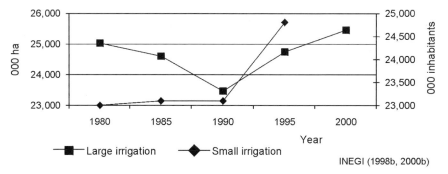

INEGI (1998b, 2000b)

Fig. 30.5. Evolution of irrigated land and rural population

The growing demand for water in areas where it is scarce has led to several cases of over-exploitation, in some cases of a seriousness that has threatened regional development. In its most immediate form this has led to a lowering of water levels (by 1-4 metres a year in some cases), rendering wells unusable, reducing the volume available for extraction, and permitting salination, the silting of land and the pollution of aquifers. Nationally, of the 340 aquifers in production, 80 have serious problems caused by over-exploitation. Of the latter, 17 suffer from saline intrusion and nine suffer from fissures and/or earth settlement, while several are polluted (Solís and Arenas 1998). Soil fertility is estimated to have been reduced by 80 percent through poor or bad management. Some soil and aquifers are affected by the presence of chromium and boron traces left by pesticides and fertilisers. Some 344,000 hectares are irrigated with domestic waste-water, almost all of it untreated. In many areas, soil and groundwater deposits have been affected by arsenic salts and leaching from waste-water.

The main obstacle to efficient water use continues to be the risk factor. Farmers prefer to over-irrigate their crops because the price of a cubic metre of water is low by comparison with the diminished potential in terms of kilogrammes of production. The same applies to the use of nitrogenous fertilisers, one kilogramme of fertiliser can mean a loss of 20 kg of production in a zone where nitrogen increases gramineous crop yields. Farmers, therefore, prefer to err on the *safe* side, overdosing on irrigation and fertilisers with consequent problems for the environment that add to the overall deterioration of agriculture (Ojeda and Peña 1998).

The recycling of water is becoming increasingly popular for different uses, one of them being waste-water for irrigation. The regulations forbid such use for human consumption, but they take no account of the impact on those who work the land. Families of agricultural workers, of which children form a high-risk group, are exposed to serious health threats. The incidence of *Ascaris lumbricoid* infections in the group most at risk, one to four years old, is six times as high among the population that uses sewage as it is among those who use rainwater (Figure 30.6). In Irrigation Districts 03 Tula and 100 Alfajayucan, both in the state of Hidalgo, over-irrigation with waste-water from Mexico City treatment overflows have raised the levels of groundwater deposits. The resultant *springs* are used by the poor of the region to meet their water needs, with consequent health problems.

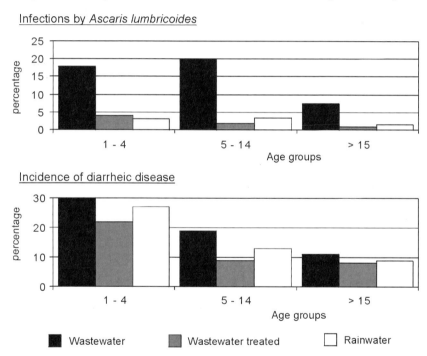

Fig. 30.6. Diseases according with water used

The risk involved in using waste-water for irrigation increases when the water from the channels is used, as it is in many cases where there is no alternative, for such domestic purposes as bathing and the washing of clothes and dishes. And the waste-water from the cities contains not only pathogens. It includes water from households, industry and high-risk centres such as hospitals, bringing a mix of chemicals, mineral elements, pathogens and organic material in general. Health problems also arrive from the consumption of products that have been irrigated with waste-water. In 1997, it was estimated that such use was restricted for 403 hectares of crops in direct contact with the soil. However, it is very difficult to know how often this measure is violated, vigilance to ensure fulfilment is insufficient.

30.1 Analysis

Inadequate management policies for agriculture have had serious economic, social and environmental impacts. Constant subsidies and a lack of interest in the maintenance of systems, practices whose origins date from the times of abundance of financial resources and land, have led to inefficiency, backwardness in technology and the operations of systems. And successive economic crises have made matters worse. At the same time, growing demand for water will lead to competition for its use in the future. Agriculture can be expected to lose out to demands for domestic consumption (which must, by law, be given priority) and industrial use. All of this runs contrary to future needs. By 2012, demand for food is expected to run at 60 million tonnes of basic grains. In order simply to maintain the current 10 million-tonne deficit, production will have to be increased by some 850,000 tonnes a year, with annual investment in water infrastructure for agriculture of 2.8 billion pesos at 1998 prices. If demand is to be met, investment will have to be of the order of 6 billion pesos a year by 2010 (Ramos 1999b).

The state's reply to this problematic has been to allow private enterprise and users' associations to finance infrastructure. Hopefully, this investment will be directed to regions where the state of existing infrastructure, soil quality and the users' organisations are capable of offering a return on the investment. One problem with such schemes is that there is a limit to the number of users who can benefit because not all have the financial resources that are needed in order to take part. Inevitably, users' organisations that do have the money will reap much more benefits than those that are cash-strapped. In addition, given the low profitability of traditional crops and the contraction of the internal market, the species that are cultivated are increasingly changing towards those that are exported. This leads to the assumption that the nation's food deficit will widen as preference is given to foreign markets. The government will, therefore, have no alternative but to subsidise either production for the domestic market or the retail prices of imports. Whichever the case, the government will have to spend large sums of money.

In general terms, there are no adequate systems of measurement or delivery of water, either in the major networks or at the level of individual land parcels. As a result, it is easy for farmers to draw off more water than authorised, or simply resort to stealing, in order to cultivate the largest expanse possible in an effort to get a return on investment. Farmers who are closest to the storage facilities have been shown to take advantage by over-exploiting the source at the expense of those who are further away (Aguirre 1998). The waste generated by low efficiency is in stark contrast with shortages and competition for the resource, and matters are made worse by the government's inability to control the situation. The net result of the transfers of ownership of the Irrigation Districts and the legal changes that have allowed lands to be sold has been to benefit economically stronger and better organised groups at the expense of the small farmers.

Private landowners and owners of communal land under the *ejido* system coexist side by side in the Irrigation Districts and Units. But they pursue different objectives, so resulting in a variety of problems in the management of infrastructure and of water itself. These include inadequate management and maintenance of the infrastructure; over-exploitation of the water by some groups at the expense of others; and monopolistic practices by some groups that have led to speculation in assigning the price of the titles to concessions. The transfer of ownership of the districts has had positive results in some cases, including increases in production and net income. But economics is not all that counts. An overall evaluation of the effectiveness of the measures that have been taken must also take account of social and environmental factors.

Waste water is frequently touted for greater use in agriculture. However, there is a lack of facilities to treat the water and of mechanisms that would ensure correct application, not to mention the social awareness of users, which probably runs at a minimum. Yet the consequences of incorrect use can be serious in the long term: major public health problems, and contamination of the soil and of surface and groundwater. The impact on both the quality and quantity of water will be obstacles to the development of the living standards of the population.

Increasing industrialization and urbanization, as well as aspirations of higher living standards, will increase food demand in the coming years at rates that outstrip population growth. Meanwhile a water crisis is developing. Radical changes must be effected in water management in the future. If not, the nation will face not only a serious shortage of water but a shortage of food that will grow in tandem.

31. Public Use

The term *public use* applies to water distributed through municipal networks to homes, businesses, industry and the municipalities' own services. By 2000, the population of Mexico was estimated at 97,361 million. Of that total, 74.7 percent lived in urban areas and the remainder in more than 190,000 rural localities of which some 150,000 had less than 100 inhabitants. Dispersion at this level creates distribution difficulties in the countryside. Of the current population, 92 percent have piped water, but the proportion drops to 62.49 percent in the countryside. (SEMARNAP/CNA 1996; CNA 1998a; Gabino and Pérez 1998; INEGI 2000b). Total extraction for public use is estimated at 13.494 km³/year, of which 69 percent comes from groundwater and the remainder from surfacewater sources. (Figure 31.1). The existing infrastructure has the capacity to disinfect 95 percent of the water supplied to the population. In addition, 2.2 km³ a year goes through some sort of process to make it fit for drinking, of 356 plants that exist for the purpose, only 287 are in operation (CNA, 2000).

Water abstracted per administrative region

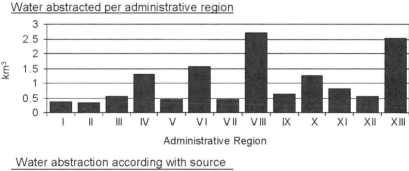

Water abstraction according with source

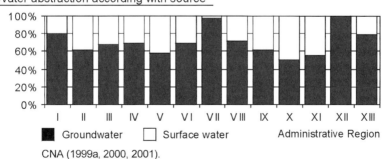

CNA (1999a, 2000, 2001).

Fig. 31.1. Water abstracted for public purposes

Volumes extracted are directly related to the geographical distribution of human settlements. The greatest volumes extracted are to be found in regions VI, VI, VIII and XIII which are those with the largest populations. The main source of supply is groundwater because of the poor quality of surface water. Table 31.1 shows the indices for provision of drinking water from 1990 to 1999. In 1990, the service covered 77.73 percent of the nation, reaching 86.4 per cent in 1999. During those years, the population grew by 1.8 million a year while 2 million more people a year were hooked up to water supplies (SARH/CNA 1994; CNA 1998b; INEGI 2000b).

Table 31.1. Water supply service evolution

Year	Population [million]	With water supply service [million]	Without water supply service [million]	Population covered [%]
1990	83.49	64.90	18.59	77.73
1991	85.13	67.20	17.93	78.94
1992	86.77	69.70	17.07	80.33
1993	88.40	71.90	16.50	81.33
1994	90.01	74.00	16.01	82.21
1995	91.16	76.74	14.42	84.18
1996	92.96	78.70	14.26	84.66
1997	94.76	80.40	14.36	84.84
1998	96.56	82.90	13.66	85.85
1999	97.36	84.12	13.24	86.40

CNA (1994c, 1997a, 1998b, 1999a, 2001), INEGI (2000a), SEMARNAT (2002)

The previous table shows that water was available to 86.4 percent of the nation. The figure may seem high but distribution among the regions was uneven, as were supplies to urban and rural areas. Among the regions, cover ranged from 66 to 92 percent (Figure 31.2).

The percentage covered refers to the number of people who have access to drinking water, but it takes no account of the quality or type of service. By 1996, drinking water was available to 20.5 million homes, of which 51.3 percent had it piped in to their dwelling, 28 percent had access outside their dwelling, 1.5 percent were supplied from a public tap, and 18.5 percent from informal systems (tanker trucks, direct extraction from the source, etc.). Access for the remaining one percent was unspecified. Taking 4.5 persons as the average household, Table 31.2 shows the water supply services distribution according with the location of tap water.

Although these figures show that some 80 percent of the population have direct access to drinking water, and only 51 percent have it piped into their home, the true figures are probably lower because the number of persons per household (4.5) is lower in middle and high-income homes. By contrast, low-income homes which lack a direct supply on average have more persons per household, boosting the numbers who lack the service. Likewise, consumption per person varies widely in accordance with socio-economic level. Daily demand might average from 80 to 550 litres per person, but there are zones where consumption is as low as 20-40 litres per person. As much as 50 percent of consumption in homes that have a garden is used outside the dwelling. Within the home, some 85 percent of water is used in showers, toilets and washing machines (Figure 31.3) (Arregin 1994; National Research Council 1995; Monroy and Viniegra 1998).

Population by administrative region

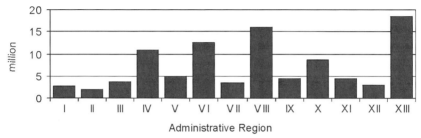

Population with water supply service

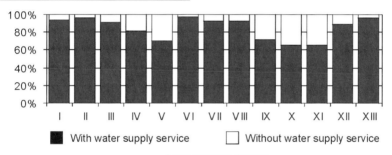

CNA (1999a, 2000, 2001)

Fig. 31.2. Population with water supply service

Table 31.2. Water supply service distribution according with the location of the tap water

Location of the tap water	Homes [million]	Population [million]
Into their home	10.52	47.34
Outside of their home	5.74	25.83
Public tap water	0.31	1.40
Informal systems	3.73	16.79
Unspecified	0.20	0.90

INEGI (2000a,b)

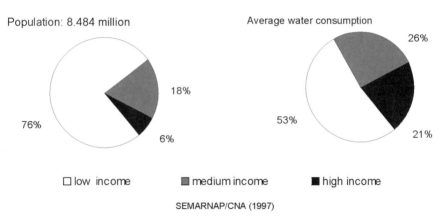

Population: 8.484 million   Average water consumption

☐ low income   ■ medium income   ■ high income

SEMARNAP/CNA (1997)

Fig. 31.3. Mexico City water consume according with socio-economic level

Towards the end of the Seventies, the provision of drinking water was one of the main social demands. In response, ambitious investment programmes were launched to clear the existing backlog and respond to the new needs of a population that was constantly growing. But despite the progress that was made, serious shortcomings remained. For decades, the Federal Government played a major role in financing and managing the systems through its Federal Drinking Water Boards. A rapid process of decentralisation that was launched in 1994 devolved functions to local authorities, private enterprise and the communities themselves. This, combined with a lack of technical-management experience and of financial resources, has generated organisms that are weak in both legislative and financial terms. As a result, despite the autonomy granted to these organisms, they continue to depend on federal and state bodies.

Staff turnover is excessive within a political culture that has people fired or moved around among different levels every three or six years, simply in order to keep promises to political allies. This has a negative effect on workplace morale and interrupts continuity. Well-trained and experienced staff are noticeably absent, often because salaries are low. Municipal planning is all but non-existent and, where it does exist, too little time is allotted to allow programmes and projects to mature. The result is that public works are undertaken without any municipal or regional framework and without any attention being paid to technical, economic, social and environmental feasibility, all with the aim of keeping promises and maintaining a political image.

Efficiency in the collection of bills is an indicator of how much of what a company invoices it really takes in as income. When it is low, it could mean that the tariffs are not appropriate (people don't pay), or that the systems of cash-collection and legislation are weak. The ability of the bodies that operate the drinking water systems to collect payment is a function of their ability to register users, check their numbers, measure the volumes distributed, and set appropriate tariffs (CEASG 1999). Measurement of consumption is limited when it exists at all, making it difficult to plan on the basis of what is being used. No records are kept of much of the water that is used because of the deficiencies in measuring consumption and the total absence of efforts to keep the register of users up to date. Coupled with inefficient billing and charging systems, the result is commercial losses (Limón 1998).

To make things even more difficult, the law forbids the authorities to cut off water supply to those who fail to pay their bills. With this as the background, income is insufficient to ensure the financial autonomy and progress of the bodies responsible. As a result, they turn to federal and state budgets and to bank credits when they need to finance necessary works. In addition, populist government policies have created a *won't pay* culture among citizens. Since the government used to absorb all financial, operational and maintenance costs, people have acquired the idea that the state has an absolute responsibility to offer the service at an extremely low cost or at no cost at all. The tariffs set by municipal councils become party political footballs, with the result that they fail to cover the real costs of the services on offer. Most are too low to meet financial, operational and maintenance costs; they are never published and increases are routinely delayed. By contrast, production of a cubic metre of water ranges from US\$1 to US\$2 (Table 31.3; Figure 31.4).

Table 31.3. Water tax according with the rate of consume and water use

City	Domestic				Commercial			
	Minimum		Maximum		Minimum		Maximum	
	Rate[1]	Tax [2]	Rate	Tax	Rate	Tax	Rate	Tax
Aguascalientes	0 - 40	2.00	>221	12.12	0 - 40	3.25	> 221	16.20
Mexicali	5 - 10	0.65	> 60	3.85	>10,000	4.14	41-100	7.72
Mexico	0 - 30	1.00	>1,500	17.25	0 - 30	4.00	>1,500	20.25
Guadalajara	0 - 17	1.20	101 - 250	6.70	20 - 40	3.30	>50,000	33.95
Monterrey	0 - 15	0.94	191 - 200	17.50	25-49	5.95	100 - 199	12.49
Chetumal	0 - 20	0.67	>61	14.80	0 - 10	2.23	>201	25.40
Puebla	0 - 10	1.52	40 - 50	2.85	0 - 120	2.00	>120	4.55
Zacatecas	0 - 20	0.88	0 - 120	14.80	0 - 10	3.52	0 - 120	14.80

Table 31.3. (cont.)

City	Industrial			
	Minimum		Maximum	
	Rate	Tax	Rate	Tax
Aguascalientes	0 - 40	3.80	> 221	16.31
Mexicali	>10,000	4.14	41 - 100	7.72
D. F.	0 - 30	4.00	>1,500	20.25
Guadalajara	20 - 40	3.30	>50,000	33.95
Monterrey	25-49	5.95	100 - 199	12.49
Chetumal	0 - 10	1.34	>1,001	31.90
Puebla	0 - 120	2.00	>120	4.55
Zacatecas	0 - 50	3.77	0 – 7,500	16.90

[1]Rate in cubic metres; [2] Tax in pesos
Source: CNA (1997a)

Tax collection – Water Utilities

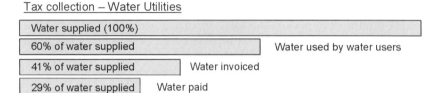

Expenditure and tax collection - National Water Commission of Mexico

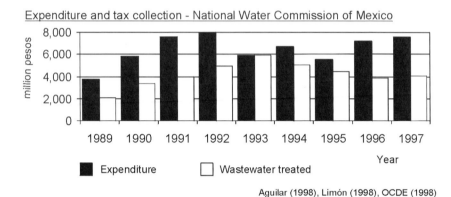

Aguilar (1998), Limón (1998), OCDE (1998)

Fig. 31.4. Water tax collection

On the regulatory front, each state has its own laws and regulations to define who has the power to provide the service and set the tariffs. In these circumstances, a national tariff structure is out of the question, not to mention the differences in costs of each service-provider and the socio-economic conditions of each regional entity.

Beset by social and legal restrictions and political power-play, the operators are all but unable to develop the ability to set tariffs that will cover their operational and maintenance costs besides paying for new infrastructure. In 1997, of the 31 state capitals and the Federal District, nine increased three types of tariff (domestic, industrial and commercial) by less than the previous year's inflation of 27.7 percent; six increased at least one of their tariffs by more than inflation; and the others froze them completely (CNA 1998b).

Operators are obliged to pay rights to use water for municipal use, but they are constantly in debt because of the failure to meet costs. As a result, in some states operators owe payments of rights amounting to more than 10 times their annual income (CEASG 1999). Under present conditions, these debts cannot be repaid, thus creating a vicious circle of recurring economic rescues by federal and state bodies (Table 31.4). Two points stand out from the table: 1) The high level of dependence on (subsidised) federal, state and municipal investment for the sector's development, compared with the (unsubsidised) resources that come from the operators and the private sector, and 2) The decline in investment from 1991 to 1996, and the subsequent recovery to 1992 levels in 1997. Charges for electricity also have a major impact on operators, on average they amount to 31-37 percent of all outgoings, and in extreme cases to more than all the operator's income (CEASG 1999).

Table 31.4. Water supply and sanitation investment (million pesos)

| Year | Subsided investment | | Non-subsided investment | | Total |
	Federal	States and Municipalities	Loans	Tax collection[1]	
1991	998	729	836	[2]	2,563
1992	1,271	626	563	[2]	2,460
1993	1,569	906	578	102	3,155
1994	1,424	427	352	127	2,330
1995	545	672	595	432	2,244
1996	1,178	346	50	171	1,745
1997	1,284	512	109	505	2,410

[1] Includ water supply and sanitation service.
[2] Figures are included in the column of loans.
Source: CNA (1997a)

Most operators are inefficient, and many are ignorant even of the infrastructure that exists. Information needed for planning either does not exist or is incomplete and badly organized. Water unaccounted for includes physical losses (burst pipes, spills) and commercial losses (underestimates of consumption, clandestine connections, fraud, errors and billing omissions, and public consumption for which no charge is made). This indicator represents the difference between the volume of water produced and the volume billed. Nationally, it has been estimated that an average of 54 percent of water by value is unaccounted for, though there are municipalities which reach 90 percent (CEASG 1999).

Macro-measurement refers to the amount of the volume of water that is captured, transported and distributed. Micro-measurement periodically quantifies the amount of water that each user consumes for billing purposes, in order to ensure rational use and maintain a suitable balance between supply and demand. Where autonomous operators are in charge, measurement systems are limited, in poor operational condition, or simply non-existent, thus creating distortions in charges made to consumers. On the other hand, the lack of reliable data on consumption makes planning difficult, though, paradoxically, when data is available nobody uses it because of the absence of anybody who knows how to incorporate it into operational and planning policies.

Leakages are another factor that militate against the proper functioning of supply systems. Types of leakage vary in accordance with what causes them: soil characteristics, poor construction, the materials used and the pressure that they have to endure, the age of the networks and the way they are operated and maintained. Studies in various cities show that losses in the supply systems range from 30 to 60 percent, with a national average of 45 percent (Arregín 1994; Aguirre and

Domínguez 1998; Bourguett and Ochoa 1998; Guitrón et al. 1998; Rangel 1998). Leakages affect finances: the water is produced and transported but cannot be billed.

Leaks in the network lead to low pressure, so the service is often offered in shifts that allow the filtration of water that can contain contaminating elements that create health problems. Not enough is done to repair leaks, and preventive maintenance is virtually non-existent. Because of cash shortages cheap, poor-quality materials are used in construction. In the long-run, though, these materials cost more because they shorten the useful life of projects and boost the repair bills as more faults appear in the system.

There is no culture of correct use of water. People in general think that drinking water should be supplied free of charge by the Government. Nor is there any awareness of the money and effort that goes into maintaining and operating the service. Installations are used incorrectly and operating costs rise as a result. In addition, more than a quarter of the urban population lives in poverty and almost half of that number in extreme poverty. In some of the major conurbations, more than half of the population is poor and heavily concentrated in satellite slum cities. At best, services for the poor are precarious. Neither the government, nor private-sector water companies, many of which face financial problems, have any incentive to extend supplies to low-income sectors, where the potential return on investment is slight. As a result, many of the urban poor have to buy bottled drinking water at prices that often are 35 times higher than those of public supplies (OPS 1994). The possession and use of water in the cities causes health problems, wastes time and creates tension and conflicts (Avila 1998; Bennett 1998).

The contamination of water sources as a result of economic activities, and sometimes by natural causes, creates serious public health risks. Studies have detected the presence of plaguicides in drinking-water wells in the countryside that exceed US-EPA recommendations 46 times over (González and Canales 1995). Eighteen systems supplied by the Yucatan aquifer have been found to contain concentrations of faecal choliforms that range from 2,000 to 25,000 NMP/100 ml (Vázquez et al. 1995). Mining activities and natural subsoil conditions have provoked arsenic concentrations in normal and chain-pump wells of 20 times the norm (Armienta et al. 1996). Leakages in the supply systems are a potential source of pollution as contaminated water seeps in. So, though potable water plants exist, the final consumer has no guarantee of quality. As well as the health risks, the costs of making water potable increase in accordance with the number of contaminating elements that are present, in some cases so many that the supply source has to be shut down. In other cases, the operators comply with the norms by adding better-quality water to dilute the water from dubious sources.

The Mexico City Metropolitan Area provides the prime example of the environmental impact generated by demand for domestic use. On average, the level of groundwater falls by 0.1 to 1.5 metres a year, but in the zones where pumping is most intensive, the net fall in level between 1986 and 1992 ranged from six to 10 metres. Over the last century, ground subsidence has been 7.5 metres on average throughout the area, causing extensive damage to buildings, drainage, roads and other elements of urban infrastructure (National Research Council 1995).

The growth in demand has necessitated imports of water from other watersheds at a high cost in both monetary terms and to the environment. The system of lakes in the Lerma Valley has virtually disappeared and the volume of the Lerma River itself has been sharply reduced as a result of the exploitation of subterranean water to supply the needs of Mexico City. As a result, the environment of the Lerma Valley has been severely damaged and conflicts have arisen over the possession of the region's water. Initially the project was meant to send $14m^3$ of water to Mexico City, but the amount was reduced to $5m^3$ as a result of the protests.

The Cutzamala system transports water over distances that range from 60 to 154 km and pumps it up to more than 1,000 m above its source, using 102 pumping stations in an operation that is highly intensive in energy use and correspondingly expensive. These stations use 1,787 million kWh/year, the equivalent of 6.05 kWh per cubic metre of water pumped, seven times as much as is used to supply other towns close to Mexico City. The Temascaltepec system was to have supplied 5 m^3, but the project has been held up by money shortages and the protests of local inhabitants. Unless the lessons of Mexico City are learned, similar problems could arise in other major cities, above all in the centre and north of the country.

31.1 Analysis

Supply systems and the institutions that operate them have severe deficiencies that derive from the lack of interest, continuity and vision in decision-making. The mentality that has pervaded the planning process for years has been to seek new sources of supply in order to meet demand, rather than look after the infrastructure and management systems that exist and try to make them more efficient. Major investments are needed to bring systems up to date, but that solution seems to be out of reach as a result of the recurring economic crises and the scarce cash flow that is generated. Some 110 billion pesos needs to be invested up to 2010 in water treatment, rehabilitation and supply, but that sum is the equivalent of the total annual budget allocated to all the states put together in 1999 and 14 times as much as the federal government allocated to the CNA in that year. Of the 7.791 billion pesos allocated to the CNA in 1999, only 3.27 billion pesos was earmarked for investment, while payments for the service were expected to bring in 4.285 billion pesos (Guillen 1999). On top of all that, many of the sources that are easiest to exploit are already in use or about to come on stream. New sources of supply will be more costly and technically difficult to exploit, not to mention new environmental and social requirements that were never imposed on the earlier projects.

In view of the government's lack of resources and capacity, the private sector has been invited to take part in the financing, operation and management of systems. But in a market economy, private enterprise can be expected to concentrate on situations where a return on investment is ensured, not the 38 percent of the population that lives in extreme poverty or the 15 percent in relative poverty. The government will have to subsidise costs if the private sector is to make a profit from supplying the poor. In other words, given the present state of the nation, private-sector participation will not solve the problem.

Another important aspect is the growing concentration of productive activities and human settlements in major cities and regions as the result of centralised planning policies. Attempts to achieve a better balance have stumbled over the lack of continuity of programmes and of finance with which to implement them. Concentration can be expected to continue in the areas that already act as magnets, the northern border and the centre of the country, presenting water suppliers with ever-greater social, economic and environmental problems. According to the National Water Commission (1997a) the strategy for increasing water supplies in the nation's main cities (the Mexico City Metropolitan Area, Guadalajara, Monterrey and Tijuana) will increasingly depend on surface sources. If this is the trend, there can be no postponement of measures to purify waste-water and preserve the quality of sources. In the light of the facts, it is clear that the provision of drinking water for the nation's future population will present a huge challenge.

32. Industrial Use

Industrial use refers to water used by industries that obtain their supplies directly from surface or groundwater then discharge it into receiver bodies. The term does not include electricity plants or industries that obtain their supplies from the potable-water network and dump their waste into the municipal sewage system. Abstraction for industrial use is estimated at 4.083 km^3/year, or 5 percent of total national extraction. Some 61 percent comes from groundwater and the rest from surface waters (Figure 32.1). Thirty-five per cent of the volume of water is used as a raw material or as a means of production in various processes. This makes quality an important factor. Groundwater is preferred because its quality is regarded as superior (CNA 2001).

Water abstracted per administrative region

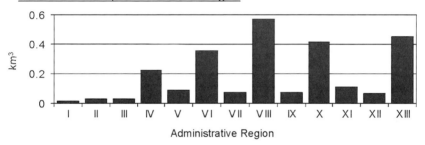

Water abstraction according with source

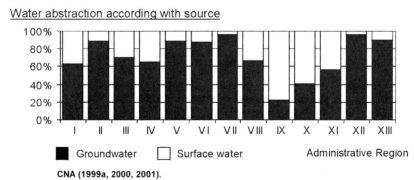

CNA (1999a, 2000, 2001).

Fig. 32.1. Water abstracted for industrial purposes

The volumes extracted bear no direct relation to the geographical distribution of the industrial activities. The states on the US border account for 58 percent of the nation's industrial parks, yet only 18 percent of the water there is extracted for industrial use because most of the industry is *dry*. The volumes of water required by industry depend on four factors: 1) The product; 2) The technology with which it is to be produced; 3) The process that has been chosen; and, 4) The efficiency of production. Only six sectors account for almost 80 percent of water consumed by industry: sugar, chemicals, oil, paper and pulp, textiles, and drinks (Table 32.1). Of the total consumed by industry, 50 percent is used for cooling; 35 percent in processes; 5 percent in boilers; and 10 percent in services (Ortiz 1997). In all supplies go to about 1,400 companies that rank as the heaviest users and those that discharge most waste-water (INEGI 1998a; CNA 2001).

Table 32.1. Main water consumers in the industrial sector

Industry	Abstraction [%]	Consumption [%]
Sugar	38.8	35.2
Chemistry	21.7	21.0
Oil	7.2	8.2
Paper and pulp	8.2	6.0
Textile	2.6	2.7
Drinks	3.3	2.4
Steel	2.5	1.7
Electric components	1.5	0.7
Food	0.2	0.2

Evaluation of water use in the industrial sector is difficult for a variety of reasons: 1) The great diversity of the different industrial sectors; 2) The size of the companies concerned ranges, within each sector, from micro to major; 3) The diversity of the technologies and the heterogeneity of the process that can be used to make the same product and within the same industrial sector; 4) The quality of the water that is available in the region; y 5) Differences, or the absence, of policies for efficient water use and management in industry. However, on the basis of the available information, some facts can be established about what water management in industry is and what it will be.

The country is estimated to have some 3.1 million economic units, of which 361,000 are in commerce, 1,242,000 in services and 1,498,000 in manufacturing. But 2,000 companies account for 96 percent of exports and 300 transnationals among them account for 50 percent (González 1999). For years, water represented no significant cost in industrial processes. This, together with the protectionist government policies that prevailed in the country for decades, led to a lack of interest in investment to make processes more efficient. The result was that several industrial sectors fell behind in technological terms.

An opening of the economy to international markets began in the last decade and the process speeded up after the North American Free Trade Agreement

(NAFTA) came into force on 1 January, 1994. At the same time, the protectionist policies have been lifted on the basis of the premise that free competition will create more efficient industrial processes, and hence companies better prepared to compete in a globalised world, with a big spin-off in social terms. Instead the growth of industry and retail sales has slowed as a result of the indiscriminate opening of markets; the technological backwardness of most industries; the recurring crises that began in the Eighties at the end of each six-year administration; the bank failures that have dried up commercial bank credits; the contraction of the domestic market because of the loss of purchasing power, and the invasion of foreign products of superior quality and lower prices. At the same time, competition in foreign markets and the slowing of growth in the world's leading economics have reduced the growth rate of the export sector. The end result is that the nation's industry, micro to medium-sized for the most part, faces a severe financial crisis (Figure 32.2).

The workers' loss of purchasing power has led to a contraction of the domestic market and a sales slump that has dealt a severe blow to non-export industry. Net sales of companies quoted on the Mexico Stock Exchange grew by only 2.1 percent in real terms between January and September 1999. Sales fell in 55 percent of economic sectors. In the metalworking and auto and auto-parts industries sales were down by 34.2 percent in real terms. Of all sales reported between January and September 1999, 25 percent of activities, notably, commerce, communications and transport, reported growth of up to 4.7 percent. Only 20 percent of sectors registered growth of 10 percent or more, as did hotels, restaurants and textiles, compared with 40 percent that were on the threshold of a sales boom in the first nine months of 1998 (Gutiérrez 1999b).

One of the sectors that was worst hit by the launch of NAFTA was capital goods, where 68 percent of companies had to close down in the face of fierce competition. Of 2,159 companies registered in 1995, only 680 survive. Competition with foreign companies has had a major impact. Companies with foreign capital are taking the lion's share of turnkey contracts that used to be the main source of income for the sector. One of the conditions that are set for companies that bid for public works contracts is that they must be able to finance the project. Given the shortage of credit, Mexican companies are unable to comply (Borjas 1999; Colín 1999).

Not only small and medium industry has felt the drop in sales. The 1998 special report on Mexico's 500 most important companies showed that only 75 percent were in the black. The problems they faced were a weak financial position and an unfavourable business environment. The report, by the research and development department of the *Expansión Business Publishing Group*, underlines that in 1998 companies were facing increased financial costs, falling demand, slowing production, less exports and growing unemployment. It adds that the 10.2 percent fall in exports for the top 500 companies could not be offset by an 8 percent growth in domestic sales. Rising domestic prices and the devaluation of the peso against the

US dollar made inputs for production more expensive. As a result, the companies reported a 7.9 percent increase in costs, far more than the growth in their turnover. Operating profits fell by 9.4 percent, from 367,434 million pesos in 1997 to 332,877 million pesos in 1998. Of the 59 activities represented in the study, only 16 reported a growth in net profits (Cappi 1999).

Foreign trade growth rate

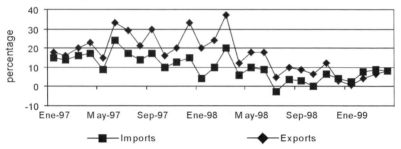

Industrial activity growth rate

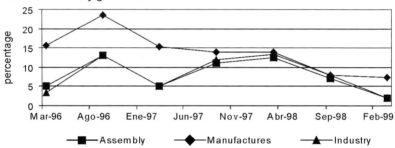

Commercial sales

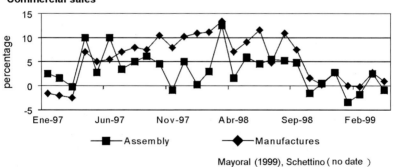

Mayoral (1999), Schettino (no date)

Fig. 32.2. Foreign trade, industrial activity, and commercial sales evolution

The lack of credits and the high interest rates that have to be paid for the few that are available have become an obstacle to companies' growth and the modernisation of industry. Central bank figures show that accumulated inflation between 1995 and 2000 was 191.55 percent, while the nominal devaluation of the peso over the same period was 183.23 percent. Over five years, the nation's leading companies paid about 125 billion pesos in debt interest alone, 60 percent more than they paid between 1990 and 1995. Their current debt burden is 30 percent higher than it was in 1994 (Sandoval 1999).

Shares in electronics, supermarkets, textiles, and the auto and steel industries grew by only 12.98 on average on the Mexico Stock Exchange amid fears that they risked defaulting on debt: 69.6 percent of their debt fell due within 12 months and they had assets of only 60 cents on every peso they owed. Volatility in the stock market is rooted in the difficulties that Mexican companies face in renegotiating credits at a time when US interest rates tend to rise and the monetary policies in Mexico are tight. Some 69 percent of Mexican companies' debt is denominated in foreign currency because the Mexican financial system has been unable to transfer sufficient resources to support productive activity as a consequence of the banking collapse that began in 1995 and persists to this day (Jiménez 1999). According to the central bank, total commercial bank lending stands at the same level as it did in 1989, marking a decade of regression in financial support (Villegas 1999).

The bad-debt portfolio of the development banks rose by 35.3 percent in the first quarter of 1999 compared with the same period of the previous year. Currently, credit unions account for 78 percent of the 1.233 billion-peso bad debt portfolio of the non-bank financial institutions. Credit unions and their like had grown to 360 in number by the end of 1994, but the crisis in the commercial and development banks slashed that figure, at present only 98 remain. *Nacional Financiera*, one of the central institutions of the development banking system, has cut back on its participation in industrial projects, particularly in those that involve small to medium manufacturing firms and mega-projects (Gutiérrez 1999c).

As a result, capital goods purchases (mainly machinery and equipment) fell from an average of 29.2 percent a month in 1998 to 8 percent in the first five months of 1999, putting a halt to the modernisation of industrial plants. Failure to renew capital goods in a timely fashion puts the nation at a competitive disadvantage and widens the gap between rich and poor companies. Even companies that have the necessary capital to bring themselves up to date fail to do so because of political uncertainty or because they want to wait until domestic demand picks up (Becerril 1999a).

In some cases, industrial plant is so out of date that private enterprise has no interest in investing to modernise it. In 1997 the government made an attempt to sell the *PEMEX* petrochemical complexes, and environmental audits were carried out to assess their condition. The results that were obtained showed that the investment needed to bring the plants up to date and meet environmental regulations was so great that private companies were sure to lose interest (Gómez et al. 1998;

Izurieta and Ruiz 1998). The nation's sugar mills use technology and processes that date from 40 years ago. Small and medium industries use, on average, 20-year-old machinery and equipment (Becerril 1999b; Sánchez 1999).

Economic difficulties are not the only factor making it hard for industries to meet environmental norms. The lack of honesty shown by industrialists also plays a part, as does the absence of the will to meet requirements. According to the CNA (2001), the volume of extraction of water is $5.1 km^3$ a year, but the true figure is reckoned to be much higher because of under-reporting, much of it dishonest. Of 64 sugar mills that were studied, 52 used more water than was authorised, in some cases twice as much.

In December 1994, more than 25,000 birds were reported to have died at the *De Silva dam* in Guanajuato (Semarnap/Profepa 1995). Inspections were made of industries that discharge waste into the public sewers in the nearby municipalities of León and San Francisco del Rincón. Later, the government and the companies signed an accord that was to move the factories to the *PIEL* industrial park, which has a water treatment plant, on the outskirts of León City. To date, not one firm has done so, on the pretext that they are waiting till the economy improves in order to have the money to make the move.

During the 1990s, Mexico became a paradise for foreign investment, attracted mainly by low wages and lax environmental regulations. But since 2000 other countries that have comparative advantages have begun to compete for foreign investment. As a result, companies have begun to move towards other regions of the world. This trend could have a severe impact on the investment programmes and modernisation plans of companies in Mexico, it might be cheaper to move to another country rather than update here.

The excessive red tape that companies have to endure in order to keep to the norms is another factor that limits their fulfilment. A few years ago, companies had to meet 800 requirements in order to keep in line with the authorities, though this has since been reduced to 170 (Becerril 1999c). At the same time, the government's need for income is so great that it is imposing taxes which, though they might help in the short term, in the medium or long term could reduce the number of industrial enterprises or restrict the growth of those that exist. On average, industrialists pay 107 percent more taxes than their counterparts in the United Staates and Canada (Colín,1999; Pazos 1999).

32.1 Analysis

The amount of water used by industry depends on the blend of goods and services that society demands, and on the process chosen in order to provide them. Water demand for industrial use can be expected to grow in future for two main reasons:

1) Growth of the population and higher living standards will increase demand for goods and services, and 2) The industrial sector's quest for more profitable activities, it takes 1,000 tonnes of water to produce a tonne of wheat, but the same amount of water can produce about US$14,000 worth of industrial goods (Kölher 1999).

Industry's water consumption is extremely heterogeneous, ranging from *dry* industries to those that use huge quantities. On the face of it, industry is not a great user of water: its consumption runs at 30 percent of domestic use and 7 percent of agricultural use. However, for marketing reasons, industries tend to locate where consumption of water runs high, mainly in and around urban centres, pitching them into direct competition with domestic users for water. Generally industries have their own water supplies (usually deep-water wells), which they themselves operate at levels of efficiency much higher than those for distribution to the public. This can give the idea that industries are privileged in receiving a good service, when the truth is that the public has no access to efficient services. The resulting ill-feeling can cause serious social tensions that spill over into disturbances such as those that have occurred in Monterrey and Morelia (Avila 1998; Bennett 1998).

In recent years, the development of industrial parks has been encouraged for market reasons and as a means of attracting investment. But many have been built in regions where natural sources of water are scarce. The workforce is growing by 8 percent a year in the northern states, four times the national average, as a result of the policies that have been followed to promote industrial development (SECOFI 1995). These are factories, most of them assembly plants, or in-bond plants, that do not need large volumes of water, but they are generating a growing flow of migration attracted by the quest for higher living standards. Some 11 million people are estimated to live in the border zone at present, and the numbers are expected to grow to 25 million by 2020. The region's population has grown so fast that it has overtaken any effort at planning. The growth has created strong demand for water supply and drainage in municipalities and states that have few resources with which to meet it. The absence of adequate industrial planning is stirring up conflicts for the possession of water and could end up as a threat to further development. If the trend continues, the impact on the region could reach dimensions that threaten to spin out of control.

In many cases, there exist technical solutions to problems of water management. However, the present state of the economy restricts the ability of companies, particularly those that are small to medium-sized, to adopt them. They involve large initial investments and most companies prefer to devote their resources to maintaining production or maintaining repayment on their debts as a means of market survival. Nor are there any fiscal advantages to be gained from investing in improvements in industrial processes. The Federal Law on Water Rights (LAN) stipulates that industries must pay from 4 to 10 pesos per m^3 of water abstracted; the exact amount depends on the region in which they are located. (Art. 223, clause A, Federal Law on Water Rights), while agricultural users pay next to nothing (Art. 224, clause I). In addition most industrialists agree that the government's deregulation programmes has had an unfavourable impact on their plant (Becerril 1999a).

While industry's use of water can lead to conflicts with other users, the main problem arises from the manner in which untreated waste is deposited in water bodies. It has been estimated that, in terms of biochemical demand for oxygen (BOD) domestic waste-water generates 1.6 million tonnes a year, compared with 1.8 million tonnes from industry. (SEMARNAP/CNA 1996). In addition, BOD of industrial origin does not necessarily represent its true impact on water bodies, industrial waste-water can also contain highly toxic elements such as heavy metals or substances classified as dangerous.

The division in Mexico's industry has grown with each year that passes. On one hand, the export sector is competitive, technologically advanced and globalised; on the other, companies that concentrate on the home market are technologically heterogeneous and highly dependent on the level of domestic consumption. While the former have boosted exports to more than 25 percent of GDP, the latter continue to provide the bulk of jobs. The financial problems of industry and the country as a whole have meant setbacks for programmes that aim to protect the environment. Targets set for the 1995-2000 National Water Programme were never met, and it will be difficult to meet those of the 2001-2006 plan. By 1999, it was estimated that 2.5 billion dollars was needed to modernise existing industry and meet environmental requirements (Ibarra 1999). Such problems highlight the absence of coordination between the environmental policies and the nation's economic situation. The lack of integrated planning is evident.

The high level of uncertainty about the future of the economy will be one of the major barriers to investment by industry in the modernisation of its factories and the installation of equipment to make water management more efficient. At the same time, constant changes in the regulations cause ill-feeling in the sector and hinder the implementation of medium and long-term plans for improvement. In many cases, industrialists prefer not to invest in the modernization of the companies because of uncertainty, investment has to be written off when changes in the environmental rules render new plant instantly obsolete. For example, NOM-001-ECOL-1996 establishes the maximum permissible limits for contaminants in waste-water discharges into national waters and other national assets. But comparing these limits with those previously established, they turnout to be much more lax, rather than tougher.

In future the state will have to establish instruments that help companies to meet requirements of environmental protection and efficient water use, through integral planning that establishes realistic targets in the short, medium and long term. The planning will have to be in concordance with the main elements that determine water use, not, as has been the case thus far, on the basis of excessively optimistic hypotheses and highly questionable suppositions.

As mentioned in previous chapters, the state's current strategy is to rely ever-increasingly on private investment to promote the various sectors of the economy. Decision-making structures have been modified to give the private sector an increasing say. The state must, therefore, strengthen its ability to ensure fulfilment of environmental norms, or the large corporations will use the threat of dis-investment so as to determine what requirements they are going to meet and how they are going to meet them.

33. Power Generation

Mexico is one of the four leading nations of the Americas in terms of electrification, together with the United States, Canada and Brazil. Generating capacity is concentrated in more than 152 power stations and 527 ancillary plants of the National Electricity System. For distribution purposes, the country has 1,850 substations with which to moderate voltage, and more than 600,000 kilometres of transmission and distribution lines. Demand for power is mainly covered by thermo power and hydroelectric plants with a total installed capacity of 35,386 MW. In 2000, electricity generation reached 189,995 GWh of which 80 percent was produced by thermoelectrics, 17 percent by hydro plants and the remaining 3 percent by geothermal plants and wind power (Figure 33.1; INEGI 1998b; CFE 2001).

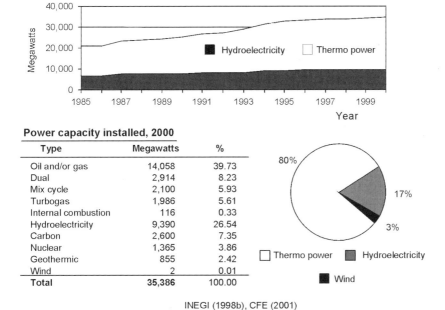

Power capacity installed, 2000

Type	Megawatts	%
Oil and/or gas	14,058	39.73
Dual	2,914	8.23
Mix cycle	2,100	5.93
Turbogas	1,986	5.61
Internal combustion	116	0.33
Hydroelectricity	9,390	26.54
Carbon	2,600	7.35
Nuclear	1,365	3.86
Geothermic	855	2.42
Wind	2	0.01
Total	35,386	100.00

INEGI (1998b), CFE (2001)

Fig. 33.1. Power capacity installed, 1985 - 2000

Hydroelectric plants use water's power to transfer it into electricity, making use of the topographical and hydrological characteristics of certain regions. Because they require storage of water for the power to accumulate, they are built in hilly regions: in our country, the Western, Eastern and Southern Sierra Madre ranges. Thermo power plants are placed in the main centres of consumption, which in our country means mainly the central plain and the north, where most of the leading industrial complexes have been developed and where, ironically, there are few natural resources, including water (Figure 33.2; Chávez 1998).

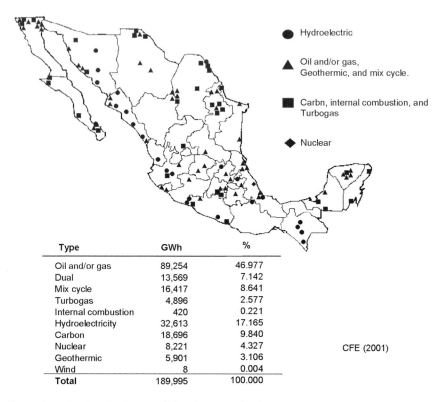

Type	GWh	%	
Oil and/or gas	89,254	46.977	
Dual	13,569	7.142	
Mix cycle	16,417	8.641	
Turbogas	4,896	2.577	
Internal combustion	420	0.221	
Hydroelectricity	32,613	17.165	
Carbon	18,696	9.840	
Nuclear	8,221	4.327	CFE (2001)
Geothermic	5,901	3.106	
Wind	8	0.004	
Total	**189,995**	**100.000**	

Fig. 33.2. Main electric plants and electric generation by type of electric plant

By 2001, installed capacity provided service to 18,682,052 users nationwide, of which 88 percent were domestic users. But industry, which accounted for only 0.6 percent of users, was the largest consumer, accounting for some 47 percent of all the power generated (Table 33.1) (CFE, 2001).

Planning to meet electricity needs is based on projections of demand for the range of sources on offer. Various sources of power are available, but they mainly boil down to thermo and hydro plants. Geographical distribution of demand and generating capacity do not necessarily coincide, but demand is met by transmission through a national grid system (Figure 33.3).

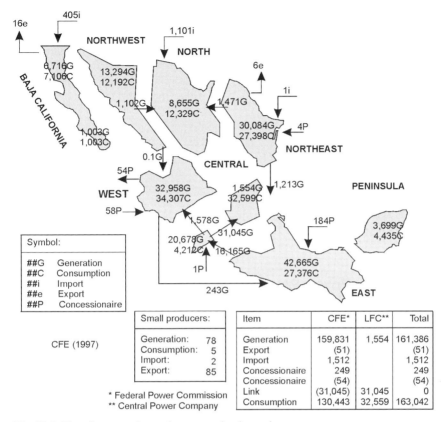

Fig. 33.3. Electric generation and consumption by region

Table 33.1. Electric sector users and sales, 2000

Sector	Sales [GWh]	%	Users **	%	Consumption [%]
Domestic	29,711	18	16'450,831	88.06	24
Commercial	8,064	5	1,912,809	10.24	6
Industrial	76,390	47	106,553	0.57	61
Services	3,947	2	119,589	0.64	3
Agriculture	7,815	5	92,270	0.49	6
Subtotal	125,927		18,682,052		
Others*	37,349	23			
Total	163,276				

* Include sales to *Central Power Company* and exportation of electricity
** Until December 31
Source: CFE (2001)

The Federal Power Commission (CFE) divides the country into zones for the administration of the National Electricity System, and these zones reveal unequal distribution of generation and consumption. The states of Veracruz, Coahuila, Hidalgo, Chiapas and Guerrero, in that order, together produce half of all the electricity that is generated. Apart from Chiapas and, to a lesser extent Guerrero, where hydroelectricity dominates, the other states have various forms of thermo power (INEGI 1998a). In terms of water, the hydro plants use 143 km^3 a year and the thermoelectrics 2 km^3 for cooling (CNA 1999a; CNA 2001).

There can be no denying that the development of society is strongly linked to energy and the ability to generate it. However, the production, transformation, storage and consumption of energy are, without doubt, some of the factors that have degraded the environment. At present, the generation of electricity is undergoing a move towards schemes that meet consumer demand, while being more efficient and hence cheaper, as well as more environment-friendly. However, in order to achieve that goal, a lot of work needs to be done with both thermo power and hydro plants.

33.1 Thermo Power Plants

Throughout the world, the Sixties was a decade of abundant cheap oil in which hydrocarbons were increasingly used to generate electricity. In those years, the Federal Power Commission gave preference to thermo power plants, particularly those that burned fuel-oil and natural gas. The number of hydroelectric projects was cut back, although the average size of those that were built increased. Thermo power projects grew both in number and the size of the generating units. Thermo power grew from 48 percent of total generation in 1960 to 70 percent in 2000.

During 1997, electricity generation used 19.736 million cubic metres of fuel-oil, 5.617 billion cubic metres of gas, 8.853 million tonnes of coal and 0.343 million cubic metres of diesel. In 1990 the nation's emissions of carbon dioxide totalled 444.489 million tonnes, of which electricity generation took the lion's share with 66.8 percent. CFE estimates indicated that, for each percentage point of the volume of electricity generated in 1997 (161,386 GWh), emissions to the atmosphere included 18,687 tonnes of sulphur dioxide, 3,451 tonnes of nitrogen oxides, 795,699 tonnes of nitrogen dioxide, 33 tonnes of hydrocarbons and 1,273 tonnes of particles. Indiscriminate use of fossil fuels can threaten the development of thermo power. According to calculations by PEMEX (Oil Consortium of Mexico), production of crude oil in 1997 was 3.022 billion barrels, with proven reserves of 47.822 billion barrels. The relation between production and reserves indicated that, at current rates of production of oil and gas, the reserves would be used up in 39 years (INEGI 1998a; INEGI 1998c).

At the same time, disposal of waste-water from the thermoelectric plants can create problems for the environment. The temperature of the waste water can break up oxygen concentrations, while heat and oxygen deficiency can harm

aquatic species that have little tolerance to either. High saline content, which results from continuous recycling and the purging of cooling systems, can change
the physio-chemical composition of aquatic eco-systems, particularly those of
fresh water. Where the wastewater is disposed of in the ground, soil can be salin-
ised and its productive potential reduced.

33.2 Hydroelectric Plants

In general, the impact of hydroelectric schemes are the same as those associated
with the construction of dams: high initial costs, long construction times, involun-
tary resettlements, modifications in the quality of water and in hydrological regimes, loss of land and living space, health problems among neighbouring communities, proliferation of aquatic weed because of the presence of nutrients, a
lowering of the security of dams as a result of attacks on structure by some of the
elements contained in wastewater, the loss of biodiversity and of arqueological
and cultural artefacts through flooding, among others (Goodland 1996).

At present, Non-Governmental Organisations (NGOs) in particular are mount-
ing strong opposition world-wide to the construction of major dams. The origin of
the opposition lies in the *black* past of planning for these projects which sparked
major controversies by taking account only of technical considerations in a vertical decision-making process. Now, however, several authors (Forsius 1993; Goodland 1996; Russo 1994; Seabra 1993) agree that hydroelectricity represents a major step forward towards renewable sources of energy. The challenge is to make
hydroelectricity respectable in the eyes of the public, and this can only be
achieved when social and environmental considerations are taken as seriously as
those of a technical nature and when society in general is involved in the process.

The benefits of hydroelectricity include the following: renewable energy; high
efficiency in converting mechanical energy into electricity (on average 85 percent,
compared with 50-55 percent for thermoelectric plants); rapid response to variations in energy demand, especially at peak hours; relatively simple technology that
is well known and easy to operate, besides being economically competitive and
environmentally clean; regulation of river flow; control and protection of land and
communities downstream from the dams; water supply for villages and towns; irrigation and the development of crops; creation of new means of communication;
development of local, regional and national economies, and development of local
tourism. In addition, costs are easily foreseeable in financial and environmental
terms, while those of thermoelectric plants are subject to the ups and downs of the
markets in hydrocarbons and to other external factors (Forsius1993; Goodland
1996; Russo1994; Seabra 1993).

33.3 Nuclear Plants

As arguments in its favour, the nuclear industry points to the environmental problems cause by the use of fossil fuels, above all the production of carbon dioxide and its possible impact on world climate change. However, it appears unlikely that the virtual moratorium on nuclear power imposed by several countries will be lifted as long as two basic technological problems remain: safety concerns will remain until a new generation of intrinsically secure reactors can be developed, and until the safe disposal of long-life highly radioactive waste can be assured (CFE 1992a)

33.4 Analysis

At present, forecasts for growth in electricity demand range from 6 percent to a maximum of 10 percent a year for the next six years (Figure 33.4). In order to meet that challenge over the period, some 13,000 megawatts of generating capacity will have to be installed, more than a third of existing capacity. Substantial investments will also have to be made in transmission and distribution systems in order to ensure unbroken supplies of sufficient quality and quantity. In all, these investments would have to be of the order of 240 billion pesos over the six years, about equal to a quarter of the nation's entire 1999 budget, and more than the total amount spent on education and social security during the year (Secretaría de Energía 1999). In 1997, the industry recorded a profit of only 200 million pesos, 7.6 million pesos less than the previous year. This decapitalisation is the result of the cancelling of the government's overdue bills which generated interest for the Federal Power Commission (CFE 1997).

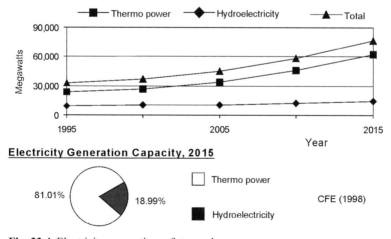

Fig. 33.4. Electricity generation – future schemes

If these investments continue to be financed as they have in the past, unbearable pressure will build up on the federal budget at a time when it faces a backlog in satisfying other needs and powerful restrictions on the availability of resources. As a result, ways are being sought to allow the private sector to participate in meeting future demand. Figure 33.4 shows that the strategy for meeting future demand relies mainly on thermoelectric plants. The Energy Ministry document entitled Proposal for Structural Change in the Mexican Electricity Industry has this to say: "Since the 1980s, electricity generation has changed substantially as a result of recent technological advances that have reduced the optimum scale of power plants and the costs of generation [...] These days, power plants are smaller and cost lest to build and bring on stream. This trend allows small companies to finance and construct new generating plants, place them wherever is most convenient, and compete freely for the opportunity to sell energy" (Figure 33.5; Secretaría de Energía 1999).

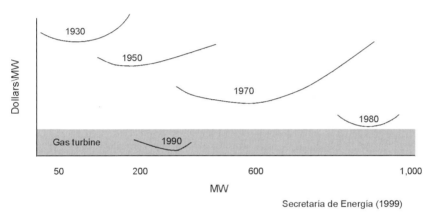

Secretaría de Energía (1999)

Fig. 33.5. Technological changes and electricity generation prices, 1930-90

Although the document does not mention it explicitly, this passage shows that the present policy of the energy sector is to promote thermo power, both in the public and private sectors. This is because thermo power cost less and take less time to build than do hydroelectric plants. But, over the long term, this strategy can have negative consequences:

- An increase in consumption of fossil fuels and consequently of the emission of gases to the atmosphere, mainly sulphur and nitrogen oxides that cause acid rain and carbon dioxide which contributes to the greenhouse effect. In addition, thermoelectric plants are sited close to consumption centres, so creating health problems for the population as a result of atmospheric emissions.
- Proximity of plants to consumption centres can also lead to competition for water with other uses, such as public supply, and safety problems linked to the means of transport for the fuels and their storage.
- Thermo power plants are being promoted because the initial investment to build them is low compared with hydro plants. But in the long run, operating costs

rise because of the continued use of fossil fuels, while those of hydro plants tend to go down (Figure 33.6). Taking into account the 39 years that Mexico has left as a producer of hydrocarbons, the price of these fuels, and with them the operating costs of hydro plants, can be expected to rise.

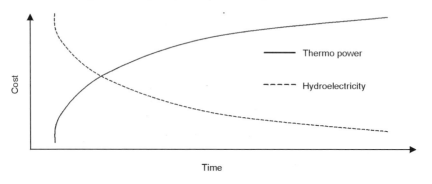

Fig. 33.6. Electricity generation costs – Hydroelectricity vs. Thermo power

As a result, alternative sources of electricity generation have to be sought in the short to medium term in order to meet present and future demand without posing a threat to national development or the environment. According to the CFE (1992b) the potential for hydroelectricity that could be development is equivalent to annual generation of 123,107 GWh, some 77 percent of what is currently generated. Dams for electricity generation, like any other major infrastructure project, are associated with social, economic and environmental costs and benefits. For a country like Mexico, where water availability and the geographical development of productive activities fail to coincide, the construction of dams could be the alternative, provided they are a step in the direction toward the use of renewable energy. If not, the country's economic development and improvements in the quality of life of its people will be stopped in their tracks.

At the same time, the guidelines for the generation of electricity are in need of a fundamental overhaul; otherwise, it will be impossible to obtain the benefits of the operation of the nation's hydraulic infrastructure, and of its hydroelectric plants above all.

34. Aquaculture

The practice of aquaculture, according to the Regulations under the Law of National Waters, is understood to be "the use of national waters for the cultivation, reproduction, and development of any species of aquatic flora or fauna" (CNA 1999b). In Mexico the bodies of national surface waters cover 3.8 million hectares, of which 2.9 million are salty water and 0.9 million are fresh. In salt or salty water the potential area for aquaculture is estimated to be slightly more than 2 million hectares; of this, 450,000 are suitable for cultivation of shrimp and 30,000 for other species. For fresh water, the potential is 900,000 hectares (SEMARNAP/CNA 1996, INEGI 2000a). The best potential fresh water is on the Pacific side in the south (Chiapas, Oaxaca, and Michoacán), in an area that comprises 49 percent of the total.

Fishing activities comprise about one percent of Gross Domestic Product (GDP) and employ about 259,000 persons. However, the socio-economic importance of fisheries is more significant at the regional or local level, given that these states have coastal areas, with coastal communities, and areas which have interior bodies of water where this activity has become a basic source of income for important segments of the population. Production in the fisheries sector between 1985 and 1997 has been unstable, but has shown reasonable recovery in the last four years, though it has not surpassed the historic record (Figure 34.1). Fishing activities have grown in recent years, including 1995, faster than the national economy as a whole. Its importance is based on its capacity to create food, employment, and to generate profits that help satisfy the needs of the population, which makes it an important instrument for encouraging regional development (SEMARNAP 1995).

Aquaculture is practiced at three levels: intensive or high yielding, at subsistence level, and rural. Nationally, there are more than 6,000 aquaculture units registered. Of these, 90 percent are rural and the other 10 percent are commercial aquaculture ventures that propagate species such as shrimp, catfish, trout, oysters, tilapia, abalone, and other less important species. These commercial operations represent 85 percent of the 20,000 hectares that are currently registered. Of these, 17,000 are for shrimp, of which 3,200 are out of operation for various reasons, including poor organisation or administration, or poor planning (SEMARNAP 1995).

It is estimated that intensive aquaculture in fresh water is practised on 2,000 hectares and uses a volume of 1.1 km³/year. In recent years rural aquaculture has received a major impetus; this activity is directed toward improving the diet of the population and generating extra income that benefits about 43,000 farm families in 2,549 communities in 514 municipalities (SEMARNAP/CNA 1996; CNA 2001). During the last 12 years, national aquaculture production has grown at an average annual rate of 2.7 percent, from 122,148 tonnes in 1983 to 171,389 tonnes in 1994, and it is expected that in the coming years this activity will continue to develop and that production will increase (Figure 34.2).

Fishery and aquaculture production

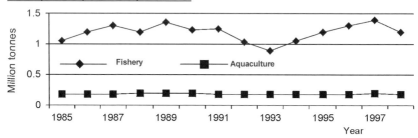

Fishery and aquaculture growth rate

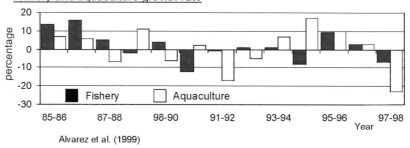

Alvarez et al. (1999)

Fig. 34.1. Fishery and aquaculture production

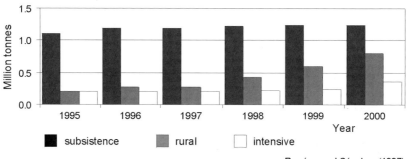

Ramírez and Sánchez (1997)

Fig. 34.2. Volume of production and projections in aquaculture activities

Of the species traditionally cultivated, tilapia, oyster and carp have the greatest social impact, accounting for 79 percent of total production, and are used almost entirely for domestic consumption. Shrimp production has grown at a rate of almost 48.9 percent between 1987 and 1997. In 1987, 286 tonnes were harvested, and production was 20,000 tonnes in 1997 (SEMARNAP/DGA 1999). In Table 34.1, one can observe the accelerated growth of this activity in the state of Sinaloa, the main shrimp producer. To meet the demand for inputs this number of shrimp farms has generated, the number of post-larva collected has had to increase, although it has not been sufficient to meet all of the current demand, which is much higher than the supply of wild larva.

Shrimp cultivation has been taken up by a large number of enterprises, both private and public, and has involved groups of *ejidos* and cooperative societies, with a corresponding generation of jobs and income. Despite this growth, the importance that shrimp cultivation represents must be underlined in terms of effective availability of lands so as to ensure that it does not affect other productive activities or other natural resources. The lack of policies for maintaining order, and for stimulating research and development, impedes a broader development of this activity.

Table 34.1. Aquaculture farms operating in the State of Sinaloa, Mexico

Year	No. of aquaculture farms	Surface [ha]	Production [ton]
1987	42	1,593	276
1988	56	3,645	901
1990	63	4,500	2,884
1991	61	4,200	3,985
1992	87	6,797	7,083
1993	90	7,200	8,725

Source: UAS (1994)

In terms of infrastructure, there are 22 production laboratories for shrimp, six for abalone, one for oysters, one for mussels, and three for lobsters; there are 39 aquaculture centres that SEMARNAT (Ministry of Natural Resources and Environment), administers and operates, and 23 others in the hands of state or municipal governments. However, there is an urgent need to improve quality and to recover production capacity in aquaculture centres, because they were only operating at 60 percent of installed capacity with production of 144.9 million fry and post-larva in 1994. It is necessary to show the productive sector the urgent need for promoting the construction and improvement of production laboratories for shrimp post-larva to meet the demand and requirements of aquaculture producers, and to counteract and avoid dependency on wild inputs which could result in over-exploitation of the resource (Table 34.2).

In general terms, the fisheries and aquaculture sector show a high level of dislocation of activities and productive operations, low levels of operating efficiency, and as a result, low productivity, both in terms of fishing fleets and processing plants and marketing. Together, this leads to low levels of consumption, especially in low-income populations, and high costs. The social and economic marginalization of the vast majority of coastal fishermen who operate small fleets is evident: there are about 70,930 small boats that comprise 95.4 percent of the national fleet and that meet the needs of national consumption.

Table 34.2. Post-larva request

Year	Requested [million]	Collected [million]
1990	650	130
1991	1,050	640
1992	1,600	900
1993	2,389	903

Source: UAS (1994)

There are significant shortcomings in several areas, in fisheries, port, and aquaculture infrastructure, especially in terms of technical support for the small fisheries fleet, and infrastructure for conservation, management of the catch, and industrial processing. The plants lack proper services, due in part to the lack of attention to conservation and maintenance, rehabilitation, and frequently, relocation of installations, equipment and machinery. The sector faces problems of credit access, high interest rates, a lack of guarantees, and restructuring of non-performing debts as a result of the country's economic situation (SEMARNAP 1995).

An analysis of official programmes for aquaculture over a period of almost 20 years shows that there is a tendency to plan for growth above what is possible for the sector, in terms of technical capacity and human and economic resources. The expectations generated in the 1988-1994 period have not been realized because of the lack of continuity in public administration, which reflects changes in policies and strategies during that period. The sector that produces fry for rural aquaculture, and the bodies of water available for aquaculture, are practically bereft of support.

With the aim of overcoming these problems, a programme has been started called Development of Aquaculture in Mexico. In June 1997, the Board of the World Bank decided to grant the Government of Mexico a credit of 360 million pesos to develop a programme to help SEMARNAP promote aquaculture. This programme will cost about 532 million pesos, including funding from the federal government and producers. The six-year project will be carried out in the states of Baja California Sur, Chiapas, Nayarit, Oaxaca, Sinaloa, Tamaulipas, and Veracruz (SEMARNAP/DGA 1999).

Consistent with the National Development Plan and the National Plan for Fisheries and Aquaculture 1995-2000, this project's goals are job creation; increasing the supply of seafood, with the goal of improving nutrition for the majority of the population; and becoming a source of foreign exchange through exporting product more competitively. The project is comprised of 14 programmes, of which 11 will be carried out directly by the government; it aims to establish procedures for simplifying administration, set out guidelines for carrying out environmental impact studies in aquaculture, set standards of sanitation, and do studies to determine the impact of aquaculture on the ecological order of regions where the projects will be carried out. Three other programmes have as a goal the development and construction of aquaculture parks. These activities will be developed jointly by the government and interested parties.

As a result of this project, it is thought that current production could be increased to 1,500 tonnes of shrimp, 900 tonnes of tilapia, and 600 million fish fry per year, which would create about 620 direct and indirect jobs in the aquaculture parks and, as a result of training programmes, at least 3,000 persons from marginal areas will see benefits and be incorporated into aquaculture as a productive activity. In terms of infrastructure, it is expected to have a central reference laboratory, two regional diagnostic laboratories (whose function will be to provide technical consulting to producers), three parks for cultivation of tilapia, one for the cultivation of shrimp, two laboratories for producing fry, and one laboratory for producing post-larva shrimp resistant to specific pathogens.

Given the increase in population and the growth of poverty, aquaculture is a viable option to meet nutritional needs, diversify productive activities in rural areas, and to be a source of foreign exchange for the nation. Aquaculture could be the key to development, and like all economic activity it has an impact on the environment and society which could turn out to be negative. In development programmes the major projects are generally taken up by large companies that have enough capital and technical experience to set up and manage them, leaving subsistence programmes to low-income sectors; those which by their nature require less capital and experience are intended to improve the standard of living for these populations. However, this type of access could turn out to be counterproductive in that it might limit the possibilities of developing from an economy of self-sufficiency to a market economy because the capital and technical consulting that are required for entry are barriers to the low-income sectors. It is necessary to seek access paths for more balanced projects of self-sufficiency or subsistence; otherwise the financial benefits of aquaculture will continue to be concentrated within a few sectors of society and the development programmes will if anything fail to contribute to lowering the levels of inequality.

An emphasis on intensive practices and exporting could create structural imbalances in the economy and local organisation, and attract the interest of large producers in regions that historically have been managed by local populations. What is more, the competition between small and large producers could create unfair productive practices in order to gain consumer preference. In some regions of the country and in some government projects, there is an almost exclusive emphasis on shrimp production with the risks that go with it; this represents an risky strategy and might force aquaculture producers to turn to other activities or to consider cultivation of other species. However, the lack of technology that is applicable to local species and conditions severely limits this activity (Fundación Friedrich Ebert 1995).

Large scale aquaculture could create problems relating to land use and social organisation, because it could displace traditional fishermen. The promise of higher development, and the hiring of workers displaced from the farms could modify the organisational structure of communities. At the same time, inadequate planning for small-scale aquaculture could create competition among producers for access to the young specimens, through sales channels, sources of apportioning the inputs, technical assistance and associated services. In addition, the borders of the locations do not generally correspond to the political borders of states or municipalities. Many bodies of water are shared and the actions of a state or municipality could affect water of neighbouring jurisdictions. Differing visions of local development in regions with shared bodies of water could create conflicts over the use of the water and impede or set back regional development

With a strategy of multiple uses for the reservoirs it is intended that communities affected by the construction of dams adopt aquaculture to compensate. However, in many cases these programmes fail because they are not accepted, and there are groups that have never had any contact with aquaculture, and where aquaculture products form little or no part of their diets. That brings out the need to consider cultural and historical reference points of these populations before suggesting aquaculture as an option for those who historically have not practiced it.

Water quality is a determining factor we must consider for the future of aquaculture in Mexico. High-yield aquaculture is based mainly on two species: oysters in the Gulf of Mexico and shrimp, as well as others such as mussels, Pacific oysters, catfish, and trout, which it was estimated in 1994 accounted for 52.6 percent of national production. However, registrations for aquaculture in 1994 reflect a decline in production, except for shrimp, mainly as a result of pollution and sanitation issues. These can incur health risks and create consumer resistance to products. Most water basins in Mexico contain some level of contamination, caused by domestic, industrial, and agricultural activities. There is a lack of infrastructure for purifying water, only inefficient means for removing contaminants, and a significant lag in the programmes aimed at purifying bodies of water because of budget shortages. The need for good quality fresh water to fill the reservoirs or to regulate salinity can create competition with other users. The prioritisation of the uses of water which favours domestic use and the relatively small size of the aquaculture sector in some regions creates a risky situation due to the possible scarcity of water or the increase in demand from other users, even when water rights are provided. This does not bode well for a future in which there must be water of the quality necessary for optimum development of aquaculture. At the same time, wastewater from the aquaculture enterprises, mainly organic material and suspended solids, can affect cultivation systems, whether molluscs, fish, or crustaceans are involved. The spread of disease from the cultivated species could be a risk if the wastewater is not treated adequately.

Inadequate planning of aquaculture can harm the environment. The outflow waters from the fish farms can change the hydrodynamics of an estuary and the lagoon systems, decrease water circulation, increase marine elements, and reduce the available surface for land use for aquaculture or land-based activities. In addition, the withdrawal of upstream vegetation, or deforestation, promotes sediment migration, increasing the build-up of sediment and decreasing the available area for cultivation. Dredging for the compacting of dikes, especially in the construction of aquaculture farms, can have severe effects, especially in fragile habitats such as mangrove swamps. Moreover, when the sediment from dredging is dumped in nearby bodies of water it contributes to the build-up, so decreasing flows within the systems, the water exchange in lagoon systems, and the surface available for catching other species. Occupying lands for development of aquaculture systems can fragment and reduce natural habitat. The withdrawal of plant covering can reduce biodiversity and create changes in the soil and water, and in the processes that bring sediments and nutrients. In addition, the introduction of species can create problems for native species and their habitats, resulting in further loss of biodiversity.

34.1 Analysis

As is the case with all economic activities, and in particular those that have the most direct links with the use of natural resources, the interactions of aquaculture with the environment are many. The importance of the fisheries sector is even greater when it is analysed from a regional perspective, because in many states and coastal communities, fishing activities have become a fundamental element in economic development and social well-being, so that care in the methods of exploitation and manipulation of fisheries resources is basic for achieving optimal use. Starting in the 1930s, institutional forces for the development of aquaculture in Mexico have been directed to the encouragement of rural aquaculture and the repopulation of reservoirs, especially the large dams. However, the results of this repopulation have still not been fully explored. Later, during the 1980s, government policies were directed to high-yield industrial aquaculture.

The trend towards volume declines in the most important species which has characterized the last four years complicates things, because the infrastructure for aquaculture production has serious problems in seed production necessary to recover the pace of growth that prevailed in 1989-1990. The picture is even more complicated if variables such as finance, sanitation, and pollution are taken into account, and this has made it difficult to bring about a full recovery of aquaculture's growth rates. Considering that most of the water basins of the nation have some level of contamination, at present there is no assurance that in the near future the infrastructure for cleaning up the basins will develop to ensure that the quality of the bodies of water will meet the minimum needs for aquaculture activity. Thus the quality of water will determine investments and aquaculture activity in the future. If this activity is to be considered as a potential option for development of dietary diversification and rural communities, as well as boosting the nation's foreign exchange earnings, it will be necessary to start by seeking mechanisms to ensure the quality of water to be used; otherwise the programme and any investments made will run the serious risk of failing to achieve the results that were hoped for.

35. Water Quality

It is a given that the availability of water, in terms of volume, is crucially important for management of the resource, as well as a determining factor in the development of society. However, there is another element that is just as important, and as such, should be given the same attention: water quality. A significant number of factors determine water quality, both natural and anthropogenic. For the purposes of this study we will only refer to the anthropogenic factors, specifically wastewater and residual solids. According to CNA data (1999a; 2001), municipal discharges, which include water of both domestic and industrial origin, total about 7.54 km^3/year. Of this volume, 5.90 km^3/year is collected through the sewage system, and only 1.46 km^3/year receives any kind of treatment. These discharges contain organic material, toxic compounds, and even pathogens. Industries that do not discharge materials into the drainage system generate 5.36 km^3/year, containing mainly toxic elements. There is installed capacity to treat just 16 percent of these discharges (INEGI 2000a). Another sector that is a heavy polluter and that is given little attention is agriculture, whose discharges contain agrochemicals and suspended particles. There are no studies on the quantities of water returned in this way on a nation-wide scale, but some studies (INEGI 1998a) estimate up to 21.20 km^3/year.

Of the total volume, the municipal sector generates 58 percent and the industrial sector 42 percent. However, this does not represent the impact of each sector on water quality. Taking as a base the Biochemical Oxygen Demand (BOD, a parameter that represents organic contamination), we may observe in Figure 35.1 that the impact of the industrial sector is higher than that produced by discharges of municipal origin. In addition, it must be considered that industrial discharges contain many toxic elements, so that their impact on bodies of water may be much higher. In the case of agricultural discharges, there are no evaluations or studies on the magnitude of the problem, either by quantity or by quality, or of the effects of contaminants discharged. Nor are there any volume evaluations on the effects of wastewaters in the soil, or on those that have filtered into the subsoil.

Another factor that contributes to contamination of bodies of water is the disposal of solid residues, including dangerous residues. It is estimated that in 1998, 30.55 million tonnes of municipal solid residues were produced (INEGI 2000a). However, an estimate from 1996 quantifies it at 31.74 million tonnes (INEGI 1998a), which illustrates the inconsistency of the sources of environmental information available. If one considers the data from 2000, it is estimated that 15.877

million tonnes were disposed of in controlled landfills, 1.007 million tonnes were disposed in uncontrolled landfills, 13.458 million tonnes were dumped into open sky trash heaps, and 0.206 million tonnes were recycled. From this we can conclude that 47 percent of solid municipal wastes are not treated in an environmentally safe way. According to information from INEGI (2000a), the residues generated by industry are estimated to be 3.184 million tonnes per year. However, SEMARNAP (1997) estimates production of 8 million tonnes per year, not including what is generated by mining activities, which is estimated at between 300,000 and 500,000 tonnes per day. It is believed that the option of disposing of dangerous residues in water basins is the general rule, considering that nearly 90 percent of the residues are in liquid, aqueous, or semi-liquid form, or dissolved or mixed in with wastewater discharges. Oils and greases together with solvents represent more than 45 percent of the total residues generated in the country. Resins, acids, and bases represent 10 percent and waste from paints and varnishes 8 percent (Martínez 1998). The spatial distribution in the generation of dangerous residues is shown in Table 35.1.

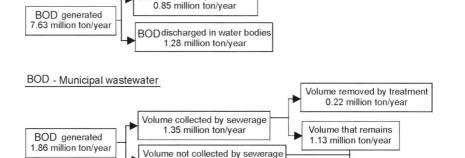

BOD - Industrial wastewater

BOD generated 7.63 million ton/year

BOD removed by treatment 0.85 million ton/year

BOD discharged in water bodies 1.28 million ton/year

BOD - Municipal wastewater

BOD generated 1.86 million ton/year

Volume collected by sewerage 1.35 million ton/year

Volume not collected by sewerage 1.35 million ton/year

Volume removed by treatment 0.22 million ton/year

Volume that remains 1.13 million ton/year

Volume discharged in water bodies 0.22 million ton/year

CNA (1999a)

Fig. 35.1. BOD generated by industrial and municipal wastewater

Table 35.1. Hazard waste generated by region

Region	States	%	Generation [million ton/year]
US border	Main industrial areas in the US border	1	0.08
North	Baja California, Baja California Sur, Chihuahua, Coahuila, Sonora, Nuevo León, Durango, Nayarit, San Luis Potosí, Sinaloa, Zacatecas, Colima, Aguascalientes and Jalisco	24	1.92
Central	Guanajuato, Michoacán, Morelos, Puebla, Querétaro, Estado de México, Tlaxcala, Hidalgo and Mexico City	65	5.2
Gulf	Tamaulipas, Veracruz and Tabasco	7	0.56
Southeast	Campeche, Guerrero, Oaxaca, Chiapas, Yucatán and Quintana Roo	3	0.24

SEMARNAP (1997)

As for the disposal of hydrocarbons, it is estimated that between three and six million tonnes per year of petroleum and its derivatives are dumped into the sea and in coastal areas, where discharges from domestic and industrial sources contribute 37 percent. In Salina Cruz, Oaxaca, the levels of dissolved hydrocarbons dispersed are about eight times higher (76.5 ppb) that what has been established as acceptable limits by UNESCO (10 ppb). The sediments in the Tonalá and Coatzacoalcos Rivers in the state of Veracruz are coastal systems that contain the highest concentrations of hydrocarbons, with 1,148,680 and 120 ppm, respectively (Botello et al. 1995).

As can be seen in the Table 35.1, the central and northern regions generate the highest quantity of residues, 89 percent of the total, since these zones have the largest number of industries. Basic and secondary chemical and petrochemical plants are the main generators of industrial residues, 40 percent of the total, followed by metalworking and basic metal industries with 10 percent, and the electrical industry with 8 percent.

In terms of sanitation, in December 1998, sewage services were provided to 87 percent of urban zones and 32 percent in rural areas, which together gives 72.4 percent of coverage at the national level (CNA 1998b; SEGOB 1998; CNA 1999a). As can be seen, there is a large shortfall in sewage services, mainly in the rural areas, and the previous numbers do not reflect the quality nor the type of service available in these areas. The following table shows the number of homes in 1997 and the type of service available for wastewaters.

Table 35.2. Type and distribution of sewage system

Sewage service	No. of homes
Mix sewage (rain and wastewater)	11,612,312
Septic tank	2,283,354
Discharge in water bodies	575,540
Without sewage service	4,856,172
Not specified	34,094

INEGI (1998b)

Of 19,361,472 homes, 30 percent discharge their waste into some body of water or lack any service, and 12 percent, even when they have a septic tank, do not know whether it has been constructed in accordance with sanitary specifications. Nor is the geographical distribution homogeneous. The largest proportion of coverage is found in the centre and north of the country; in the southeast the service is grievously deficient (Figure 35.2).

In December 2000 there were 1,018 municipal systems with an installed capacity of 75.9 m^3/second, of which 793 were operating with an installed capacity of 45.9 m^3/second; treatment efficiency varied from 95 percent to less than 30 percent, with an average of about 60 percent (CNA 1998b; Parra and Sarchini 1998;

CNA 2001). In terms of treatment of wastewaters from industry, the number of plants in 1998 was 1,354 with a design capacity of 29,321.81 litres/second equivalent and an operations flow of 21,951.48 litres/second. Of the total number of plants, 468 meet the Specific Conditions of Discharge (SCD), with an operations flow of 7,298.74 litres/second. On the other hand, 880 plants with an operations flow of 14,652.74 litres/second do not meet the SCD (CNA, 1998c; INEGI, 2000a).

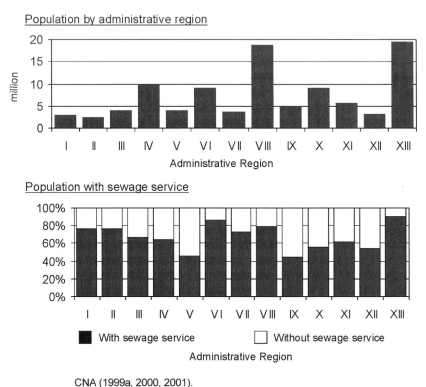

CNA (1999a, 2000, 2001).

Fig. 35.2. Sewage service spatial distribution

Studies carried out by the National Water Commission in 218 water basins, covering about three quarters of the territory in which 93 percent of the population lives, along with 72 percent of industrial production and 98 percent of the irrigated areas, shows that most water basins have varying degrees of contamination, as shown by the presence of organic, industrial, and/or agrochemical wastes. It shows that the central region has low water quality in its receptor bodies; in the majority of the water basins of the country water is classified as being of medium quality. Just 20 water basins generate 89 percent of the total organic discharge in the country in terms of SCD, and this area is comprised of: the Valley of Mexico, Lerma, San Juan and Panuco (Figure 35.3; CNA, 1994c; CNA, 1999a).

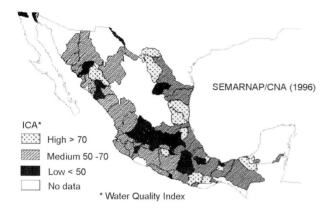

Fig. 35.3. Water surface quality (1975 – 1992)

The presence of nutrients from the discharge of wastewaters has encouraged the presence of aquatic weeds. They have infested about 68,000 hectares in 268 bodies of water, 10,000 km of canals, and 14,000 km of drainage ditches. This problem in turn encourages the development of insects and other organisms that are vectors for illness, limiting the development of recreational activities and affects fishing and navigation. It also causes an excessive loss of water for evapotranspiration, lowering the capacity of the reservoirs, impedes the efficient use of the hydraulic works, and increases their operating costs (SEMARNAP/CNA 1996).

The over-exploitation of aquifers has created contamination problems through marine intrusion in the following aquifers: San Quintín, Maneadero, San Vicente, San Rafael, San Telmo, Vicente Guerrero, and Camalú, in Baja California; Santo Domingo, San José del Cabo, and La Paz, in Baja California Sur; Caborca, Hermosillo, and Guaymas, in Sonora; and Veracruz City in Veracruz. It has also increased problems through the concentration of salts in the aquifers in the Guadiana valley in Durango, in the Aguascalientes Valley in Aguascalientes, and in the La Laguna Region (Figure 35.4) (SEMARNAP/CNA 1996; CNA 1999a; CNA 2001).

Moreover, discharges of wastewaters and inadequate disposal of dangerous residues have created problems of filtration of contaminants in the following aquifers:

- Yucatán (Mérida): Concentrations of nitrates with values higher than 50 mg/l and fecal coliforms higher than 2,000 NMP/100ml (Cabrera et al. 1997; CNA 1996; Pacheco et al. 1997; Villasuso 1996);
- Zacatecas (Calera, Chupaderos, Guadalupe, Bañuelos, and Benito Juárez): Concentrations of arsenic of 0.335 mg/l, due to migration of this element from overpumping (Castro et al. 1997);
- Hidalgo (Zimapán): Concentrations of arsenic, from values below the standard (0.05 mg/l), to levels that exceed it by 20 times (Armienta et al. 1996);

- State of Mexico (Tlalnepantla): A study carried out on 23 wells shows that in nine the water is classified as not potable, in seven the quality is medium, and the rest are of good quality. The parameters that exceed the standards are: chlorides, lead and nitrates (Frausto and Guise, 1996);
- City of Guadalajara (Tesistán-Toluquilla): A study carried out in 64 locations shows that waters from the aquifer contain high concentrations of boron (0.8-14.9 ppm), arsenic (0.08-5.5ppm), fluorine (18.51 ppm), and STD (620-1,208.8 ppm) (Gutiérrez and Sánchez 1996);
- Guanajuato (León): A study of wells for supplying potable water for the city of León shows that the waters of the aquifer contain high concentrations of chlorine (800-1,000 mg/l), sulphates (800-850 mg/l), sodium (800-850 mg/l), and nitrates (10 to 25 mg/l); in the first 15 cm of soil, concentrations of up to 250 mg/kg of chromium were found, and in areas that have been irrigated for 40 years chromium was detected in varying concentrations, at depths of up to 80 m (CNA 1996);
- Hidalgo (Mezquital): The constant increase in the flow of wastewaters for irrigation has begun to dominate the groundwater flows. As a result, it has changed the original aquifer and new areas of discharge and springs have appeared. This new water resource is being used as the only one for public storage of water for about 500,000 persons, as well as for industrial and agricultural uses (CNA 1996);
- Sonora (Yaqui Valley): In six of eight wells studied for public storage, five organochloric pesticides have been detected: HCH, Dieldrin, Endrin, DDE, and DDT. The concentrations vary from 1.6×10^6 to 5.2×10^3 mg/l (González and Canales 1995);
- There have been contamination problems in aquifers located in the states of Aguascalientes and San Luis Potosí, as well as in Celaya and Salamanca, in Guanajuato (SEMARNAP/CNA 1996).

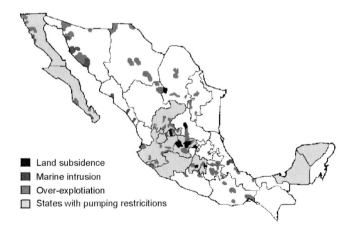

■ Land subsidence
■ Marine intrusion
■ Over-explotiation
□ States with pumping restricitions

Fig. 35.4. Main aquifer problems in Mexico

In all of these cases the parameters mentioned are found to be above the standards for storage of potable water (NOM-127-SSA1-1994).

In Mexico contamination has had a severe impact on public health. The shortfall in purification services and of health have resulted in the resurgence of Third World illnesses such as cholera; between 1991 and 1997 about 50,000 cases were reported, with about 500 deaths. However, it is thought that the true numbers would be much higher due to the low quality of information and the fact that not all cases are reported because of the lack of medical services, especially in the poorest areas. Gastrointestinal infections resulting from consuming contaminated water represent the second highest cause of infant mortality in Mexico (278 of every 100,000), although this has been decreasing. In 1965, 68,000 cases were reported; in 1972 there were 3,680 cases. The cause of this incidence of illness in one-third of the states is due to pollution of fecal material in the water, and it affects above all the poorest section of the population (Figure 35.5). Costs associated with the health effects due to gastrointestinal infections, currently one of the most serious problems in the nation, have been estimated at US$3.6 billion (OPS 1994; Saade 1997; WHO 1997; UNAM 1997; INEGI 1998a; OECD 1998; WHO 1998).

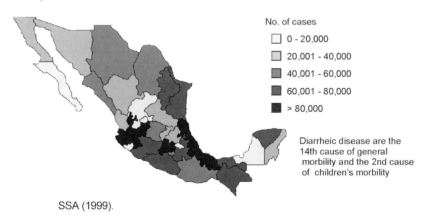

No. of cases

☐ 0 - 20,000

☐ 20,001 - 40,000

■ 40,001 - 60,000

■ 60,001 - 80,000

■ > 80,000

Diarrheic disease are the 14th cause of general morbility and the 2nd cause of children's morbility

SSA (1999).

Fig. 35.5. Diarrheic disease cases by State, 1997

If we consider that illnesses like cholera develop through consumption of water contaminated with faecal material; the increase in annual spending to prevent this could be thought of as an indirect indicator of the inefficiency of sewage and purification systems. Since 1991 the CNA has had to carry out operations every year to prevent cholera, and Table 35.3 shows the year-to-year development of this. This situation has brought about a need to devote funds to improve water purification systems, money which could have been spent on other activities if there had been adequate planning and management of the purification infrastructure.

The use of wastewaters in irrigation is a growing practice, and it is estimated that 344,000 hectares have used municipal wastewaters, much of which has not been treated. It is supposed that contaminated water should not be used for irrigation of products for direct consumption or that are in contact with the soil; however, it is difficult to determine the extent to which this standard is violated. In 1995, 161 hectares of restricted cultivation irrigated with wastewaters were reported, and in 1997 the number had increased to 403. The use of wastewaters in combination with poor irrigation practices could also create problems of salination; in some irrigation districts up to 40 percent of the irrigated areas are affected by this.

Table 35.3. Cholera prevention operatives in Mexico, 1991-1997

Activity	Year						
	1991	1992	1993	1994	1995	1996	1997
Localities included in operatives	-	1,077	1,385	1,678	5,174	3,780	3,594
Chlorination of water sources	-	29,767	14,397	34,530		59,929	73,358
High risk places disinfected	-	604	1,038	1,221	42,846	4,212	12,725
Infection focus whitewashed	-	14,247	6,638	25,570	141,254	57,615	59,269
Chlorine monitoring	-	4,869	7,081	8,519	58,293	329,101	25,586
Operatives developed	4	2	10	7	69	109	111
Inhabitants benefited (thousand)	450	1,102	6,156	2,695	6,596	6,700	7,522

Data: CNA (1994c; 1997a; 1998b)

The uncontrolled discharge of wastewaters has caused hydroecological emergencies. In December 1994, there was a massive die-off of birds in the De Silva Dam, municipality of San Francisco del Rincón, State of Guanajuato. This dam is fed with rainwater and wastewaters from the sewage systems of León and San Francisco del Rincón, and from petroleum, food, textile, and chemical industries located in the two cities. Mortality was more than 25,000 aquatic and land birds representing 30 migratory and native species (SEMARNAP/PROFEPA, 1995). As a result, there was heavy pressure from international organisations on the federal government to take measures for the improvement and preservation of the areas where these birds spend part of each year. It is a concern that with the passage of time these events are going to occur with increasing frequency. In Querétaro, more than 8,000 migratory bids died in the Tequisquiapan dam. In Zacatecas 8,000 fish died in the El Pedernalillo dam, and in Espejo de Lirios Lagoon in the State of Mexico more than 12,000 fish died. Similar incidents occurred in La Virgen dam in San Luis Potosí, Tres Palos Lagoon in Guerrero, Nautla Lagoon, Vega de Alatorre Lagoon, Grande and Chica Lagoons in Veracruz. In Baja California, in Ojo de Liebre Lagoon six whales died as well as several dozen turtles. In Baja California Sur three whales died in Magdalena Bay and four other whales in Los Cabos. In Sonora the mortality of whales rose to 25 at the port of Topolobampo, Piedra

Island, and Jitzamuri. In Tamaulipas, 20 dolphins died on the beaches of Bagdad, Matamoros. In Nayarit, eight dead whales appeared on the beach of Guayabitos.

Another factor that deteriorates the quality of water is the presence of suspended solids from erosion. According to SEMARNAP, 60 percent of national territory suffers from severe erosion. This is not only altering the productive potential in places where it occurs, but it is also having a negative impact in the areas where sediment builds up in water storage areas and drainage canals, contaminating lakes and rivers, and bodies of water in general. This is caused by a lack of integrated planning for agricultural activities (irrigation and seasonal), and livestock and forestry operations. Urban concentrations encourage the rapid deterioration of ecosystems due to the immoderate land clearing, woodcutting, overpasturing, reduction of vegetation cover, and exposure of surface soil to water and wind erosion (PNUMA 1993; INEGI 1998a; INEGI 2000a).

Currently, with the redesign of the National Network for Monitoring Water Quality (RNMCA), the main network has 403 permanent stations, of which 215 are located on bodies of inland water, 45 in coastal zones, and 143 in aquifers; there also are 244 semi-permanent or mobile stations in the secondary network, of which 227 are located on inland waters and 17 in coastal waters. In addition, for detection and control of specific problems, there are studies that require seasonal stations. The annual information, generated through about 100,000 analyses carried out from 4,000 samples, is managed by the National System of Information of Water Quality (SNICA), which uses the data of the RNMCA, incorporating national inventories of discharges, purification plants, and municipal and industrial waste water treatment plants, as well as information related to contingencies and hydroecological emergencies (CNA 2001).

Even with the installed infrastructure there are serious shortcomings; the time that elapses from taking samples to the integration of the final report is about six months, and there are no reports in real time; the sample sites do not have infrastructure or installations that allow adequate sampling; the laboratory equipment is obsolete and the installations in many cases are improvised and do not meet even minimum standards of security; the personnel is not properly trained in the management and lacks knowledge to carry out the analyses; the information is not available to the public; and to date no CNA laboratory has been accredited with the Secretary of Commerce and Industrial Development in accordance with Article 2 and Article 69 of the Federal Law of Metrology and Standards (Diario Oficial, July 1, 1992), so that the results obtained from the laboratories lack official validity.

35.1 Analysis

Discharges of wastewaters, domestic, industrial, and agricultural, as well as the disposal of solid wastes and dangerous materials in bodies of water have serious environmental, economic, social, and health consequences. With the increase in population and economic activities, the volumes of wastewaters, solid wastes, and dangerous materials can be expected to continue to grow. This will pose a huge challenge in terms of safe management of these residues.

Generally when there are references to pollution through discharges of wastewaters, the reference is to municipal and/or industrial discharges; agricultural discharges are almost never mentioned. This sector contributes the largest volume of wastewaters, with a high content of agrochemicals; the complexity of analysis of this type of discharge is due to the fact that they are spread out geographically and intermittent in time. The problem in Mexico is no different from that of other countries with extensive agricultural lands, but most experts agree that the best way to prevent contamination from agrochemicals is adequate management in their application. In Mexico there are no programmes to verify adequate management of agrochemicals; federal authorities lack tools and resources to verify with users what type and form of agrochemicals are being used. It is estimated that in 1995, 3,299,300 tonnes of fertilizers were sold (INEGI 1998a), but there are no data for insecticides.

The continuous degradation of bodies of water will make it increasingly difficult and costly to find adequate storage basins. Most surface waters in the country require extensive treatment to be used as water for human consumption and for industrial activities. Groundwater, the main source of potable water, is not always free from fecal coliform or other substances dangerous to health. A sample of water quality in border cities shows the impact of *maquiladoras* (assembly plants). In Nogales, in a public water source which supplies several communities beside a chrome-plating plant contained high levels of trichloroethylene (TCE), 20 percent above the standards for potable water permitted in the United States (Moreno 1995). The cost of cleaning up the water may be expected to increase due to the distance from new sources of water, construction of transport and potabilisation infrastructure.

The use of waste waters in irrigation spawns two main problems: 1) The relationship with the health of farmers and their families, and 2) The problems associated with pollution of soil and of surface and groundwater. The scarcity of water and the increase in competition for sources can be expected lead to increasing use of wastewaters in irrigation. Proper management of this can create benefits: lowering the use of agrochemicals because wastewater already contains nutrients to increase soil fertility, a reduction of volumes of pure water used in irrigation, and a decrease in extraction from sources. However, poor management could create serious problems, for example: pollution of soil and of surface and ground water, reduction of fertility and production in soils because of problems of salination and pollution, harmful effects on the health of farmers, a deterioration of the infrastructure that was designed to function with pure water and now uses mainly wastewaters, health problems among populations that consume products irrigated with wastewaters, and limits to the type of plants that may be cultivated.

The serious shortcomings in infrastructure for sewage and purification presents a major financial challenge, given the scarcity of funds, and political interference with the way that decisions are taken on them. For water cleanup in the Mexico City metropolitan area, in 1997, a programme was proposed to start construction of wastewater plants in Texcoco Norte, Coyotepec, El Salto, and Nextlalpan, with a combined capacity of 74.5 m^3/second, including extraordinary expenses for the rainy season (CNA 1997b). The investment needed for these projects was of the order of one billion dollars, which would have been financed with a loan from the Interamerican Development Bank (IADB) for 365 million dollars, and a further 410 million dollars from the Foundation for Overseas Development of Japan. The rest was to be financed by the government of Mexico City. To meet this expenditure, the Mexico City government requested authorisation from the local congress to borrow 7.5 billion pesos to get the project started. The local congress only authorized 1.7 billion pesos. The paradox is that the loans from the other financial institutions had already been approved. In 2002, the CNA must pay 15 billion pesos to the IADB for a line of credit that was never used.

As for industry, the current economic crisis and the unrestricted opening to international markets has placed small and medium-sized businesses at a serious disadvantage, and set them back. The vast majority are operating at 50 percent of installed capacity and prefer to invest in resources for maintaining production rather than comply with environmental standards. This situation does not appear likely to improve in the future, and we can conclude that the industrial sector, especially small and medium-sized industry, will have serious problems complying with legislation relating to discharges of wastewaters. In the case of large industries, the lack of federal funds makes private investment increasingly necessary, and this has obliged decision-making organisations to change. As private capital gains ground, fears are growing that the lack of government investment is going to result in failure to apply the legislation, and in concessions that are outside the legal framework. It is hoped that one day industry will be the motor of the national economy, but a solid regulatory apparatus is needed to allow for the establishment of environmental regulations and compliance with them; without this condition we can confidently state that no framework of standards will be adequate to improve the quality of the bodies of water.

Use of wastewaters will be a necessary condition in the future for any integrated management plan for water resources, and this implies compliance with the minimum conditions that ensure that there will be no negative effects on the health of the population or the ecosystems. It will be necessary to increase budgets to overcome the shortcomings in purification, create new infrastructure, and maintain what is already in place. It will be necessary to involve more directly those responsible for discharges into water bodies, making them responsible for treatment of their effluents and for provision of the resources needed to verify and follow up on these actions. Society should become conscious of the need to pay for services of purification, and to establish the economic mechanisms needed for compliance. In the future the quantity of water available will be a function of two basic factors: 1) Efficient use of the resource, and 2) Control of pollution in water bodies. Both factors will allow the provision of extra resources for new demands for water, as well as for public health, economic, social, and environmental benefits. The federal government does not have any choice but to invest in water purification and to establish the mechanisms and actions so that society can be involved in this task. Otherwise, the storage of water, the health of the population, and the maintenance of the ecosystems are likely to be severely compromised in the future.

36. Regulatory Framework

Mexico has a long tradition of water management that goes back to prehispanic times. However, modern water policy in Mexico has its roots in the Constitution of 1917. From that a body of laws, regulations, and institutions has developed over the past 80 years to define the limits of intervention of the authority, as well as the rights and obligations of private and public individuals and organisations that make use of national waters.

After the revolution, the need to increase food production and to populate extensive regions of the area bordering on the United States led to the creation of the National Irrigation Commission and the promulgation of the Irrigation Act of 1926. Agrarian reform was the touchstone for post-revolutionary governments, and agrarian policies and water planning developed together. Irrigation Districts and small irrigation units developed all over the country. In 1934 and 1936 the Law of Water as National Property and its regulations were passed (Ortiz 1997; Herrera 1998).

In 1946, the National Irrigation Commission became the Ministry of Water Resources, and for the first time, responsibility for water development was concentrated in a single organism. In addition to the water-for-agriculture infrastructure, new multipurpose projects were built. Demand for water increased substantially as a result of the confluence of policies designed to strengthen the national economy with special emphasis on industry. At the end of the 1940s and the early 1950s, the first Water Basin Commissions were set up for the most important rivers of the nation (Grijalva, Papaloapan, Coatzacoalcos, Balsas). The goal of these commissions was to encourage regional development based on the implementation of large-scale water projects. Later, at the start of the 1960s, rehabilitation projects were developed on a large scale in irrigation districts with the goal of increasing agricultural productivity. Plans were also developed for the transfer of water among basins to expand the irrigated areas in the northeast region, and to ensure future storage of water for Mexico City. In 1956 and 1958, the Law and Regulations for Exploiting Subsoil Waters were published. From this date forward the country began to regulate the extraction and use of groundwater (Ortiz 1997; Herrera 1998).

In 1972 the Federal Water Law was passed, specifying for the first time the functions and responsibilities of the federal institutions, especially those from the Ministry of Water Resources and, for the first time, establishing regulations governing residual waters. In 1975 the first National Water Plan (PNH) was spelled out, with the intention of integrating water policies and the economic and social development objectives of the country into a coherent whole. It defined regions by hydrological criteria, incorporated the social and economic variables most relevant to the process of water planning on both a national and regional scale, and established scenarios with a time horizon of 25 years. Other objectives contemplated in the PNH were to achieve a better population distribution and to decrease levels of soil and water contamination. The PNH of 1975 was the first attempt at an institutionalised process that aimed at integral planning for the exploitation of water resources. In 1976 the National Commission of the Water Plan was established to start executing the plan and carrying out the respective changes and updates.

In 1976 a series of new institutional arrangements were made within the federal government. The Ministries of Water Resources and of Agriculture were merged into the Ministry of Agriculture and Water Resources, and the government decided to focus mainly on the problems of the agriculture sector. This became a serious limitation for moving forward with the strategies and actions of the PNH. Later, in 1985, new difficulties surfaced for implementing the PNH: the National Commission of the Water Plan was changed into the Mexican Institute of Water Technology (IMTA). This resulted in the break-up of the work team and the redirecting of the objectives. As a consequence, the PNH never could be put into practice as originally conceived. Paradoxically, this situation arose just when the problems of the water sector began to require more effective government action.

Another defining moment for the water sector came in 1982, when the taxes for use and exploitation of national surface and groundwater were placed within the Federal Law of Rights. This law went through two basic reforms: the introduction in 1986 of taxes with respect to the regional availability of water, and in 1991 charges were imposed for the discharge of residual waters containing pollutants. Finally, as the problems related to water continued to grow, in 1989 the National Water Commission (CNA) was created as an organisation of the Ministry of Agriculture and Water Resources. In 1992 the Law of National Waters (LAN) was passed, and in 1994 the regulations for the LAN were promulgated. Under the LAN, the CNA is the only legislative authority for national water issues. It came into being almost 15 years after water management was centralised into a single body.

36.1 Legal Framework for the Management of National Waters

Article 27 of the Mexican Constitution establishes that all water resources are public property under the control of the federal government. Rights are granted for use of water for 50 years, and may be revoked if there are changes in the use to which the water is put or as a result of management practices that are contrary to the public interest. Article 115 of the Constitution (amended in 1985) establishes that the municipalities are responsible for providing potable water services, sewage, and purification, with the assistance of state governments if the municipal government requests it due to lack of financial or technical resources. In December 1992, the national Congress approved the Law of National Waters, which regulates the exploitation of the resource and the safeguarding of its quality and quantity. It is the only general law that refers to water management, and it obliges the federal government to formulate and execute the National Water Programme. The goals of the LAN are to regulate the exploitation and use of national waters, control its distribution, and maintain its quantity and quality to achieve its sustainable development. To do this it has established four relevant areas that must be considered in the implementation of national water programmes: 1) strengthening the institutional capacity for water management; 2) decentralization of functions; 3) the use of economic tools to measure efficiency of water use; and 4) involvement of the private sector in the financing of water infrastructure (Herrera 1997). Each state has its own law of potable water and purification, aimed at establishing the conditions of supply of the corresponding services, and establishing fees for these services.

The Regulations for the Law of National Waters (RLAN), published in 1994, establish the procedures and administrative details for the application of the law. Specific time periods, administrative actions, and procedures for conflict resolution are included in the regulations. The Federal Law of Rights on Water Matters (LFDA) establishes the legal framework and the mechanisms under which the federal government may establish fees for water use and for residual discharges into the bodies of water in accordance with predefined parameters. The LFDA stipulates: i) the principle on which water shows economic value according to its availability, and ii) the principle under which those who pollute must pay. The law was updated every year, and after 1999 it was updated every six months. The law fixes usage taxes and fines for polluting for different categories of usage according to the availability of water in a specific zone. It is interesting to note that, paradoxically, given that agricultural use is historically the largest consumer of water it has been exempt from the payments of taxes for water use.

The General Law of Ecological Equilibrium and Environmental Protection (LGEEPA), in the amendments of 1996, establishes the levels of competence for the federal, state, and municipal governments relating to environmental issues. It also makes obligatory the carrying out and presentation of the corresponding environmental impact study for all actions required; promotes change in favour of an integrated system for permits; and creates an inventory of all environmental emissions, including discharges of effluent into the water. The General Law of Public Health stipulates the quality standards that must be met in water destined for human consumption.

36.2 Institutional Framework

In the past, people and economic activities have developed inversely to the distribution of the nation's water resources. The growth of the economy and population has accelerated, but there has been a total absence of long-term strategic planning and vision, and a lack of institutions capable of an integrated management of the water infrastructure. As a result, by the end of the 1980s many problems had emerged: over-exploitation of aquifers, costly transfers of water among watersheds, serious levels of pollution in the main water basins of the nation, a deterioration of the water infrastructure, and a growing number of conflicts among users over who controls the resources. In response to this, new strategies for the water sector have been articulated: 1) Development of the infrastructure needed to eliminate the existing inadequacies and to satisfy growing demand; 2) More efficient use of water; and 3) Decreasing pollution levels. Water policy was designed taking into consideration the strategies developed in the National Water Plan of 1975, with the necessary adjustments for the new social, economic, environmental, and political circumstances. The first step towards the implementation of this policy was the creation of the National Water Commission of Mexico (CNA).

The CNA was created by Presidential Decree on January 16, 1989 as part of the then Ministry of Agriculture and Water Resources. The CNA became the only federal body with the authority to administer national water resources. In 1994, with the new federal administration, the CNA was transferred from the agriculture sector to the then Ministry of the Environment, Natural Resources and Fisheries (SEMARNAP), and later, in 2000, it became part of the Ministry of Environment and Natural Resources (SEMARNAT).

This change reflects the importance the federal government accords the management of the environment and the use of natural resources of which water forms a part, and strengthens the exercise of authority so that the regulation of water usage is not subjected to the rules of several departments.

To carry out its responsibilities the CNA is divided into six key areas: administration, water administration, construction, operations, planning, and technical. In addition it has the following support functions: rural programmes and social participation, communications, internal control, legal affairs, review and liquidation of financial credits, potable water and purification, and water planning. During the last administration the regional managers, who report to the general managers, were reorganized and their numbers increased from six to 13, located in various states of the republic, based on strict hydrological criteria (See Chap. 28). State managers were established, except in states where the offices of the state managers and the regional managers coincided. The state managers report to the regional managers. The CNA has the support of the Mexican Institute of Water Technology (IMTA), which is responsible for providing technical assistance and training for the CNA. Its activities include research, development, and technology transfer, as well as development of training programmes within the water sector. The responsibilities of the CNA are shown in Figure 36.1.

Fig. 36.1. National Water Commission responsibilities

The CNA has a technical board whose chairman is the head of SEMARNAT. Board members include the heads of the Ministries of Finance; Agriculture, Livestock, Rural Development, Fisheries, and Food (SAGARPA); Social Development; Health; Energy; and Control and Administrative Development (CNA 2001). When the board considers it convenient, it may invite heads of various departments and organisms of the federal public service and state and municipal representatives, as well as users. The technical board meets regularly every two months, or at any time the chairman calls a meeting. The powers of the technical board are as follows (Art. 11 of the LAN):

- To be familiar with the policies and measures for the planning and actions coordinated among the federal departments which have interests in water policy,
- To understand the matters that are submitted for its consideration relating to water management and the revenues, assets and resources of the CNA,
- To be familiar with the programmes and budgets of the CNA, to supervise their execution, and to be familiar with the reports which the general director submits,
- To propose terms under which the credits required by the CNA may be secured,
- To agree on the creation of the Water Basin Boards.

The general director is responsible for following up on the decisions adopted by the technical board and providing sufficient and timely information on the CNA projects which require coordination and support from the departments with ties to the board (Art. 14, section XIV of the LAN). From the foregoing, it may be assumed that the technical board is the real decision-maker for Mexico's water policy. There are two implications: 1) decisions taken in the water sector will not be reversed, and, 2) important decisions are not necessarily taken within the sector.

36.3 Principles of Mexico's Water Policy

The following is a brief description of the elements on which water policy in Mexico is based. Although they are in no particular order of significance, each plays a role in the development of policy.

36.3.1 Water Planning

The National Development Plan for 2001-2006 is the basic six-year planning tool for the Federal Executive, and provides the principles, objectives, and strategies that guide activities for the period. It is the guiding document for all activities in the federal public administration. Mexico's water policy falls within the National Water programmes. In accordance with the LAN it is the responsibility of the National Water Commission to develop the National Water Plan, to keep it up to date, and to ensure that it is executed. The National Water Plans are designed to ensure the availability of water to satisfy the needs of the people and to encourage the development of economic activities in a way that is compatible with the environmental capacity of each region. Table 36.1 shows the objectives and quantitative goals of the National Water programmes of 1995-2000 and 2001-2006.

Table 36.1. 1995-2000 and 2001-2006 National Water Plan

National Water Plan 1995-2000	
General Objectives	Goals
• To diminishes the lack of access to water in poor population, • To increase the sanitation activities in those river basins with high level of pollution, • To strength the legal security on water rights, • To establish adequate water prices according with economic and environmental criteria, • To foster the public participation in the water planning process, • To foster the decentralization of functions and responsibilities from central level to local authorities, • To encourage efficient water management in agriculture, municipal, and industrial sectors.	• To increase the water supply service for rural population from 13.8 to 18.8 million, and to increase the sewage service from 5.5 to 15.1 million, • To increase the water supply service for urban population from 62.8 to 68.8 million, and to increase the sewage service from 56.0 to 60.6 million, • To keep the present level of water purification in at least 95% increasing the water purification capacity from 2.21 to 2.37 million m^3 per year, • To increase the water treatment capacity from 0.536 to 2.586 million m^3 per year, giving priority to those river basins with high level of pollution, • To increase land irrigated in 1,040 km^2 and to rehabilitate 8,000 km^2.

Table 36.1. (cont.)

National Water Plan 2001- 2006	
Objectives	Goals
• To foster efficient water management in the agricultural sector, • To increase the quality and quantity of the water supply, sewage and sanitation services, • To achieve the integrated water management in river basins and aquifers, • To promote the technical, administrative and financial development of the water sector, • To consolidate the public participation in the planning water process, and to promote the water culture, • To diminish the effects and risk of droughts and flooding in the population.	• 89% of population with access to water supply systems, • 78% of population with access to sewage systems, • Treatment of 65% of wastewater collected, • 23% of irrigated land with efficient irrigation technology, • 25 River Basin Councils working with its own administrative system, • 41 Technical Groundwater Committees working with its own administrative system • 100% of titles and water permits verified (volume of abstracted water, wastewater discharged, and standards of quality), • To collect 39,497 million pesos (2001 figures), from taxes on water rights.

SEMARNAP/CNA (1996), CNA (2001)

The National Water Plan for 2001-2006 is the first one that mentions the elements that historically have formed part of water planning. For the first time it includes what is called the Vision of the Water Sector in Mexico to 2025, and mentions the regional long-term water programmes for 2001-2025. It does not spell out objectives, strategies and actions for achieving the long term plan, but at least it breaks with the tradition of limiting the goals of the sector to five years. It may be interpreted as a first attempt to establish a long-term planning system. Another important element is the explicit recognition of the importance of the participation of users and society in general in the process of water planning.

36.3.2 Decentralisation of Functions

Historically, federal organisms responsible for the administration of national waters have controlled regulatory, administrative, financial, operating, and construction functions, as well as the promotion of water development, all of which have been carried out through a centralised structure. The authorities and local bodies have been very restricted in terms of power to promote water development. The users are not organized and they have only been the beneficiaries of the services and the infrastructure. This has impeded the expeditious and efficient resolution of the complex problems of the sector. As part of the strengthening of federalism, the CNA has encouraged the decentralization of functions, programmes, and federal resources in favour of the state and municipal governments and the organized users; and supports the establishment and consolidation of state water commissions.

The CNA has transferred programmes of potable water and purification for rural areas; clean water; control of aquatic weeds; and potable water and sewage in urban areas, to the states to be carried out by their governments and departments. To increase the effectiveness of this process, the CNA supports the adjustment of the legal framework of the states that request it, to provide impetus to initiatives that allow them to create state water commissions, so that these new organisms can assume responsibilities currently under the CNA. Water-for-agriculture programmes for the rehabilitation and modernisation of irrigation districts; land development; efficient use of water and electricity; and integral use of the water-for-agriculture Infrastructure are joined with the programmes of SAGARPA in the Alliance for the Countryside (Alianza para el Campo). To encourage integral and regional agricultural development, these federal funds are transferred to the Agricultural Development Trust Funds, to be used directly by the users. To encourage efficient water usage, the infrastructure for irrigation districts has been transferred to the users. In July 2001, a total of 3.3 million hectares was transferred to 525,000 users, of whom 387,000 are members of *ejidos* and 138,000 are small farm owners organized into 444 Citizens' Associations and 10 Societies of Limited Responsibility. The land transferred represents 98 percent of the total area of the 82 irrigation districts. Due to actions carried out and the increase in fees for irrigation surfaces, the level of financial self-sufficiency of the districts increased from 43 percent in 1989, when the transfers were begun, to 69 percent in 2000. The delivery of the funds from the CNA to the state and municipal governments and the users was of the order of 468 million pesos in 1996; 769 million pesos in 1997; 850 million pesos in 1998; 1.254 billion pesos in 1999; 1.269 billion pesos in 2000; and for 2001, 1.28 billion pesos are budgeted. The contributions of the states, municipalities and users doubles this investment (CNA, 2001).

36.3.3 Economic Tools

The strategy Mexico has adopted to streamline water management rests on a delicate balance between government regulation and market mechanisms. One of the main goals of the latest legal reforms has been to establish this balance. Water management is now based on two principles:

- First, a license or concession is necessary for each public or private user of national waters, and it is also necessary to secure a permit for the discharge of residual waters into bodies of water or into the soil.
- Second, those who benefit from the use of the waters, or who use the bodies of water to dump residual waters must pay for: a) the management and maintenance of the storage areas, and b) the re-establishment and the improvement of water quality, in proportion to the quantity of water consumed or the quantity of contaminants present in the residual waters.

The LAN establishes the conditions which must be met in order to transfer concessions and permits from one user to another. Recent changes in the legislation contemplate the commercial exchange of concessions, although in most cases these would be linked to land tenancy. In accordance with the pre-established rules, the regulation of water markets was foreseen, especially for the transfer of water rights in irrigation districts and other collective systems, and where industrial and urban demand compete for water used for agriculture. The LAN establishes the Public Registry of Water Rights (REPDA) which it is to be managed by the CNA. All water rights must be registered in the REPDA, including modifications, extensions, suspensions, terminations, and transfers. The CNA has the power to revoke the rights if the usage of water or the volume of water used or discharged differs from what is allowed in the permit. The LAN gives the CNA the role of arbiter and conciliator to resolve conflicts, though the prevailing principle is that whoever possesses the rights has the last word. The LAN also establishes economic incentives with the goal of improving efficiency levels in water usage. The basis for consolidating the financial system for water in Mexico is established through: 1) more participation by users in the financing of the water infrastructure, and 2) a system of tariffs, taxes and incentives for water usage, as well as for the discharge of residual water.

36.3.4 Water Rates

Federal investment in the sector in recent years shows a trend toward stabilizing at levels below what is desirable. As a central point in the strategies for achieving efficient, fair, and environmentally friendly water usage, the concept is growing of water as an economic resource and not as free merchandise. From this point, water policy promotes a juncture of policies and measures designed to improve and strengthen the finances of the user system with the goal of achieving financial self-sufficiency in the short to medium term, mainly through a quota system which will allow the system to recover the total cost of the services provided (CNA 2001).

It is estimated that by increasing charges, a system would be consolidated which would allow the financing of programmes and actions within the water sector. The same charges would provide a better flow of resources as a counterpoint to the credits obtained from the development bank, or alternatively, to form part of the financial packages with the participation of state and municipal governments, users, and beneficiaries, and private initiative. The fiscal legislation also establishes the costs and tariffs which must be charged to users of water services provided by the federal government, with the goal of recovering all of the costs of operation, conservation, and maintenance related to the supply of water to centres of population, industry, or irrigation districts. On the other hand, the Law of Contribution for Improvement of Federal Public Works for Water Infrastructure is being used as the instrument for recovery of federal water infrastructure investments that directly benefit individuals or companies.

36.3.5 Participation by the Private Sector in Financing Water Infrastructure

In the past, investment in infrastructure was a responsibility of the public sector. Until just a few years ago, the government was the main, and in some cases the only, sector involved in the construction, operation and maintenance of infrastructure. However, the growing need for investment to satisfy the demand for goods and services has surpassed its ability to finance projects from its own resources and external credits. To provide more development in the potable water, sewage, and sanitation subsector, and to help resolve the problem presented by operators, the LAN supports the participation of the private sector in terms of using its technical experience, applying leading edge technology and its financial solvency. In addition, with the participation of the private sector, continuity of services is assured in spite of political changes that might take place in the state and municipal administrations.

In recent years there have been several examples of participation by private companies in the administration of potable water, sewage, and sanitation services in Mexico, in medium-sized and large urban centres, including tourist centres, with service concessions and contracts. There are instances in Aguascalientes, Cancún, the Federal District, Puebla, Navojoa, and Saltillo. The participation of the private sector in the financing of infrastructure can take different forms. The simplest is a service provided by a private operator, but other forms include concessions; a variety of schemes for building, operating, and transferring infrastructure; self-help and the issuing of bonds (Aguilar 1998). Recently 2 billion pesos of funding was approved for the programme for the Modernization of Water Management Groups (FINFRA 2). This programme was developed with BANOBRAS (National Bank for the Development of Infrastructure), and is directed mainly toward helping private water managers who serve localities with populations of more than 50,000, which represent slightly more than 50 percent of the population of the nation (CNA 2001).

36.3.6 Public Participation in Planning: The River Basin Councils

Before the LAN was launched, water administration was developed exclusively by the federal government through a vertical decision-making structure. It gave little or no priority to environmental issues or to the participation of users or affected groups. As a result of the unsustainable environmental situation of Mexico, in 1998 the General Law of Ecological Equilibrium and Environmental Protection (LGEEPA) was passed. It established as an obligation the presentation and authorization of environmental impact studies for water works (Art. 28 of the LGEEPA), and provided for the participation of society in the planning, execution, evaluation, and monitoring of environmental policy and natural resources (Art. 157 of the LGEEPA). With the passage of this law, the Water Basin Boards were established under Article 13 of the LAN.

To facilitate the coordination of water policies and programmes among the three levels of government in Mexico – federal, state, and municipal – and to co-ordinate the objectives, goals, strategies, policies, programmes, projects, and actions between the federal water authority and the duly accredited water users and the diverse groups and organisations, the LAN contemplates and orders the establishment of River Basin Councils, giving exclusive competence for setting them up to the technical board of the CNA (Art. 13 of the LAN). Article 15 of the LAN regulations defines who should comprise the water basin boards (Fig. 36.2). The broad objectives of the boards are the following: 1) To organize the various uses of the water; 2) To organize the water basins, waterways, and bodies that receive the water so as to prevent contamination; 3) To promote and encourage the recognition of the economic, environmental, and social value of the water; 4) To conserve and preserve the water and land of the water basins; and 5) To ensure efficient use of water.

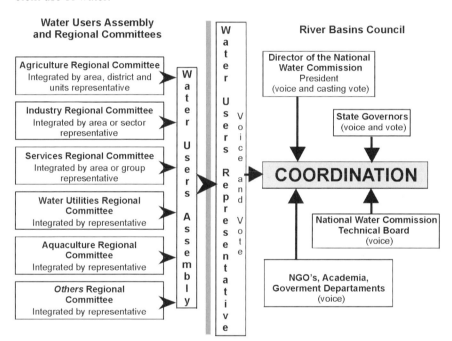

Fig. 36.2. River Basin Council organization

The creation, installation and follow-up of the water basin boards are carried out in three phases, which are called gestation, installation, and operation. Not all of those involved with the board are given voting power, only spokesmen for the users, officials from the state governments, and the chairman, who has a casting vote (CNA 1997b; 1998a). It should be noted that users of national waters who participate in the water basin boards must be accredited by the CNA, on the basis of their concession titles or permits that legitimise their usage. Third parties interested in participating must be constituted as organized groups and recognised by the commission.

According to the CNA (2001), as of September 2001 there were 25 water basin councils in place, as well as 6 water basin commissions, 4 water basin committees, and 47 technical committees for groundwater. Although the three last-mentioned groups are not considered in the LAN, the experience in the approval of the water basin councils demonstrates the need for them as support groups. The gestation and consolidation of these organisations is currently underway. During the last three and a half years, there have been 1,463 meetings. The information that has emerged from the previous process was one of the main inputs in the elaboration of the national water programme 2001-2006.

36.4 Analysis

Water policy in Mexico has been developed over 80 years, and during that time it has been amended and redirected in response to development objectives that have been imposed on the nation at various times. The planning of the water usage of the country has traditionally been carried out from the point of view of supply, since the government, in its work of ensuring the survival of the nation over the long term, has unilaterally provided the water as demanded by society and the different economic sectors. In this sense the institutions and laws for managing water resources until the decade of the 1970s were focused mainly on creating the necessary conditions for the exploitation of water resources and the development of the water infrastructure. The result of this unilateral action was that by the beginning of the 1970s the system was beginning to feel the effects of over-exploitation of water resources, and their contamination. The first laws relating to the environment were passed in the mid-1940s, at which time the Law of Conservation of Soil and Water was passed. Subsequent legislative measures provided more breadth to environmental policy, but the government's priorities were clearly based on the growing economy and industrialisation, and generally ignored the effects of inefficient natural resource use and pollution.

Starting in the 1970s, there was a realisation of the need for more broad-based planning to make water policy compatible with the goals of national development. It was realized that integral planning should link economic and social development with the availability of resources, and should endeavour to improve the environment by lowering pollution levels. With this in mind, the National Water Plan was developed in 1975. However, the economic crisis at the beginning of the 1980s obliged the government to revise its investment programme and reorganize its finances, and as a result more than 400 projects were abandoned. Later, when the Ministry of Agriculture and Water Resources was created in 1985, the water sector was reoriented to focus on the agriculture sector, and the commission in charge of the PNH was transformed into the IMTA with the function of providing technical assistance, developing technology and training in the water subsector, and abandoning its planning function. In terms of environmental issues, even though the first legislation dates from the 1940s, it has only been in the last 20 years that this has gained importance. However, environmental policy in Mexico has suffered from the cyclical variations of the economy and the subsequent changes in priority in terms of environmental issues. Also, during the last 10 years, the federal institutions responsible for environmental protection have changed several times. This has meant that environment policy in Mexico has been characterised by a lack of continuity in its programmes and actions (OECD 1998).

The political and financial uncertainty and the high turnover of the bodies in charge of water administration have been reasons why water policy in Mexico lacks long-term strategic planning among the institutions. This has resulted in inappropriate and fragmented strategies. In addition, the planning process has been based on overoptimistic assumptions, and this in turn has led to objectives that in practice are difficult to meet. For decades there has been an absence of coordination among institutions and even within institutions, and this has created duplication of work in some cases and an absence of any action in others. The lack of funds for water administration or pressures from political promises taken on by each administration have in many cases resulted in planning that was done hurriedly, with a lack of technical elements, and with social and environmental aspects either undervalued or simply not considered. The lack of political will at high levels and pressures from strong economic groups have meant that the legislated mandates have not been completely fulfilled. The development of laws and regulations has been promoted and decreed at times when problems have been extreme. This is probably due to lack of vision, incorrect or partial understanding of national water situation, bureaucratic inefficiency, and inadequate coordination between the legislature and the institutions in charge of water management.

In recent years, especially after 1994, Mexico has created new institutions and instruments to develop a more efficient political atmosphere, and applied a consistent strategy to achieve sustainable economic growth. This has been in response to pressure from international financing groups (the Interamerican Development Bank, the World Bank, the International Monetary Fund), the international treaties to which Mexico is signatory, and recognition by the government that the environmental problem could generate social problems with high political costs. Even so, in a nation that faces rapid population growth, migration to the cities, with 30 percent of the population in extreme poverty and another 30 percent in relative poverty, to achieve sustainable development as established in the National Development Plan of 2000-2006 presents economic, social, and environmental challenges that are extremely daunting.

In terms of administration of water resources, currently the National Water Commission of Mexico looks after water requirements and sanitation for socio-economic development in Mexico, through the support of bodies responsible for providing infrastructure and services for urban, industrial, and agricultural uses. It is also responsible for control of the water system through the construction and operation of works to protect the system against natural disasters and emergencies. Its responsibilities also include the administration and custody of the resource. It is recognised that in recent years, the CNA has introduced significant changes in water management, in response to the accelerated growth of demand, increasingly scarce sources of stored water, accumulated deficiencies in coverage and quality of services, insufficient financial resources, and growing competition for water among users. Despite the importance of the actions taken, and even when there has been an important advance in establishing an integral legal framework for water administration, shortfalls persist in important areas, as well as lack of definition and gaps in standards.

As a result, there are organisational deficiencies, even at high levels of management. According to the present law, the Director General of the CNA is appointed by the President of the country (Art. 12 of the LAN) and it is his responsibility to report to the President as well as to the technical board. In 2000, the CNA became part of SEMARNAT, and hence, at least in theory, the CNA Director must report to the head of SEMARNAT; however, the law does not spell this out. The heads of SEMARNAT and the CNA both report to the President, and so, in practice, the CNA acts independently and SEMARNAT has little or no control over its activities, policies, and programmes. The idea of locating the CNA within SEMARNAT so that the two would coordinate their activities in the protection of water resources has not worked in practice. Another example is that the legislation relating to water discharges shows significant inconsistencies. The criteria for evaluating discharges are under three different laws. The LGEEPA uses ecological criteria in classifying receptor bodies of water; the LAN considers only the classification of receptor bodies of water; and the LFDA has established a list of receptor bodies of water without any technical consideration. While the LGEEPA and the LAN recognize the Mexican Official Standards (NOMs) and the Specific Conditions for Discharge as regulatory instruments, the LFDA recognizes the Specific Conditions for Discharge as the only instrument for controlling wastewater discharge. None of the laws has priority over the others in terms of implementation. It is not clear to users which parameters they should be using: Specific Conditions for Discharge or NOMs, because there is no hierarchy among the laws (Tortajada 1999b).

In terms of decentralisation of functions no positive results are expected over the short term. The municipalities are responsible for the sanitation and the water supply services. Many municipalities also have regulations and environmental offices adapted to their needs. Since mayors cannot be re-elected when their three-year terms end, the lack of continuity, and the absence of planning and responsibility diminishes the efficiency of municipal administrations. The small size and scarce resources of many municipalities also impede their performance on environmental issues (OECD 1998). There is also a strong influence on local regulations from federal bodies, and that limits their capacity for action. On June 3,

1998, the NOM-002-ECOL-1996 was published in the Official Gazette (Diario Oficial de la Federación), establishing the maximum permissible limits for contaminants in the discharge of residual waters into sewage systems. Even when legislation establishes that the municipalities will be in charge of sanitation services and consequently should be in charge of fixing the maximum permissible limits of discharges as they relate to local conditions and the capacity of each municipality, NOM-002-ECOL-1996 limits this power. Of course there is always the possibility that federal and state bodies might help the municipalities.

It is recognised that more than good administration is needed to resolve the environment problem. Nevertheless, with the political and financial uncertainty from each change in administration, each administration has established unrealistic goals, in many cases, goals that do not match the political, economic and social needs. On Jan. 1, 1997, the 43 norms that were set up as specific limits for the discharge from industrial plants were replaced with NOM-001-ECOL-1996, which establishes the maximum permissible limits for pollutants in the discharge of wastewaters into all national waters. This NOM establishes a system which is the same for all users, and depends on the type of receptor body of water. In addition to the quality goals for the discharges, where the population is greater than 50,000, the deadline for compliance was Jan. 1, 2000. So far there has been no study as to whether this NOM has been fulfilled or whether the goal has been reached. The federal government has not made any statement on it, but it is reasonable to suppose that the goal has not been met, and the most obvious case is that of the Federal District. In 1997, as part of its water management strategy, four large water treatment plants were built as part of the overall project for purifying water in the Valley of Mexico. However, as mentioned in previous chapters, this project was never completed (see Chap. 35).

The participation of users was recently incorporated into the framework of regulations. In terms of the administration of water, the water basin boards have been established, and users participate through them, in a way that can be taken into account in the establishment of policies and strategies for adequate water resource administration on a regional basis. Some of the biggest obstacles that have limited the development of the boards is the lack of experience by the users in getting together, organizing themselves, and taking decisions. This is a result of more than 70 years of vertical and centralized decision-making structures. Some of the reasons are: the lack of knowledge of the economic, social and environmental value of water, the failure to understand water policy and the administrative process for water because of the scarcity of information flowing from government to society, the authorities' lack of experience in management of these organisations, the lack of information about the current state of water resources on either a regional or national scale, and the lack of financial autonomy.

Currently the federal government has started a process of orientation and organisation for users with the goal of helping them understand the functions and responsibilities of the water basin boards, to establish strategies and directives for the internal organisation of the committees and assemblies of users, and to involve them in the decision-making process and the assigning of responsibilities. On the other hand the CNA is developing internal rules that will guide the operation of the boards. In this way it is hoped that these organisations will consolidate and acquire sufficient maturity and experience to begin operating as it was hoped they would during the first years of the decade. In keeping with the goals established it was hoped that by the year 2000 the water basin boards would have been in place and operating, and that they would have financial autonomy. The failure to meet this goal is evidence of the complexity of the process of encouraging public participation. The lack of experience with this new form of organisation for taking decisions, from the standpoint of both government and users, has turned into a learning process for both groups, in the course of setting up the basis for improved board operations.

One of the main links that sustains the strategies for preserving resources and lowering pollution levels is the application of economic instruments. These are defined in the administrative and regulatory mechanisms with fiscal, financial, or market-driven characteristics. Through them users take on both the benefits and the environmental costs of their economic activities. The economic strategy is based on two principles: 1) water user pays, and 2) water polluter pays. This establishes the following elements: 1) a system of taxes and fiscal incentives that recognize the economic value of the water and is designed to stimulate efficient use of the resource; 2) a system of concessions through which legal security is granted to users, for the use of the water and the disposal of sewage; 3) creation and maintenance of a registry of concession titles for adequate control of the volumes provided and the transactions involved; and 4) establishment of market mechanisms for the exchange of concessions.

The economic strategy has barely begun and it is still necessary to create the necessary conditions for it to operate adequately. The unfinished business includes the following: there are not enough users working within the system; the steps required must be simplified to encourage participation; the Public Registry of Water Rights has flaws and is not totally reliable; there is not enough water and weather information to determine the availability of water; in some water basins the volume of water assigned is greater than estimated availability; only 63 percent of users pay for their rights; the payments system must be improved, especially with respect to legal and fiscal issues related to uses of the water; the rights are not economically rational and favour agricultural use over industrial and public use; conditions for the exchange of titles and efficient assigning of them must be improved; and most title exchanges are effected outside the legal framework (CNA 1997b).

International experience (OECD 1998) shows that economic instruments can be useful tools for more efficient use of resources. In the case of Mexico, this strategy has good possibilities for contributing to more efficient management of water; however, the economic characteristics of the country exhibit spatial conditions that must be considered in the application of these tools. The existence of a dual economy is recognised, in which a dynamic and competitive system coexists with a sector that is broad but presents serious shortfalls. It is expected that the sectors that are more developed economically and technologically will not have problems and can adapt to the application of the economic instruments. However, the lagging sectors may find it difficult and this could be a limiting factor for their development. It is necessary to establish appropriate tax differentials, taking into consideration issues of equity and poverty, or else, far from becoming an instrument to promote the efficient use of water, the system could turn into a limiting factor for the development of broad sectors of the national economy.

The water programmes for 1995-2000, and for 2001-2006 promote the participation of the private sector for the development of the water infrastructure, and for its maintenance and operations. The big lag is in the broad sectors of economic activity and the regional disparity in the country, and this has been looked upon as an opportunity to create a growing environmental industry that closes the gap. The application of this strategy should recognize three important factors: 1) As mentioned in the previous paragraph, there is a dual economy, one dynamic and the other lagging; 2) All companies, no matter what business they are in, seek to provide services and obtain revenues and profits; and 3) Someone has to pay for the services.

In the case of industry, the agriculture sector, and urban centres, which by its nature has the capacity to pay for the services the environmental companies offer, it is sure that these sectors will have the benefits of the advises of the environmental companies. On the other side, the economic shortages of broad sectors of the economy will be a serious obstacle for obtaining the services of environmental companies. This could actually worsen the gap in the dual economy. Companies capable of obtaining environmental services will benefit from fiscal incentives, improving their productive processes, modernising their industrial plant, gaining more acceptance for their products in a society that is beginning to enjoy products generated through environmentally friendly processes, and they will be more competitive in the internal market and may be able to sell in the international market, with environmental certification for their products and services, including ISO 14000. On the other hand, the poorer sectors will have serious difficulties taking advantage of these benefits. In the case of urban centres or populations that are not attractive to private investment, the government will have to continue to subsidise the services or promise to pay the private sector to attract the needed investment. In addition, the quality of service will depend heavily on investment for the services to be obtained.

It must be recognised that those responsible for water management in the federal, state, and municipal governments have a huge challenge in making the water resource management strategies work. However, this responsibility is not totally theirs. Users and society have a large stake in the fulfilment of these strategies. It will be necessary to effect a cultural change in society and make it a partner in the activities related to the efficient management of water. In the past, a common practice of non-fulfilment of the legislation has developed, and the prevailing attitude is to ask why they should follow the rules, and to seek loopholes in the rules or the administration so as not to comply. Many environmental consulting businesses have specialised in this. It will be necessary to develop a new sense of responsibility, and of personal and professional ethics. For this the government will have to gain the confidence of society through transparent management of its processes and in the fulfilment of the commitments it has taken on. The past decades, starting in the 1970s, have been characterized by recurrent economic crises, inappropriate financial strategies that have left 50 percent of the population in poverty, scandalous corruption at high levels, and non-fulfilment of social promises. The government apparatus has an enormous challenge if it is to regain the confidence of society, and this is the necessary condition for any strategy of resource management to function.

It will require considerable sustained time and effort for the measures adopted to be firmly established in the management of water resources. There has been a concerted effort to establish a legal framework that permits efficient water administration. However, there are also some serious obstacles: it will be essential to assure the financing, to create a cultural change both within the government and in society so that the decentralization will be effective, to progressively build an institutional capacity at the state and municipal levels, and these governments must gain more financial autonomy to manage their environment. Environmental information and educational programmes must be expanded. The results emanating from the application of legislation must be assessed on a continuous basis, with the goal of closing the gaps and modifying the law where necessary and, most important, ensuring the continuity of the reforms on a national scale.

37. Discussion

Mexico has an average annual rainfall of 780 mm, of which about 27 percent becomes runoff that totals 410 km^3 per year. The annual renewal of aquifers is estimated at about 63 km^3 per year, of which 48 km^3 is natural and the other 15 km^3 is related to irrigation management. In addition, there is an estimated stock of 110 km^3 of water in the aquifers, and this is considered as a non-renewable resource. There is broad variation in regional climate, from the topical forests with more than 3,000 mm of precipitation per year to the deserts with less than 50 mm. The variation in runoff is even more extreme, from 2 km^3/km^2/year in the wettest areas to practically zero in the deserts. In the arid regions, precipitation and runoff are very erratic, with extreme seasonal differences. In these regions the precipitation comes during a period of 2 to 4 months and is related to hurricanes and tropical storms, which can be very intense and can cause severe erosion. Runoff is related to precipitation and many basins are dry during the off season.

Historically, human settlements and activities are greatest in the arid zones, that is, 76 percent of the population is located in an area which captures 20 percent of the precipitation, 90 percent of the irrigation, 70 percent of the industry, and which generates 77 percent of GDP (Figure 37.1). One quarter of the population lives in regions more than 2,000 metres above sea level, where just 4 percent of the runoff occurs. Fifty percent of the runoff occurs below 500 meters.

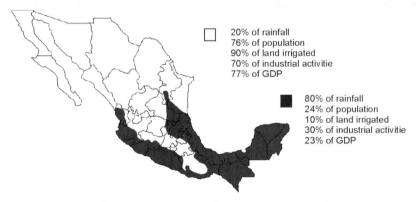

20% of rainfall
76% of population
90% of land irrigated
70% of industrial activitie
77% of GDP

80% of rainfall
24% of population
10% of land irrigated
30% of industrial activitie
23% of GDP

Fig. 37.1. Spatial distribution of water availability, population, and economic activities

This situation may appear contradictory, but it has been possible only due to a continuous generation of infrastructure and over-exploitation, mainly of ground water. Over-exploitation is an important problem in the arid and semi-arid regions of Mexico where most of the population and industry are located, and where irrigation comprises the greatest use of water. Two million hectares of land are irrigated, 55 million people live in cities, and more than half of industrial production in the country depends on groundwater to satisfy its needs. In many cases, other sources of water are not available, or are too expensive. The lowering of aquifer levels creates higher costs of pumping, or wells are abandoned, or new ones must be drilled, and the quality of the water is degraded. Over-exploitation is a non-sustainable condition, and because important regions of the country depend mainly on groundwater, in the future strict controls will be needed, based on adequate planning, knowledge of the condition of the resource, and commitment by government and users.

According to the National Population Board (CONAPO), by the year 2020 the population of Mexico is expected to be about 130 million, and will likely be concentrated in the center of the country and the northern states, mainly in the municipalities in the US frontier area. Taken together, these regions will have about 57 percent of the population (Figure 37.2). Gross Domestic Product is expected to continue growing at an annual rate of 3 percent.

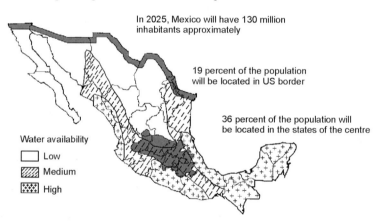

Fig. 37.2. Projection of the spatial distribution of the population for 2025

Population growth, economic expansion, and the tendency to concentrate in certain regions will increase the demand for water. In this situation conflicts over water can be expected to increase between urban and rural areas, between neighboring cities, and more frequently, between neighboring states or regions. According to the CNA, the projected demand for water by 2020 will be 92.42 km³/year (Paz 1999), compared with the natural availability of water (496 km³ per year). It would appear that Mexico will not have problems of scarcity in the coming years, with the annual availability per capita of 3,600 m³/year. That is more than the worldwide average (2,500 m³/capita/year), and far above the minimum recommended for avoiding stress on the water supply (1,000 m³/capita/year).

However, from a regional standpoint the situation is very different. In the areas with arid or semi-arid climates the demand is already above, or is about to rise above, availability. With the new administrative regional division, the availability of water and the projected demand, Regions I, II, VII and XIII are reducing, or will reduce over the next 20 years the natural availability of water, and Regions VI and VIII are already at the availability limits (Figure 37.3).

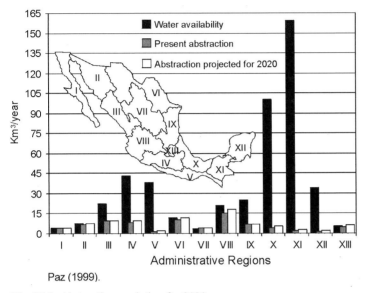

Paz (1999).

Fig. 37.3. Abstraction evolution for 2020

Table III.2 of Chap. III shows the estimates of naturally available water per year per person for each of the 13 regions, including surface water and groundwater. In Regions I, VI, VII, and VIII it is barely above 1,000 m³/capita/year. In Region XIII it is 289 m³/capita/year, and demand in this region must be satisfied by importing from the Lerma and Balsas river basins. The current rate of population growth in the aforementioned regions indicates that they will have less than 1,000 m³/capita/year. The increasing technical and financial difficulties to develop new sources of storage mean Mexico faces serious problems of water management in the future, with the situation in some regions comparable to countries with severe water scarcity.

The traditional way in which the government has tried to resolve storage problems has been through a continuous search for new sources and very high investment for the creation of water infrastructure. However, through time this focus has encountered a growing number of technical and financial problems. The water sector not only must face the challenge of developing new infrastructure for the satisfaction of future demands, at the same time it must allocate sufficient funds to maintain and make more efficient the existing infrastructure.

In the urban public sector it is possible to observe quantitatively the magnitude of this problem. According to studies carried out by the CNA (Paz 1999), in the most probable scenario for 2020 it is expected that the rates for potable water and sanitation will be 98 and 95 percent for urban zones, and 80 and 75 percent for rural zones. This gives us the following:

Table 37.1. 2020 Scheme for water supply and sewage services

Population (1997): 94,277,632			
Rural: 24,951,467		Urban:69,326,165	
Population with water supply service	Population with sewage service	Population with water supply service	Population with sewage service
15,834,200	7,819,790	65,097,269	60,480,146
Population (2020): 130,000,000			
Rural: 33,800,000		Urban:96,200,000	
Population with water supply service	Population with sewage service	Population with water supply service	Population with sewage service
27,040,000	25,350,000	94,276,000	91,390,000

SEMARNAP/CNA (1996), CONAPO (2000)

To achieve the objective it will be necessary to incorporate each year services of potable water and sewage for between 0.51 and 0.8 million additional residents in rural areas, and between 1.33 and 1.4 million new residents in urban areas. Comparing these numbers with the average population increase between 1990 and 1999 (see Figure 37.4) we can see the following:

The average annual population that benefited from services over the last decade has been the following: 1.63 and 1.22 million in urban areas and 0.57 and 0.51 million in rural areas, for water supply and sewage services. To obtain this average 1995 was excluded because the previous year, 1994, was a presidential election year and generally investments for services increase after an election. As a result, 1995 is not representative of normal investment within the sector. If we compare the average annual population between 1990 and 1999, with forecasts for 2020, we can observe that with a continuation of the trend only in the water supply service, both for urban and rural areas, the goal will be met. The scenario will likely require important investment flows. But there is one factor that the numbers do not show: the quality of service. As was shown in Chaps. 35 and 31, service for potable water is not necessarily a tap water in the house, and sewage services do not necessarily imply a connection with a drainage system. Similarly, these numbers do not indicate the quality of water being received, the shift system that is imposed in some zones, the amount of water lost in leaks, the number of people who become sick from waterborne diseases either from the water they receive or by living close to bodies of water that have become open sewers, the increasing pollution in bodies of water, the man-hours lost carrying water to houses where there are only communal sources, and the business done by water truckers to meet the needs of the people.

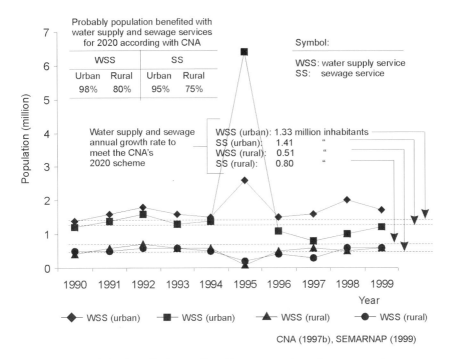

Fig. 37.4. Population benefited annually with water supply and sewage services

This probable scenario presented by the CNA for 2020 may be less favorable if we consider the following factors: 1) the federal government does not have a large enough financial base to provide the services; 2) external or internal financing from private banks and development banks, especially in construction, has been continuously falling in recent years (Figure 37.5); 3) collection of fees for services at the federal, state, and municipal levels has been far below the necessary investments (see Fig. 37.4); 4) the revenues of 63.2 percent of the population come to less than two minimum salaries, 30 percent of the population lives in conditions of extreme poverty, and 20 percent in conditions of relative poverty; and 5) private initiative has been reluctant to invest large amounts of money in potable water and sewage services. This indicates serious obstacles to obtaining financing.

One of the objectives of recent governments has been to reduce poverty. To the extent that development programmes meet this goal, there will be a direct impact on demand for water, and more infrastructure will be needed. The government recognises its lack of financial and institutional capacity to generate this infrastructure, and this has opened the door for private initiative to participate in the financing and operation of the programmes. However, there are two factors that add to uncertainty in this strategy: 1) private initiative is sometimes more interested in profits than in service to society; and 2) the poor financial situation of the government and of a large section of the population. With the failure to provide sufficient economic incentive and the high risk involved in investing under these conditions, it is to be expected that the private sector is reluctant to participate, when it has more secure alternatives for investment. This places the government in a critical situation, in that it may be obliged to seek the funds needed to finance and offer the services itself, or to subsidise the participation of private investors. As mentioned above, this will not solve the problem under current conditions.

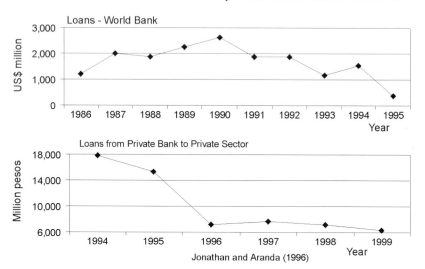

Fig. 37.5. Internal loans and foreign credit evolution

To show more clearly the implication of failure to develop and maintain the water infrastructure in the coming years, the following scenario has been developed on the basis of the following assumptions: a) currently there is national coverage of 85 percent of the water supply needs and 72 percent of sewage services, on average; b) in the last 10 years the average growth rates of services have barely kept pace with population growth; c) the percentages of coverage per state remains constant, to establish on a regional basis where the most serious coverage problems exist; d) although the foregoing paragraphs referred to quality of service, for the purposes of this exercise, they will be assumed to be optimum. Table 37.2 shows the results of the projection of coverage per state for potable water supply and sewage.

If the growth rates of services continues as they are now, by 2020 the system will have to: 1) maintain the storage and sewage infrastructure that currently exists and provide service to 83 and 69 million persons respectively; 2) generate new infrastructure for water storage and sewage for 26 and 24 million persons; 3) 20 and 35 million persons will not have services of potable water storage and sewage; and 4) huge backlogs will persist in the poorest states, such as Oaxaca, Guerrero, Chiapas, Hidalgo, Puebla and Veracruz. In addition, the quality and the continuity of service will have to be taken into account.

Table 37.2. 2020 water supply and sewage services projection

State	Population [thousand]				
	Total	With water supply service	With sewage service	Without water supply service	Without sewage service
Aguascalientes	1,509.9	1,472.7	1,409.0	37.1	100.8
Baja California	4,717.2	4,057.6	3,556.1	659.6	1,161.0
South Baja California	684.5	614.8	504.5	69.7	180.0
Campeche	1,122.6	874.6	653.3	248.1	469.4
Coahuila	2,850.1	2,683.9	2,158.9	166.2	691.2
Colima	733.3	697.8	684.1	35.5	49.1
Chiapas	5,207.0	3,389.9	2,718.3	1,817.1	2,488.7
Chihuahua	4,271.5	3,902.9	3,359.7	368.6	911.8
Federal District	8,815.2	8,566.4	8,563.1	248.7	252.0
Durango	1,508.1	1,346.0	972.3	162.1	535.9
Guanajuato	5,748.7	5,096.7	4,047.5	652.0	1,701.2
Guerrero	3,906.7	2,520.7	1,803.4	1,386.0	2,103.3
Hidalgo	2,806.1	2,222.7	1,573.3	583.3	1,232.7
Jalisco	8,576.7	7,787.7	7,638.4	789.0	938.2
México	19,142.0	17,455.4	15,910.3	1,686.5	3,231.6
Michoacán	4,587.4	3,944.3	3,162.9	643.1	1,424.5
Morelos	2,653.4	2,384.7	2,142.6	268.6	510.7
Nayarit	1,102.6	946.1	818.5	156.5	284.1
Nuevo León	5,553.0	5,225.6	4,901.5	327.4	651.5
Oaxaca	3,890.0	2,589.2	1,621.3	1,300.8	2,268.7
Puebla	6,372.4	4,992.6	3,587.8	1,379.8	2,784.5
Querétaro	2,074.2	1,841.5	1,387.5	232.7	686.7
Quintana Roo	2,034.5	1,803.0	1,540.2	231.6	494.3
San Luis Potosí	2,741.6	2,008.6	1,460.7	733.1	1,280.9
Sinaloa	3,126.1	2,733.5	2,090.3	392.7	1,035.9
Sonora	3,216.5	3,001.7	2,346.4	214.9	870.1
Tabasco	2,813.2	1,825.3	2,299.3	988.0	513.9
Tamaulipas	3,655.6	3,234.5	2,387.8	421.1	1,267.8
Tlaxcala	1,366.9	1,303.6	1,029.3	63.1	337.5
Veracruz	8,630.9	5,350.4	5,199.3	3,280.5	3,431.6
Yucatán	2,350.9	2,003.6	1,142.7	347.3	1,208.2
Zacatecas	1,364.7	1,125.3	788.8	239.4	575,9
Total	129,133.5	109,003.3	93,459.1	20,130.1	35,681.8

SEMARNAP/CNA (1996), CONAPO (2000)

One of the main goals in the political agenda for the coming years will continue to be water and sewage services for broad sectors of the population. Failure to do this could create more social tension and unrest, with high political costs. The need for a new water infrastructure is recognised, but so too must the need for adequate planning, administration and non-intensive use of technology. This could increase the availability of water in quantity and quality, where and when it is required, and in some cases without new infrastructure.

The future development of the country will be a function of the capacity to make the resource available to those sectors that need it, in the quantities and quality that is required. However, the construction of water infrastructure will be much more complex, because the cheapest and most accessible sources have already been, or are being developed. Environmental and social factors, land use, and relocation of settlements have reduced the possibility of large new developments. They have been made more complicated or the time required for their completion has increased. Other alternatives for increasing water availability, such as importing water from other basins or desalinisation are not economically viable. However, it is possible to move toward the water supply goals using mechanisms such as: 1) improving systems of water management; 2) existing infrastructure is sometimes underutilised because of a lack of maintenance, lack of complementary infrastructure, or because it is badly operated; 3) improving demand management; 4) using alternative sources of supply such as rainwater; 5) using alternative non-intensive technology for energy use, financial resources, and easy maintenance; 6) encouraging the aquifer recharge, naturally or artificially, as a means of preserving them; and 7) treating wastewaters for recycling or reuse.

Technological aspects and the development of infrastructure will certainly be the determining factors for water management in the next few years. However, throughout this work we have observed that the solution of problems related to water is not exclusively dependent on generating water infrastructure or on issues of technology development. If there is a will to make water management sustainable it will be necessary to give special attention to a factor that historically has been neglected: the human factor, which we consider to be the determining factor for water management. As a result of this study, we present below eight factors which we consider to be important determinants for improving Mexican water management, and where we can observe that, in the end, the development of the human factor through education, knowledge, social participation, working together, tolerance, respect, and ethics will be the elements that play the most important role over the next few years in water management.

37.1 Water Availability

It is expected that in the next few years demand for water will increase for two reasons: 1) the population increase; and 2) the increase in living standards of the population. Together these factors will require an increase in food production, more energy generation and more goods and services. We should recognize that the volume of water each region has is finite, and so the increase in per capita demand will make the future availability of water decrease. According to official statistics, only in Region XIII is availability per capita below 1,000 m^3/year (see Table 28.2), which as has been mentioned is the minimum quantity needed to avoid pressure on the system. However, official figures refer to gross volumes of runoff.

These figures do not take into account that a large quantity is not economic or technically available for the following reasons: 1) there is not the infrastructure for storing it in some regions; 2) in the north of the country precipitation is so small that it does not create runoff that can be captured; and 3) mixing with polluted waters means that part of the precipitation or runoff is not available or its potential usage is limited.

In recent years the environment itself has taken on importance as a water user. The National Water Law considers the environment as a factor which should be assigned flows or minimum volumes for its preservation, something that would limit volumes for usage. Consequently, with the principle that the development of society cannot occur in a deteriorating environment, it must be considered as a user more than simply as a vehicle to deliver sufficient water, with the quality needed for its maintenance. This means makes the environment a new competitor added to existing water users, and all the conflicts for water will intensify, especially in those regions that are already in short supply. In addition, the process of administering water will be more complex, since it will have to determine the minimum quantity and quality required by the ecosystem to maintain itself. However, the bodies responsible for the administration of water must understand that care of the environment is not a luxury but a necessity, and the availability of water depends not only on the quantity of rainfall and runoff, but also on its capacity to regulate.

Finally, the phenomenon of global climatic change must be considered. According to the latest report of the Intergovernmental Panel for Climatic Change (IPCC 1997), in which the regional problem is considered, in Latin America, global warming will aggravate the periodic scarcity that already occurs in arid and semi-arid zones. For social and economic reasons, developing countries will be especially sensitive to these increased periods of scarcity. As well, according to estimates of the IPPC, there are indications that the problems of flooding are going to increase in many temperate and wet regions, and it is recommended that the design of reservoirs and dikes be reviewed. There are few studies in Mexico on the possible effects of global warming.

For the foregoing reasons it is expected that the economic, technical, and environmentally available volumes will be different from those being managed at present, and consequently the availability per person that is officially managed will be less, or perhaps more. It must be recognised that the hydraulic network has serious deficiencies and the number of stations does not allow the delivery of all of the volume of rainfall and/or runoff that occur. On the other hand, the lack of studies on the effects of climate change on the availability of water in the country makes future estimates more difficult. As a result, it will not be surprising if the volumes that are available naturally in some cases are not accurately estimated.

37.2 Wastewater Treatment

The use of surface and groundwater has increased over the years, and this has been accompanied by a continual degradation in the quality of water. The discharges of residual waters coming from industries, urban centres, and agriculture for the most part receive no treatment and are often highly contaminated. The degradation of quality of groundwater could also worsen as a result of the concentration of salts or the intrusion of saline water into aquifers close to bodies of water that contain a high amount of salt. Water contamination is especially prevalent in the centres and north of the country.

For decades sanitation has been the neglected sector that has received less attention and funding. The strategy to solve problems of water scarcity has been based on finding new sources of water rather than to establish water treatment programmes for recycling. In previous chapters it has been mentioned that the appropriate use of treated water could improve availability and eliminate the need to find new sources of water or develop new water infrastructure. In recent years the CNA has carried out activities to control the problem of contamination. Nevertheless, the reticence of the government to face this problem and put it in a proper perspective has meant that the actions for sanitation have not been seen as urgent. In general the results relating to the improvement of the condition of ecosystems have only been considered for the medium and long term. In many cases there are no results visible to public opinion, so that spending money in this area is not politically convenient. Interest in investing in purification is proportionate to the benefit that results, and the lack of funding obliges the government to invest in programmes that are more politically feasible in the short term. On the other hand, for users who generate residual waters there are various factors: scarce funds, which they would rather spend on maintaining productive systems rather than in environmental areas; an absence of interest in contributing to improvement of the environment; a lack of political will to force users to comply with the law; the fact that at times it is more cost-effective to pay for the discharge than to treat residual waters; the lack of capacity of institutions to carry out inspections; and corrupt practices among inspectors.

Proper treatment and management of residual waters could provide the following benefits over the medium to long term: a) an increase in the availability of water for those users who do not require high quality; b) an exchange of clean water for treated water between agriculture and domestic users; c) combined with the natural capacity for filtration of the soil, treated water could be used to stabilise aquifers that are currently overused, or at least it could reduce the level of overuse; d) reduction of the cases of water-borne illnesses; e) preservation of ecosystems; f) an increase in the potential uses of bodies of water; g) a decrease in conflicts among users due to scarcity of water; h) protection of storage sources.

There are various actions that, developed properly, could provide incentives for treatment, re-use, and recycling of water. Economic tools, through taxes, could help to direct prices so that it would be cheaper to treat water than to discharge it without treatment, or cheaper to recycle water than to use water from water supply systems. It would be best to introduce this strategy gradually so that operators could adapt their rates and society would not resent a sudden escalation of water rates. In the case of industry it would also be necessary to allow a lapse of time so that they can change their processes or renew their technology without undermining their productive capacity. This strategy should be accompanied by strengthening the capacity for inspection, in a way that ensures the activities of the operating bodies and industry.

Given that residual water treatment is attractive to different sectors of society it is necessary to develop alternative technologies that allow for the reduction of treatment costs. The federal and state governments could stimulate research at the universities and technology centres, through their own resources or strategies that link the schools with industry. Economic stimuli could be established through tax relief for industries or water utilities that establish agreements with research centres undertaking research on developing treatment systems. There are other benefits to be obtained with the funds that are transferred to the schools: for example, improving installations, acquiring equipment, providing professors and students with research projects, developing human resources, compensating for the lack of budget from which most schools suffer, increasing revenues for the academic community. But one must not lose sight of the so-called soft technologies which can provide a good option considering their low costs and ease of operation. This technology could be directed at rural populations, which do not have access to systems that require high investment. However, care must be taken not to fall into the trap of thinking soft technologies should serve only poor populations. In highly developed countries such as Japan, people in cities receive incentives to install equipment to collect rainwater and use it in their houses. In this way we can see how taxes plus easily managed technologies can improve water management practices.

According to the regulations, the institutions responsible for water management have to apply themselves to develop the minimum elements through which better to administer water quality. This work includes: classifying water bodies; developing a complete and reliable inventory of discharges; establishing specific conditions for discharges by users; improving inspection systems for verifying compliance; establishing water quality goals backed up by cost-effectiveness, taking into account current economic conditions and the seasons; providing continuity of plans and programmes; ensuring that laboratories are up to date, as well as equipment and certificates at the central and regional levels; updating and redesigning the National Network of Monitoring Water Quality to make full use of its resources; developing information systems on water quality; and putting this information within the reach of society as a strategy for the creation of environmental consciousness. The foregoing calls for creation of special technical teams for management of residual waters, elements such as treatment, re-use, environmental impact legislation, economies of water, health implications for the use of residual waters. There should be training for this work, and not just the technical or engineering aspects.

The government should be tough in ensuring compliance with the regulations. In many cases the lack of compliance is not due to an inadequate legal framework but rather to the lack of will and training to enforce compliance of the law. At present our politicians do not pay enough attention to the improvement of the quality of the water resources of the country, possibly because this aspect of water resources management does not pay dividends, politically speaking, under current circumstances. Nevertheless, it is necessary that our politicians understand that the treatment and re-use of wastewaters provide a huge opportunity for increasing water availability in many regions of the country. Treating water is just as important as finding new sources.

37.3 Efficient Use Of Water

In Mexico there are high losses of water. Agricultural systems operate at about 40 percent efficiency, and municipal systems at about 60 percent. This means that for agricultural usage, for 60 km^3/year extracted, only 24 km^3/year is used effectively. In domestic usage, of the 9.54 km^3/year that is consumed, 5.7 km^3/year is used. That constitutes a total loss of about 40 km^3/year. The efficient use of water is determined mainly by four factors: 1) the type of water infrastructure; 2) the condition of the water infrastructure; 3) operations policies or programmes; and 4) owners of use and consumption. One of the decisive factors over the next few years for improving the availability of water will be to make efficient use of the water being extracted.

Given the increase in population and in industrial activity, it is expected that the agriculture sector will have to give way as a result of the pressures of increasing demand for water. As a result, it will be in this sector where the main efforts should be made, in part because it is the biggest consumer (76 percent of water abstracted), and in part because it is the sector with the largest losses. Failure to proceed with implementation of efficient usage practices in agriculture that would free up large volumes of water for other users could result in severe conflicts between users and even within the agriculture sector. It is true that the backwardness of the agriculture sector is complex and difficult to manage. The groups that comprise the sector are heterogeneous and their interests conflict. State planning has lacked any vision or continuity, and the same political groups have been hindered for want of adequate development of the same. The decline of international prices of basic grains has placed national producers in a difficult financial situation, and this has caused the government to continue subsidising this sector because of its social and economic impact. Even allowing for the foregoing, it is necessary to recognise that current water usage practices in agriculture are not sustainable. The technology exists for making water usage more efficient through trickle systems, sprinklers, and microsprinklers. The technical capacity and knowledge exists to maintain the infrastructure and improve irrigation practices. The critical points for carrying out these actions are: finance and the will of the groups to fulfill the promises.

It is time the government began to reconsider its policy of not charging the agriculture sector for water, a policy which has responded more to political questions than to real action for developing the countryside. If it were to charge 0.5 pesos for each cubic meter used for irrigation, the annual levy would be about 30 billion pesos, or about three times the annual budget of the CNA. The main obstacle for this is that the revenues from taxes enter the Treasury as general revenues and do not return directly to the sector that generates them. The solution would be simply to return the contribution of agriculture as direct investment to the sector, since this would allow for improving the sector and encourage producers to get used to paying taxes while seeing the benefits they can generate. Government and the interested parties must develop alternatives that allow them to improve the sector's efficiency, while taking special care to find the right balance between investments and the time needed to achieve efficient use and the needs and requirements of the producers. To raise efficiency to 55 percent, which is still low, would permit savings equivalent to all of the water that is consumed in the nation for public and urban usage.

Even as the agriculture sector puts out a major effort to improve water usage practices, the rest of the sectors should not be excluded. One result that will accrue from special care over the next few years is the type of infrastructure that will develop and the kind of investments made. To meet the demands for water in Mexico City, the Temascaltepec Project is being developed. It consists of a dam with a curtain 120 meters high and a length of crest of 743 meters. It will have a pumping station with a capacity of up to $12m^3$/second, and the water will be carried from the Valle de Bravo Dam through a tunnel 18.75 km long and 3.5 meters in diameter. The depth of the tunnel will vary between 16 and 700 meters. According to official figures the initial investment for the project is calculated at 501.9 million dollars. The project will bring 5 m^3/second of water to the Metropolitan area of Mexico City. In contrast, Mexico City loses $18m^3$/second as a result of leaks in its network. Surely a redirection of investment from the Temascaltepec Project would generate more benefits if it were applied to projects such as plugging the leaks, capturing rainwater, systems for the aquifer recharge, and treatment and re-use of residual waters. The traditional focus of solving water scarcity problems through a continuing search for new sources no longer works. More creativity is needed in the search for alternatives for efficient water use, as well as investment and management of financial resources in the sector.

37.4 Culture of Water

The need to create a culture of awareness of the environment is recognised, and the need in this case is for awareness of the need for water management. For this reason the government has launched several campaigns regarding proper use of water, trying to involve society in the task of achieving efficient usage. However society's response has not been what was hoped for, mainly because of two main factors: lack of information and lack of confidence in the government.

On the first point we could say that an uninformed society is a non-participating society. If we establish the premise that citizens' participation is basic for the resolution of problems, information must flow to society so that its members can be conscious of the economic, social and environmental value of water. In this way, to work or to be informed on the real problem of the resource will allow people to be conscious and to create the space through which those who consume water can change. It is not possible to change society's mentality when it is not allowed to meet the problems head on, or if it only sees small facets of the problem and, what is worse, the magnitude of the problem is minimised for political reasons. There must be enough information, and it must be of sufficient quality, and above all honest and transparent, and it must be presented in a way that even non-specialists understand. Thus, NGOs, the media, and education centres have a great opportunity to link themselves to the huge effort that would signify water management in the new millennium.

Permitting access to information creates an opportunity to bring society and the institutions of government closer together. In this way society will also be interested in knowing its institutions better, how they operate, who works within them, and what their objectives and functions are. It is logical to think that society will only work in conjunction with institutions that are known and never with those that are unknown. So government institutions must abandon their old practices of being uncommunicative, and allow information to flow. There have been advances in this over the last few years. Although the General Law of Ecological Equilibrium and Environmental Protection (LGEEPA) makes the right to information explicit, the same law also establishes items where information can be denied. This creates a situation in which society has the right to ask but the government does not have the obligation to respond.

The second point is extremely critical in establishing a healthy society-government relationship. Mexico is a young nation, with 57 percent of the population between the ages of zero and 25 years. This implies that a broad segment of the population was born in the 1970s, at the same time as the beginning of a period of successive financial and governmental crises which, even today, we have not completely overcome. The poor management of finances has caused a pervasive weakness in living standards of the population, purchasing power has contracted so that access to goods and services has been limited, and 63 percent of the economically active population earns barely twice the minimum wage. As well, we have a continuous series of scandals of corruption and financial fraud at high political levels, which have been revealed to the public. This situation has generated an atmosphere of resentment in the population and especially among the young, who feel defrauded by a system that has not responded to their needs and has limited their access to the opportunity for a higher standard of living. We should remember that this segment from zero to 25 years of age will be the main actors in society in the next 20 years. The groups that achieve power will have to start to change the way in which they act. If they hope to gain social participation they will first have to regain from society confidence in government institutions, and they will only be able to do that through ethical, responsible, and transparent management of these institutions. Put another way, government institutions will have to start responding to the needs of society and not of special interest groups.

37.5 Capacity Building

Four factors have determined the direction of the formation of skills in the water sector: 1) the concentration of functions at central level has meant that training has also been concentrated. As a result there has been a lack of solid institutions in the states and municipalities. The weakness of the local operations has become an obstruction to the process of water management as the system decentralises. 2) Historically the development of water infrastructure has been carried out on the basis of technical and financial considerations, so that the formation of skills has been directed mainly toward the engineering aspects, leaving aside other important elements for water planning, such as social and environmental factors. 3) The number of courses may be an indication of the importance given to training, but it is not a reliable indicator of its efficiency. In the water sector most courses last just one or two days, enough to provide only a superficial review, and a thorough analysis of the topics presented is lacking. 4) The availability of funds for carrying out training programmes is a determining factor for their success. However, experience shows that when funds are in short supply, as has been the case in recent years, training is one of the first budget items to be cut.

Conditions in the country are changing rapidly and the water sector must assimilate and adapt to these changes just as quickly. New decision-makers are joining the process of water planning, and new players are starting to become relevant in decision-making. Those with the responsibility of water administration will have to give more importance to training if they want to succeed in overcoming the challenges of the water sector in the future. It is clear that the solutions of the past will not be appropriate to the new conditions.

The process of training in the coming years must be applied not only the federal bodies, but must be extended to regional and state bodies, and to users. These groups will not be able to assume their new responsibilities adequately if they lack the knowledge to perform. Training must be based not only on the technical and financial aspects, but must also recognize that we are getting involved in a complex field, and that water management involves countless political, economic, social, cultural, historical, and environmental factors, which interact in a dynamic process. This underlines the need for an integral and interdisciplinary focus that allows a better evaluation of the human problem with respect to water. Not only will it be necessary to have good technicians, it will also be necessary to build the necessary skills to incorporate economic, social and environmental variables into the process of planning in the water sector.

37.6 Development of Legal Frameworks and Applicable Regulations

The environment problematic touches on so many sectors that different government ministries have been given the task of establishing plans, programmes, and actions for managing it. However, this has been done from the narrow perspective and interest of each of these ministries. The legal framework and the institutions responsible for water management are relatively new and the magnitude of the problem has made it necessary to develop diverse legal instruments and administrative mechanisms in a short period of time. The result has been that the legislation is inconsistent within the different regulatory instruments, and this is evident in the lack of coordination among government ministries and even within ministries. Often the personnel of a ministry do not know the ministry's general goals, or even the structure and functions of each branch. It is necessary to establish both vertical and horizontal coordinating mechanisms so that the formulation of the law and its implementation will be more effective. This means the law must be applicable, and must be consistent with the economic, social, and political circumstances of each country in a way that ensures that it is an instrument that allows the meeting of the integral management objectives of water and the environment.

In the case of water management, the conservation of the resource through an appropriate distribution of users is one of the principles under which current water policy operates in Mexico. The Public Registry of Water Rights (REPDA) was set up to make this principle operational. It has a variety of objectives, including: ensuring the concession of water rights, approving transfers, providing data on the volumes used and discharged in each region, encouraging the exchange of rights within the law, authorizing discharges, improving the establishment and calculation of taxes, and establishing and administering the special reserve zones.

A register of users is a basic element for adequate administration of the resource. However, the REPDA has inconsistencies because of the way it was first set up. At the end of December 1995, the CNA reported 26,000 registered users under REPDA (SEMARNAP/CNA 1996), and in June 1999 it reported 265,000 (CNA 1999a). This suggests that in 42 months there were 239,000 new users registered. To achieve that 190 users would have had to be registered every day, something that is not probable given the slow and complicated administrative processes within REPDA, and the CNA recognises this (CNA 1997b). For decades users carried on with little or no government supervision, creating a culture in which they believed they had no responsibility to report on the way they were using the resource. It is not possible that in three years the thinking of 239,000 users had changed. It is estimated that there are actually 368,000 users, but the exact number is not known.

So that the REPDA can consolidate and operate as a key element, and one that allows the correct assignment of volumes among users as a function of the water capabilities of each region, it will be necessary to develop some aspects that certainly would contribute to the fulfillment of the objectives: more flexibility in the registration process; more clarity in the procedures for calculating taxes for water use and discharges; attention to the taxes in a true economic rationale; the end of favouring one use over another; and establishing tax differentials according to users' ability to pay. For an adequate understanding of the resource, it is necessary to have technical elements, including: information on the availability of the resources on a regional basis; adequate water evaluations; performance of aquifers and regional planning studies; reliable data bases; adequate hydrological models for the establishment of the potential impact on third parties; hydrological information for the control of inflows and outflows of dams; and data for improving management of the water infrastructure during periods of extraordinary rainfall and droughts. Failure to proceed on this basis would place in doubt the strategy of the CNA for the administration of water rights where the REPDA is the backbone of the system.

An adequate process of calculating and collecting taxes could be an incentive for more efficient water management, but for this strategy to work it would be necessary to develop two important elements:

1. A balance between taxes and costs. Establishing an appropriate balance between taxes for use or discharge of water and costs for services of supply or treatment could promote conservation, efficient use of water, and the reduction of pollution. However, for this to operate the law would have to be really applied, with sanctions to users that do not comply and benefits to those who take actions to make more efficient use of the water. Supervisory and inspection activities by the authority would be extremely important for meeting the goals, and these must be carried out in an ethical and responsible way.
2. Relation between levies and financing. Currently there is no direct relationship between tax levies and the financing of water administration activities. As was mentioned above, revenues generated by charging taxes go into the Treasury as general revenues and later the CNA negotiates its budget with the Finance Ministry. Once it gets the tax revenues, the federal government assigns the budget to the states, which in turn assign it to the municipalities, although there could be direct negotiation between the federal and municipal governments. This situation limits the development of the water sector at both federal and

situation limits the development of the water sector at both federal and local level. To establish a direct relationship between levies and water administration could open the door to the possibility of financing administrative activities and creation of infrastructure. The lack of budget in the water sector is not only due to the existence of the custom of not paying, it is also because of the diversion of funds generated from the water sector to other activities considered more important. In addition, if taxes could be applied directly in the regions where they are collected to reinforce the water administration systems, this could increase the willingness of users to pay because they could see the benefits in their regions, and the benefits for administration.

Special care is necessary in the implementation of economic instruments because without proper regulation and monitoring of the same, under a market system, it could degenerate into monopoly situations on water rights, with economic coercion of some groups over others. If the economic aspect is fundamental, care must be taken not to fall into exclusively economic conceptions in water management, as if that were the only alternative.

37.7 Participation of Society in the Water Planning Process

The LAN identifies the water basin boards as the appropriate units for the administration of the resource and the bodies through which public participation can be brought into the processes of managing and planning the water sector. The CNA describes the boards as "plural, effective, participatory, and democratic organisations [...] to advance more efficiently in the direction of sustainable development" (CNA 1998a). However, there are elements within the water sector that limit the opportunity for public participation.

In the first place, these organisations have emerged from a process never before used in the history of Mexico, and are contrary to the centralised process for decision-making and legal structure designed for them. The federal bodies have had to develop the judicial tools and initiate the legal reforms that allow for the creation of a space through which decision-making is opened to sectors outside of the federal level. However, this process is far from being complete. The 1917 Constitution, with a 1983 amendment, only empowers the municipalities to supply water, sewage, and sanitation services (Art. 115 of the Constitution), and leaves in the hands of the federal government issues relating to the exploitation, use, distribution, and control of what is defined as national waters (Art. 27 of the Constitution). The LAN grants exclusive powers to the Federal Executive to legislate for and administer national waters (Art. 4 of the LAN), and establishes that the participation of users and individuals will be promoted by the Federal Executive only in terms of carrying out and administration of the works and the water services (Art. 5 of the LAN). The water basin boards are established as bodies of coordination (Art. 13 of the LAN), and as such, are empowered exclusively for issuing recommendations to the government bodies and users. The water basin boards do not have authority to issue any official regulations or to exercise legal or judicial action, and do not replace any authority or organisation (Castelan 2000).

Second, because of the accumulation of responsibilities, functions, and resources, in the central bodies of the Mexican public administration, there are no equivalent solid institutions at the state level. The weakness of the local bodies has become an obstacle. Paradoxically, when decentralisation began to catch on in Mexico as an alternative, it was not noted that for the most part the states lacked – and in most cases still lack – the physical, institutional, and human infrastructure to taken on the new responsibilities that were envisioned (Merino 1995). In addition, the specific forms of control, the party discipline between governors and municipal presidents that the central model generated and the lack of knowledge and experience in issues relating to water resources, have resulted in a lack of interest and apathy, especially among municipalities, in participating and proposing alternatives. For decades the participation of the state and municipal bodies in the planning and decision-making processes was little or nothing, so that the development of local capabilities was never developed. Ironically, now that there are spaces for their participation, in many cases they do not know what to do, how to act, or what to propose, a situation that is magnified by the limitations imposed by the law.

Third, it says that "The River Basin Councils [...] are the modern current expression of the new ways of integrating water issues [...] and a way foreseen in Mexican law for society to participate in the definition and orientation of the tasks of water activities" (CNA 1998a). However, an analysis of the structure of the boards reveals two things: 1) it is not possible to speak of a true user representation because the diversity of characteristics, needs, and interests they represent is not considered, nor are the differences in needs of the small and large producers; nor that the spatial distribution problems are different for those located in the upper part of the basin than for those in the low or medium levels. The economic coercion that the powerful groups can bring to bear on the others profoundly distorts the processes of decision-making and of electing representatives. Paradoxically, even when the users are part of the group, their interests do not necessarily represent those of their groups. Ultimately they operate as a defence process for special interest groups, not social groups, and are occasionally antagonistic; and 2) the participation of society, non-governmental organisations, and education institutions or research centres, and of other government bodies are brought in at the invitation of the CNA, if they consider it convenient (Art. 15, Section III of the Regulations of the Law of National Waters). From this we can see that water basin boards are not open forums for broad participation. The LAN confers a high level of discretion to the CNA in deciding who participates and who does not. This allows it to determine the balance of power, and in this way direct the decisions toward objectives or results that have been established by other bodies.

Fourth, in relation to the users and, mainly, society, there is a total absence of a culture of participation. There is no experience in the processes of organisation and much less in public participation. There is a lack of understanding of the economic, social, and environmental value of water and of the planning and administration processes in the water sector, and there is no access to information on water sector planning. This brings up a question: how can society become part of the planning and decision-making process when there is a total absence of understanding about the resource and of how to participate? If the process of opening up decision-making has hardly begun and there are already deficiencies, we cannot help but realize that they represent steps toward a democratic society. However, important as these spaces are, it is just as important to educate society so that it can participate properly in these processes. Otherwise the process will only create the same structures for decision-making as before, where a few hold the knowledge and the technical information for discussing and making proposals, while the others must accept them due to their inability to establish alternative proposals.

It may be observed that even with the reforms made by law, the space which society occupies at decision-making levels continues to be very small and, ultimately, decisions continue to be made at the federal level. The participation of citizen groups in society is becoming more important every day in terms of water resources. The big challenge is to integrate the participation of diverse sectors of society with diverse and occasionally antagonistic interests, such as the level of complexity that increases when new players arrive to participate. The foregoing makes necessary a rethink of what constitutes public participation both horizontally (among different groups within society), and vertically (between users and the different levels of government). Citizen participation must be given more respect than in the traditional forms of participation and with a high degree of professional ethics from which participation is coordinated, because the user will take part only to the extent that the organism responsible for water responds to its needs. Citizen participation should generate considerably more social benefits.

In terms of decentralisation, the water basin boards could be a good alternative for integral management of water. However, they must have real autonomy for regional planning and real decision-making capabilities. The vertical and centralized system for planning and taking decisions will have to be gradually abandoned, and this also includes a change of ideology at the highest national levels which for years have decided the way the country is going. Put another way, the Mexican political system must start to learn how to share power.

37.8 Effective Planning

Although Mexico's water problems are not recent, it was not until the early 1970s that they started to have enough weight to enter the national political agenda. The disorder and lack of planning in the use of water resources began to create difficulties and, above all, conflicts among users. This situation created the need for an instrument that would allow the organising of the uses and establish rules at the national and regional levels. In about 1971, the Ministry of Water Resources (SRH) prepared the terms of reference for what would be the fist National Water Plan (PNH), which was prepared for the World Bank, with the intention of seeking financing and technical support. A Bank commission discussed the terms with the regional representatives of the United Nations Development Programme (UNDP), which decided in favour of the project. This was how a tripartite commission was put together for carrying out the PNH. The Mexican government and the World Bank provided the funds and technicians and the UNDP provided technical assistance. It was also considered relevant to have an independent body to formulate the plan, and so in 1972 the Planning Commission was created, presided over by the head of the Ministry of Water Resources and staffed with personnel from the different federal departments related to water management and national and international consultants. This commission acted on a semi-independent basis from the Ministry of Water Resources (see Figure 37.6). The long term goal of the PNH was to formulate and bring into play a systematic process for planning and management of water resources that would allow appropriate selection of programmes, policies, and projects that would contribute to achieving the economic development goals of the country.

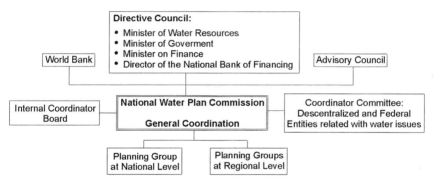

Fig. 37.6. 1975 National Water Plan organization

The work officially began on Sept. 1, 1972, and took two years and a half to complete. Thus was begun an intense interinstitutional work programme. Between 1972 and 1975, there were four meetings of the management board to establish work policies. The board consultant reviewed and discussed numerous reports and on five occasions directed the alignment for formulating plans and programmes. The coordinating committee held 40 meetings to review progress and to issue comments and observations. During the same time there were 25 coordinating meetings with the SRH; 16 of them were with the general directors and the other nine with regional directors. Also, the plan was submitted for consideration by deputies and senators in Congress. To ensure that the plan would be accepted, there were meetings with the state governments, users, and interested groups from the private sector, and there was wide coverage in the news media. Finally in

January 1976 there was a formal presentation by the PNH to the President of Mexico. Some of the results are summarised below (Herrera 1997):

- Dividing the country into regions on hydrological criteria;
- Estimating the regional availability of water through hydrological balances;
- Regional and national socio-economic diagnosis, including traditional development sponsors;
- Specific studies on the agriculture sector, since it was the largest consumer;
- Establishing specific objectives, goals, and programmes, including structural and non-structural measures, the electrical sector and urban centres; increasing service coverage; management of water in international basis; managing ground water; and security of water infrastructure;
- Criteria for changes in pricing policy seeking financial self-sufficiency within the sector;
- Programme for the training of personnel connected with the water sector;
- Programme for the collection, updating, and generation of information on issues related to water management, with special emphasis on the quantity and quality of water; and
- Financial programme for the implementation of the plan.

The projections, results, and conclusions of the PNH enabled the establishment of a group of quantitative goals for the year 2000:

- Food self-sufficiency by doubling the irrigated area. In 1976 the irrigated area was 3.2 million hectares (INEGI 1999b);
- Increase in aquatic production from 56,000 tonnes in 1970 to 316,000 tonnes;
- Water supply services to 95 percent of the urban population and 70 percent of the rural population;
- Sewage services for 80 percent of the urban population and 63 percent of the rural population;
- Increasing hydroelectric generation from 40 billion KWH to 500 billion KWH;
- Providing communities with populations of more than 2,500 with wastewater treatment systems;
- Maintaining pollution levels in bodies of water at or below those of 1970.

On March 26, 1976, a presidential decree created the National Water Planning Commission (CNPH), with the goal of executing the plan. In 1977 the PNH was revised with the goal of evaluating results obtained and updating the plan as part of the water planning process.

The necessary works for the PNH demanded a heavy output of human and financial resources. In 1975 the CNPH had 82 staff, of whom 40 percent had postgraduate degrees or a high level of specialisation. The costs of designing the PNH in December 1975 were estimated as follows: The Mexican government contributed US$5.8 million and the UNDP provided US$1.5 million. The work consumed 216,000 man-hours of foreign consultants' time and 1,166,400 man-hours of national specialists' time (Herrera 1997).

In 1976 the federal administration changed and with it the orientation of the nation's development goals. The ministries of Water Resources and of Agriculture were merged into the Ministry of Agriculture and Water Resources (SARH), with the goal of unifying the actions of government directed at resolving the growing problems of the agriculture sector. Under this new system the CNPH became part of the planning branch, while many of its functions were transferred to the water infrastructure branch. The main work of the CNPH is now directed at resolving problems in the agriculture sector, although it has some freedom to work in other areas of water policy. The goal of staging a review and updating of the PNH in 1977 was never carried out realised because it was thought that the benefits would be marginal in comparison with implementing the strategies established in 1975. The new institutional system imposed serious limits on the implementation of the PNH, causing plans and programmes for water management to fall apart, and much of the information generated was lost in the never-ending structural changes.

In 1983 the Ministry of Urban Development and Ecology (SEDUE) was created, and activities related to water quality were transferred there. In 1985 the functions of the SARH, including the CNPH, were grouped under the water infrastructure branch. In 1986 the CNPH became the Mexican Institute of Water Technology (IMTA), with the goal of reinforcing the 1975 recommendations of the PNH, with respect to technological development and training of personnel. In 1989 the National Water Commission (CNA) was formed as a body of the SARH, with exclusive powers to legislate and regulate water issues. Finally, in 1994, the Ministry of the Environment, Natural Resources and Fisheries (SEMARNAP) was created, with the CNA located within it with the goal of making coherent the activities of water management and the national objectives of environmental protection.

Over 25 years water management activities passed through four Ministries, and in 15 years they have been split up among two or more departments. This has created chaos in water management for lack of interinstitutional coordination, differences between the specific objectives of each ministry, lack of continuity in government structures, and dispersal of functions. This has been a serious barrier in planning processes with an integral focus and a long reach within the water sector, and the attainment of objectives. There is a planning process in the water sector, but the uncertainty about the continuity of the political structures and in recent years the economic uncertainty with each change of administration has distorted the planning processes, generating instruments and situations in which no one benefits from the development of the water sector, and which have become a burden for the resolution of water problems. Some of these aspects are presented below:

1. **Lack of consistency among objectives, goals, and time frames for their fulfillment**. The water policy spelled out in the water programme 1995-2000 set as its main goal the assurance of availability of water to satisfy the needs of the

population and to encourage the development of economic activities compatible with the environmental capabilities of each region. To do this, the programme sets out five quantitative goals (see Chap. 36). Of these, four refer to the creation of infrastructure for water supply, sewage, and sanitation, and only one refers to the broadening and rehabilitation of irrigation land zones. The result is that the goals proposed only refer to the generation of water infrastructure and do not include aspects such as: integration of water basin councils, decentralisation of functions, capacity building in the water sector, participation by the private sector, implementation of economic instruments, and training of human resources. The main goal is extremely broad and surpasses many of the quantitative proposals of the programme. In other words, even when the goals are met the main objective will not be close to being reached. Thus, we cannot believe that six years are enough to overcome the inertia and solve a problem that has been around for more than 50 years.

2. **Extravagance of economic resources and experience**. The lack of continuity in plans and programmes, and the sharp changes in national development strategies with each change of administration, create an important loss of time, money, and experience. Unfortunately, in the political arena there is a philosophy that originated long ago that no longer holds true, that each administration must reinvent the nation and must create diagnoses, plans, programmes, and executive projects that fulfill the new political directives and interests of the group in power. Thus in each administration significant amounts of money and man-hours are expended to establish new lines of action while the work of the previous administration is for the most part lost. Moreover, the restructuring of the institutional web results in a rotation of personnel, changes in management bodies, and a splitting up of work groups, so that the experience learned during the previous administration is diluted or lost, and the new groupings must spend time learning what they must do and how to do it. Generally water projects mature during the medium to long term, so that it is impossible for a single administration to develop a project and see its benefits. This, together with political uncertainty or because some other project becomes the latest fad, means there is a lack of institutional programmes to evaluate projects after they have been completed. Thus an impressive accumulation of information and experience is lost and is not applied to the planning processes.

3. **The absence of local proposals in the planning process**. Mexico is a complicated mixture of climates, topography, economic activity, society, and culture. The priorities, goals, and problems that face each region are different. This requires regional planning, in which the local players, based on their knowledge of their needs and the availability of resources, establish the mechanisms for appropriate management. The central planning and long-distance management that have prevailed for decades has denied this reality. In recent years there have been efforts to encourage regional planning, but the central vision of the federal officials always prevails, and this is evident in the fact that regional plans or programmes are often no more than a carbon copy of the central plans. This is not regional planning. As discussed above, the lack of training at the local level has been an obstacle for the development of their own plans. However,

we think that the real problem lies in the inability of federal officials to delegate power and decision-making authority to local officials. This is based on the argument that they might lose control, although what *loss of control* might mean is never made clear.

4. **The *lack of funds* argument.** Since the 1980s a new factor has arisen that has limited the planning process and made it uncertain: the recurring financial crises. Despite this, it could be argued that resources exist, given that Mexico is the 13th largest economy in the world. However, Mexico's political system has been traditionally reactive and not proactive. As a result, the system has developed a new way of planning: planning for crisis. This means that when a crisis begins one sector gets attention and money, while other sectors are ignored. For example, we have the highway rescue and the bank rescues, and now we have the rescue of the sugar sector, which has absorbed a considerable quantity of financial resources from other areas, mainly from funds that had been intended for social programmes. From this we must total the resources lost as a result of the lack of continuity of programmes and institutions. This creates not only poor management of the nation's natural resources but also terrible management of financial resources.

Planning puts into practical effect the strategies and actions the aims of a nation, and as such it is an indispensable element for development in any country. The sudden changes in development strategy with each administration should reinforce the notion that there is no clearly defined long-term national project in Mexico. As mentioned above, planning reflects the interests of the group in power at a point in time, and as such is limited to what can be done in six years. After that, others will decide what to do.

If what is wanted is consistency with the idea that broad-based national development is in tune with the principles of sustainability, the first thing to do will be to establish the different actors involved in the national scheme in an exercise that matches interests and objectives, and form a social pact in which the principles are set forth regarding what the national project should be. Once this has been done, the project must be publicised effectively throughout society with the intention that it be internalised and translated into action. It does not matter what has been done in the past, because each plan or programme established in the past emerged from an exercise similar to the one described above. The question then becomes: What is the difference this time? Consider the following: 1) for decades it has been only the political, academic and private sector elite that has been involved in this process; 2) information management has been closed; and 3) there is no mechanism that ensures the continuity of policies for the long term. Thus it is clear that these are the points that must be worked on. Whoever participates in this exercise must have a high sense of ethics and responsibility to fulfil the commitments.

37.9 Commentary

It is clear that the focus that prevails in the management of water resources is far from the principles of sustainability. The traditional foci that for decades appeared to solve water-related problems have been shown to have many weaknesses. Mexico faces a situation unlike any other in its history. The challenge for the future is great: to break the inertia that has formed over decades is not an easy task and in the next few years difficult decisions must be taken if problems are to be solved. The main challenge is how to manage water resources within the limitations. To think that the situation will be salvaged only through the investment of millions of dollars would be from the outset to set an impossible challenge. As long as sustainability is defined in these terms in high circles of world power, the concept becomes one more element of pressure on developing countries.

We think we have not passed the point of no return, and there is a possibility of giving things a different spin. The key is a change of attitude not only among the managers of water resources in the country, but also on the part of society. Otherwise the development and improvement of living standards of the people will simply not happen. This change of attitude must be supported by a real and extensive knowledge of the problem, the elements involved, the way in which the possible solutions are related; the development of a new set of values where humans and their relationship with the medium and its similarities occupy the centre, instead of macroeconomic indicators; and a new ethical sense where respect, joint responsibility, and sense of commitment will make it possible to establish a new social contract through which our problems will be solved together. In the long term, Mexico has no other choice.

38. Final Comments

Mexico's rapid growth, starting in the 1930s, created a need for accelerated development of water infrastructure to satisfy the growing demands of a growing population. The development of millions of hectares of irrigated land, dozens of industrial centres, providing thousands of services and the maintenance of important urban centres, has been made possible by the overuse of water resources, unjustified investments, and high environmental and social costs. There have been positive results, almost exclusively due to technical and financial aspects of the planning of water infrastructure, and this has created a large gap in several areas of the sector.

In general there are many shortcomings: the gap has been broadening between those who have access to water and those who do not; over-exploitation and pollution of surface and groundwater continues; the lack of water infrastructure, the lack of maintenance, or the lack of an adequate territorial organisation, and the effects of droughts and floods is increasingly severe; there is inefficiency and significant losses of water in the storage of potable water and purification; there is a lack of capacity in local and regional bodies in charge of water administration; and lack of a water consciousness in society. These are some of the problems Mexico faces with respect to water. The government needs to recognise and incorporate this into its political agenda. It is simply not possible to conceive of national development if the resource base on which development is based is not maintained. To proceed in any other way is to place at risk the development and social stability of many regions of the nation.

Through the analysis that has been presented, it is clear that the argument that the shortcomings of the water sector are caused by a lack of money lacks validity. Development goals based on purely economic considerations, together with the search for broad short-term financial benefits, have meant that the water sector has become a provider of resources without any thought given to the environmental or social effects that could result. This is totally at variance with the notion that national development should be constructed on the foundation of preserving natural resources and equal distribution of the benefits.

Instead of water being a strategic element for any development, the government has made it the basis for a series of client relationships with different sectors and different actors in society. The result was a total absence of users in the operation, maintenance, improvement and efficient use of water. As the years passed, the notion grew that water services provided by the government should be offered at no cost, or at very low cost. The government did not want to put a brake on this because so long as the economy of the 1940s, 50s, and 60s generated positive economic benefits, it was relatively easy to keep on offering free goods and subsidies to users, and at the same time to hide the deficiencies that began to develop in the water sector. Thus the image of the State as Benefactor grew, to the benefit of a certain political class.

Later, with the problems of the national economy in the 1970s, deficiencies that had begun to develop years earlier became obviously. In the same decade a new element appeared on the national scene: the recurring economic crises. From that point on, the main concern of the State was issues like the balance of payments, the trade balance, the external debt, the government deficit, and inflation. A plethora of macroeconomic indicators and concepts began to creep into the speeches of politicians. The argument grew: economic growth and orderly finances will provide the financial resources to lead to the protection of the environment, a more efficient use of natural resources, and a better distribution of wealth. According to the World Bank, Mexico is the 13th largest economy in the world; its achievements and economic discipline have enabled it to be accepted into the World Trade Organisation and the Organisation for Economic Cooperation and Development. If the arguments for this were justified, today we would be seeing an improvement of the standards and quality of life of the people, and of the nation's environment. If one were to make a cautious analysis of the data presented by SEMARNAT, INEGI, INE, CEPAL, CNA, and other organisations, and to evaluate the historic trends, it would become obvious that not only have things not improved, they have become worse. The question arises: Do economic growth and orderly finances equal improvement of the quality of life and of the environment? The evidence suggests that it does not, because there is no lineal connection. Certainly, financial capacity and investment are important factors for developing the water sector; that is why the government urgently needs to involve users and the private sector in generating and maintaining the water infrastructure. Paradoxically, the sector's problems began to emerge when there was a relative abundance of economic resources; nonetheless, there was also a total lack of vision and capacity to manage the sector, as well as an atmosphere of few ethics in the nation's political spheres.

The future is sure be complex, given existing shortcomings and future demands, new challenges imposed by globalisation, a society more critical and sensitive to social and environmental issues, technologies that in some cases will not be environmentally friendly, new paradigms for the management of natural resources, new structures and political networks, and actors who for decades did not take part in water planning. All of the foregoing will create situations that have never been seen before, and untested and novel solutions must be designed. For a nation like Mexico, where economic activities and the availability of water are unevenly distributed, the development of capacity in the water sector, among the users and in society, is a key condition for better water resource management.

If society wants to meet the goal of ensuring the availability of water to satisfy the needs of the population and to encourage the development of economic activities in a way that is compatible with the environmental capacity of each region, planning and water resource management must be done on a basis of having available better scientific understanding and technology, in such a way that development will not face political obstacles, so that poverty can be eradicated and the environment protected. This requires a total change in the thinking of those responsible for the water sector: the social, environmental, political, cultural, and economic factors must be genuinely considered in practice and not only in speeches, as has been the case in the past.

In the final analysis, our actions and not our words will be most important factor in solving water problems in the future. We must define objectively what the potential is and what the problems are, so that we know how to manage them efficiently within a reasonable time frame. The origin of the problem is well known, as are the solutions. The fact is that while the technological problems are relatively easy to solve, the political, economic, institutional, and social problems often present the greatest difficulties. In seeking technological alternatives, it will also be necessary to seek alternative ways of thinking based on ethics and responsibility. To think that only though the investment of millions of dollars can a situation be salvaged, is simply to create an insurmountable challenge. So long as sustainability continues to be defined in financial terms the concept will only become one more element of pressure on the country. The real question for Mexico is how to manage the water sector so that the economic, social, and environmental benefits are maximised and the costs minimised.

References

Aguilar A (1998) Financiamiento de Programas Hidráulicos. In: Arreguín FI, Donath E, Escalante C. Fernández A, Fuentes O, García N, Gutiérrez E, Hernández E, Luna H, Martín A, Rodríguez JA (eds) XV Congreso Nacional de Hidráulica – Memorias . Asociación Mexicana de Hidráulica – Instituto Mexicano de Tecnología del Agua, México, vol. 3, pp 1177-1182

Aguirre E (1998) Dificultades en la distribución de agua en distritos de riego que tienen superficies en dos o más estados de la República. In: Arreguín FI, Donath E, Escalante C. Fernández A, Fuentes O, García N, Gutiérrez E, Hernández E, Luna H, Martín A, Rodríguez JA (eds) XV Congreso Nacional de Hidráulica – Memorias . Asociación Mexicana de Hidráulica – Instituto Mexicano de Tecnología del Agua, México, vol. 3, pp 67-72

Aguirre A, Domínguez E (1998) Breve Historia del Abastecimiento de Agua en la Cuenca del Valle de México. In: Arreguín FI, Donath E, Escalante C. Fernández A, Fuentes O, García N, Gutiérrez E, Hernández E, Luna H, Martín A, Rodríguez JA (eds) XV Congreso Nacional de Hidráulica – Memorias . Asociación Mexicana de Hidráulica – Instituto Mexicano de Tecnología del Agua, México, vol. 3, pp 537-544

Alvarez P, Ramírez C, Orbe A (1999) Desarrollo de la Acuacultura en México y Perspectivas de la Acuacultura Rural, http://www.red-arpe.cl/paper/doc_04.pdf

Armienta M, Rodríguez R, Ongly L, Villaseñor G, Lathrop A, Mango H, Aguayo A, Ceniceros N, Cruz O (1996) Origen y Transporte del Arsénico en el Acuífero de Zimapan, Hidalgo. In: Asociación Latinoamericana de Hidrología Subterránea para el desarrollo (ed) 3er. Congreso Latinoamericano de Hidrología Subterránea. San Luis Potosí, México, vol 3, pp 101-107

Arregín F (1994) Uso Eficiente del Agua en Ciudades e Industrias. In: Garduño H, Arreguín-Cortés F (eds) Uso Eficiente del Agua. Comisión Nacional del Agua, México, pp 61-88

Avila P (1998) Agua, Poder y Conflicto Urbano en Morelia. In: Avila P (ed) Agua, Medio Ambiente y Desarrollo Sustentable. El Colegio de Michoacán, México, pp 202-214

BANAMEX-ACCIVAL (1999a) Indicadores Económicos. Examen de la Situación Económica de México 881:176-177.

BANAMEX-ACCIVAL (1999b) Indicadores Sociopolíticos. Examen de la Situación Económica en México 879:91

Becerril I (1999a) Inadecuado Ambito Económico para el Pleno Desarrollo Industrial. El Financiero press note 2/Jun, México

Becerril I (1999b) Temen empresarios invertir en ampliación de plantas. El Financiero press note 14/July, México

Becerril I (1999c) Desaliento e incertidumbre en el sector industrial. El Financiero press note 26/July, México

Bennett V (1998) Housewives, Urban Protest and Water Policy in Monterrey, Mexico. Water Resources Development 14:481-497

Borjas S (1999) Extranjeros comen el mercado de la rama de bienes de capital. El Financiero press note 22/July, México

Botello A, Villanueva S, Ponce G, Rueda L, Wong I, Barrera G (1995) La contaminación en las zonas costeras de México. In: Restrepo I (ed) Agua, Salud y Derechos Humanos. Comisión Nacional de Derechos Humanos, México, pp 53-122

Bourguett V, Ochoa L (1998) Consideraciones sobre la Reducción de Pérdidas de Agua Potable. In: Arreguín FI, Donath E, Escalante C. Fernández A, Fuentes O, García N, Gutiérrez E, Hernández E, Luna H, Martín A, Rodríguez JA (eds) XV Congreso Nacional de Hidráulica – Memorias . Asociación Mexicana de Hidráulica – Instituto Mexicano de Tecnología del Agua, México, vol. 3, pp 615-624

Cabrera A, Pacheco J, Coronado V (1997) Presencia de Organismos Coliformes Fecales en el Agua Subterránea de una Granja Porcícola en el Estado de Yucatán. In: Federación Mexicana de Ingeniería y Ciencias Ambientales A. C. (ed) Memorias Técnicas - XI Congreso Nacional de Ingeniería Sanitaria y Ciencias Ambientales. FEMISCA, México, vol. 1, pp 111-119

Cappi M (1999) Endeble situación financiera de las grandes empresas mexicanas. El Financiero press note 20/July, México

Castelán E (2000) Los Consejos de Cuenca en el Desarrollo de Presas en México. (Final report by the Third World Centre for Water Management, Mexico for the World Commission on Dams)

Castro, A. Iturbe R, Martínez JL, Marín L (1997) Características del Agua Subterránea en el Ciudad de Zacatecas y Alrededores. In: Federación Mexicana de Ingeniería y Ciencias Ambientales A. C. (ed) Memorias Técnicas - XI Congreso Nacional de Ingeniería Sanitaria y Ciencias Ambientales. FEMISCA, México, vol. 1, pp 263-270

CEASG (1999) Diagnóstico de Organismos Operadores de Agua Potable, Alcantarillado y Saneamiento 1995-1997. Comisión Estatal de Agua y Saneamiento del Estado de Guanajauto, Guanajuato, México

CFE (1992a) La Electricidad en México. Comisión Federal de Electricidad, México

CFE (1992b) Hacia el Siglo XXI. Comisión Federal de Electricidad, México

CFE (1997) Informe Anual 1997. Comisión Federal de Electricidad, México

CFE (1998) Capacidad Instalada de Centrales Eléctricas en las Regiones del Sistema Eléctrico Nacional: Escenario Diversificado. Comisión Federal de Electricidad, México

CFE (2001) Informe Anual 2000. Comisión Federal de Electricidad, México

Chávez RT, Gama F, Mata V (1998) Evaluación y Control de Acuíferos en Explotación para Abastecimiento de Agua a Centrales Térmicas de la Comisión Federal de Electricidad. In: Universidad de Chapingo (ed) I Seminario del Uso Integral del Agua. Universidad de Chapingo, México, vol 2, pp: 4_4_1 - 4_4_8

Cifuentes E, Blumenthal U, Ruiz-Palacios G (1995) Riego Agrícola con Aguas Residuales y sus Efectos sobre la Salud en México. In: Restrepo I (ed) Agua, Salud y Derechos Humanos. Comisión Nacional de Derechos Humanos, México, pp 189-202

CNA (1993a) Presas en México: Volumen I. Comisión Nacional del Agua. México. 348p

CNA (1993b) Presas en México: Volumen II. Comisión Nacional del Agua. México. 339p

CNA (1993c) Presas en México: Volumen III. Comisión Nacional del Agua. México. 337p

CNA (1993d) Presas en México: Volumen IV. Comisión Nacional del Agua. México. 267p

CNA (1993e) Presas en México: Volumen V. Comisión Nacional del Agua. México. 313p

CNA (1994a) El Agua y su Aprovechamiento Múltiple. Comisión Nacional del Agua, México

CNA (1994b) Programa Nacional Hidráulico, 1994. Comisión Nacional del Agua, México

CNA (1994c) Informe 1989 - 1994. Comisión Nacional del Agua, México

CNA (1994d) Presas en México: Volumen VII. Comisión Nacional del Agua. México. 273p

CNA (1994e). Presas en México: Volumen VIII. Comisión Nacional del Agua.México.271p

CNA (1994f) Presas en México: Volumen IX. Comisión Nacional del Agua. México. 286p

CNA (1994g) Presas en México: Volumen X. Comisión Nacional del Agua. México.278p

CNA (1994h) El Agua y su Aprovechamiento Múltiple. Comisión Nacional del Agua. México.

CNA (1996) Los Recursos Hídricos Subterráneos y la Disposición de Aguas Residuales Urbanas: Interacciones Positivas y Negativas. OMS-PNUMA-GEMS/OPS-CEPIS/ODA-BGS, Querétaro, México

CNA (1997a) Situación del Subsector Agua Potable, Alcantarillado y Saneamiento a Diciembre de 1996. Comisión Nacional del Agua, México

CNA(1997b) Estrategias del Sector Hidráulico. Comisión Nacional del Agua, México

CNA (1998a) Los Consejos de Cuenca en México, Definiciones y Alcances. Comisión Nacional del Agua, México

CNA (1998b) Situación del Subsector Agua Potable, Alcantarillado y Saneamiento a Diciembre de 1997. Comisión Nacional del Agua, México

CNA (1998c) Inventario de Plantas de Tratamiento. Comisión Nacional del Agua, México

CNA (1999a) Compendio Básico del Agua en México. Comisión Nacional del Agua, México

CNA (1999b) Ley de Aguas Nacionales y su Reglamento. Comisión Nacional del Agua, México

CNA (1999c) Diagnóstico Hidráulico y Lineamientos Estratégicos en la Cuenca Lerma – Chapala. Comisión Nacional del Agua, México

CNA (2000) El Agua en México: Retos y Avances. Comisión Nacional del Agua, México

CNA (2001) Programa Nacional Hidráulico, 2001 - 2006. Comisión Nacional del Agua, México

Colín M (1999) Persisten fallas estructurales en exportaciones manufactureras. El Financiero note press 2/August, México

CONAPO (1992) Síntesis del Estudio Binacional México – Estados Unidos sobre Migración. Consejo Nacional de Población y Vivienda, México

Fabre L (1999) La pobreza, a Prueba de Programas Sexenales. El Financiero note press 2/June, México

Forsius J (1993) Rapporteur's report on session 1: Regulatory challenges facing hydropower today. In: International Energy Agency (ed) Hydropower, Energy and the Environment: Conference Proceedings. OCDE, Sweden, pp 9-11

Fragoza DF (1998) Descargas Intermitentes, una Alternativa para el Uso Eficiente del Agua de Riego. In: Universidad de Chapingo (ed) I Seminario del Uso Integral del Agua. Universidad de Chapingo, México, vol 2, pp: 2-16-1

Frausto O, Guiese S (1996) Hidrogeoquímica del Sistema de Pozos del Municipio de Tlalnepantla, Estado de México. In: Asociación Latinoamericana de Hidrología Subterránea para el Desarrollo (ed) 3er. Congreso Latinoamericano de Hidrología Subterránea. ALHSD, San Luis Potosí, México, vol 3, pp 87-91

Fundación Friedrich Ebert (1995) La acuicultura en el Desarrollo Regional. Fundación Friedrich Ebert – Desarrollo Ambiente y Sociedad, S. C., México

Gabino R, Perez F (1998) Prospectiva para Agua Potable, Alcantarillado y Saneamiento. In: Arreguín FI, Donath E, Escalante C. Fernández A, Fuentes O, García N, Gutiérrez E, Hernández E, Luna H, Martín A, Rodríguez JA (eds) XV Congreso Nacional de Hidráulica – Memorias . Asociación Mexicana de Hidráulica – Instituto Mexicano de Tecnología del Agua, México, vol. 3, pp 1001-1008

Garcés C, Marañon B, Scott C (1997) Mexico, Irrigation Sector Profile. International Irrigation Management Institute, México

Gómez M, Izurieta J, Saldaña P (1998) Los Estudios de Clasificación de Corrientes como Herramienta para Evaluar el Impacto de las Descargas de Aguas Residuales. In: Arreguín FI, Donath E, Escalante C. Fernández A, Fuentes O, García N, Gutiérrez E, Hernández E, Luna H, Martín A, Rodríguez JA (eds) XV Congreso Nacional de Hidráulica – Memorias . Asociación Mexicana de Hidráulica – Instituto Mexicano de Tecnología del Agua, México, vol. 3, pp 259-266

Goodland R (1996) The environmental sustainability challenge for the hydro industry. Hydropower & Dams 1:37–42

González CA, De León MB, Fuentes RC (1998) El Drenaje Agrícola en México. In: Arreguín FI, Donath E, Escalante C. Fernández A, Fuentes O, García N, Gutiérrez E, Hernández E, Luna H, Martín A, Rodríguez JA (eds) XV Congreso Nacional de Hidráulica – Memorias . Asociación Mexicana de Hidráulica – Instituto Mexicano de Tecnología del Agua, México, vol. 3, pp 121-128

González V (1999) El TLC no garantiza la eliminación de las crisis sexenales: analistas. El Financiero press note 28/May, México

González R, Canales A (1995) Contaminación por plaguicidas en el acuífero del Valle del Yaqui. In: Restrepo I (ed) Agua, Salud y Derechos Humanos. Comisión Nacional de Derechos Humanos, México, pp 203-220

Gorriz C (1996) Irrigation Management Transfer in Mexico. World Bank, Washington, D.C.

Guerrero G (1999) Mensaje del Director General de la Comisión Nacional del Agua. In: Arreguín F, Herrera C, Marengo H, Paz GA (eds) El Desarrollo de las Presas en México. Asociación Mexicana de Hidráulica, México, pp 221-223

Guillen G (1999) Paga el usuario $1.60 por m^3 de agua; cuesta $8.0. El Universal note press 23/March, México

Guitrón A, Saavedra JC, Hernández I, Méndez S (1998) Programa de Mejoramiento de la Eficiencia del Servicio de Agua Potable de la Zona Hotelera de Cancún, Q. Roo. In: Arreguín FI, Donath E, Escalante C. Fernández A, Fuentes O, García N, Gutiérrez E, Hernández E, Luna H, Martín A, Rodríguez JA (eds) XV Congreso Nacional de Hidráulica – Memorias . Asociación Mexicana de Hidráulica – Instituto Mexicano de Tecnología del Agua, México, vol 3, pp 633-640

Gutiérrez C, Sánchez L (1996) Hidrogeoquímica del Sistema Acuífero Tesistán – Toluquilla, Jalisco – México. In: Asociacion Latinoamericana de Hidrología Subterránea para el Desarrollo (ed) 3er. Congreso Latinoamericano de Hidrología Subterránea. ALHSD, San Luis Potosí, México, vol 3, pp 19-34

Gutiérrez E (1999a) El Sector Social, privado de blindajes y perspectivas El Financiero press note 28/May, México

Gutiérrez E (1999b) Estancamiento en ventas internas; crecieron 2.1 por ciento real hasta septiembre de 1999. El Financiero note press 12/November, México

Gutiérrez E (1999c) Aumenta la banca de desarrollo el monto de recursos para respaldar intereses vencidos. El Financiero note press 22/June, México

Herrera C (1997) National Water Master Planning in México. In: Biswas AK, Herrera-Toledo C, Garduño-Velasco H, Tortajada-Quiroz C (eds) National Water Master Plans for Developing Countries. Oxford Press, India, pp 6-53

Ibarra M (1999) Los programas de protección ambiental, afectados por la crisis. La Jornada 10/November, México

INEGI (1996a) Perfil Sociodemográfico, XI Censo General de Población y Vivienda 1990, México 1992; y Conteo de Población y Vivienda 1995. Instituto Nacional de Estadística, Geografía e Informática, Aguascalientes, México

INEGI (1996b) XI Censo General de Población y Vivienda, 1990; y Conteo de Población y Vivienda, 1995. Instituto Nacional de Estadística, Geografía e Informática, Aguascalientes, México

INEGI (1998a) Estadísticas del Medio Ambiente – México 1997. Instituto Nacional de Estadística, Geografía e Informática, Aguascalientes, México

INEGI (1998b) Anuario Estadístico de los Estados Unidos Mexicanos, 1997. Instituto Nacional de Estadística, Geografía e Informática, Aguascalientes, México

INEGI (1998c) El Sector Energético en México. Instituto Nacional de Estadística, Geografía e Informática, Aguascalientes, México

INEGI (1999a) Estadísticas Históricas de México I. Instituto Nacional de Estadística, Geografía e Informática, Aguascalientes, México

INEGI (1999b) Estadísticas Históricas de México II. Instituto Nacional de Estadística, Geografía e Informática, Aguascalientes, México

INEGI (2000a) Estadísticas del Medio Ambiente – México 1999. Instituto Nacional de Estadística, Geografía e Informática, Aguascalientes, México

INEGI (2000b) Agenda Estadística. Instituto Nacional de Estadística, Geografía e Informática, Aguascalientes, México

INEGI (2001) Niveles de Bienestar por Entidad Federativa, http://www.inegi.gob.mx/difusion/espanol/niveles/jlynivbien/entidad.html

IPCC (1997) Impactos regionales del cambio climático: evaluación de la vulnerabilidad: Resumen para responsables de políticas, http://www.ipcc.org

Izurieta J, Ruiz A (1998) Simulación de la Calidad del Agua en el Río Cazones, Veracruz. In: Arreguín FI, Donath E, Escalante C. Fernández A, Fuentes O, García N, Gutiérrez E, Hernández E, Luna H, Martín A, Rodríguez JA (eds) XV Congreso Nacional de Hidráulica – Memorias . Asociación Mexicana de Hidráulica – Instituto Mexicano de Tecnología del Agua, México, vol 3, pp 249-258

Jaimes S, Robles B, Sánchez J, Iñiguez M (1998) Sistema para la Evaluación Económica de Proyectos de Infraestructura Hidroagrícola. In: Arreguín FI, Donath E, Escalante C. Fernández A, Fuentes O, García N, Gutiérrez E, Hernández E, Luna H, Martín A, Rodríguez JA (eds) XV Congreso Nacional de Hidráulica – Memorias . Asociación Mexicana de Hidráulica – Instituto Mexicano de Tecnología del Agua, México, vol. 3, pp 417-424

Jiménez B (1999) Podría aumentar la volatilidad de corto plazo en la BMV, por la combinación de factores políticos y financieros. El Financiero press note 24/June, México

Jonathan F, Aranda J (1996) Decentralization & Rural Development in Mexico, Center for U.S.-Mexican Studies. University of California, San Diego

Kölher R (1999) El Uso de Agua Reciclada para la Irrigación Agrícola. In: Fideicomiso de Infraestructura Rural y Riego Compartido (ed) 4to Simposium Internacional de Fertiirrigación, FIRCO, Jalisco, México, pp 40-56

Limón H (1998) Modernización del Sector Hidráulico en México: Autofinanciamiento del Sector. In: Arreguín FI, Donath E, Escalante C. Fernández A, Fuentes O, García N, Gutiérrez E, Hernández E, Luna H, Martín A, Rodríguez JA (eds) XV Congreso Nacional de Hidráulica – Memorias . Asociación Mexicana de Hidráulica – Instituto Mexicano de Tecnología del Agua, México, vol. 3, pp 461-468

MACRO (1995) Realidad Económica de México 1995. Editorial Iberoamericana, México

Martínez H (1998) Efectos a la Salud por Aguas Residuales. In: Universidad de Chapingo (ed) I Seminario del Uso Integral del Agua. Universidad de Chapingo, México, vol 1, pp: 3_8_1 – 3_8_10

Mayoral I (1999) Crecimiento Muy Heterogéneo. El Financiero press note 10/July, México

Merino M (1995) Algunos Dilemas de la Descentralización en México. In: Foro del Ajusco (ed) Desarrollo Sostenible y Reforma del Estado en América Latina. Colegio de México – Programa Universitario de Medio Ambiente y Desarrollo, México, pp: 167-190

Monroy O, Viniegra G (1998) Manejo Sustentable en la Ciudad de México y Zona de Influencia. In: Universidad de Chapingo (ed) I Seminario del Uso Integral del Agua. Universidad de Chapingo, México, vol 1, pp: 3_11_1 – 3_11_11

Moreno J (1995) El Agua en Sonora: Escasa, Mal Utilizada y Contaminada. In: Restrepo I (ed) Agua, Salud y Derechos Humanos. Comisión Nacional de Derechos Humanos, México, pp 221-258

National Research Council (1995) El Agua y la Ciudad de México: Mejorando La Sustentabilidad. National Research Council – Academia de la Investigación Científica, A.C. – Academia Nacional de Ingeniería, Washintong, D.C.

Novelo G (1998) Participación de FIRA en el Uso Racional del Agua y la Energía en el Sector Rural. In: Universidad de Chapingo (ed) I Seminario del Uso Integral del Agua. Universidad de Chapingo, México, vol 1, pp: 2_1_1 – 2_1_13

OCDE (1998) Análisis del Desempeño Ambiental, México. Organización para la Cooperación y el Desarrollo Económicos, México

Ojeda BW, Peña PE (1998) La Calendarización del Riego en Tiempo Real: Una Alternativa Hacia el Uso Eficiente del Agua. In: Universidad de Chapingo (ed) I Seminario del Uso Integral del Agua. Universidad de Chapingo, México, vol 1, pp 2_6_1 – 2_6_9

Oliva C (1999) Estado Actual de las Presas, Breve Reseña. In: Arreguín F, Herrera C, Marengo H, Paz GA (eds) El Desarrollo de las Presas en México. Asociación Mexicana de Hidráulica, México, pp 33-37

OPS (1994) Las Condiciones de Salud en las Américas. Organización Panamericana de la Salud, Washington, D.C.

Ortiz G (1997) La Política del Agua en México en el Marco del Desarrollo Sustentable. Ingeniería Hidráulica en México, pp 25-36

Pacheco J, Cabrera A, Gómez A (1997). Especies nitrogenadas en el agua subterránea de un área con actividad agropecuaria en el Estado de Yucatán. In: Federación Mexicana de Ingeniería y Ciencias Ambientales A. C. (ed) Memorias Técnicas - XI Congreso Nacional de Ingeniería Sanitaria y Ciencias Ambientales. FEMISCA, México, vol. 1, pp 120-128

Parra H, Sorchini H (1998) Situación Actual del Recurso del Agua y Demanda Potencial de Agua Tratada en el Valle de México. In: Arreguín FI, Donath E, Escalante C. Fernández A, Fuentes O, García N, Gutiérrez E, Hernández E, Luna H, Martín A, Rodríguez JA (eds) XV Congreso Nacional de Hidráulica – Memorias . Asociación Mexicana de Hidráulica – Instituto Mexicano de Tecnología del Agua, México, vol. 3, pp 1267-1272

Paz G (1999) El Panorama del Agua en México. In: Arreguín F, Herrera C, Marengo H, Paz GA (eds) El Desarrollo de las Presas en México. Asociación Mexicana de Hidráulica, México, pp 9-26

Pazos L (1999) Hablando en Plata, Más impuestos a las maquiladoras. El Financiero press note 17/July, México

PNUMA (1993) Aspectos Técnicos Relevantes para las Negociaciones de la Convención Internacional para Combatir la Desertificación en América Latina y el Caribe. Programa de las Naciones Unidas para el Medio Ambiente, México

Ramírez C, Sánchez V (1997) Una propuesta de diversificación productiva en el uso del agua a través de la acuacultura. In: Universidad de Chapingo (ed) I Seminario del Uso Integral del Agua. Universidad de Chapingo, México, vol 3, pp: 1_9_1 – 1_9_7

Ramos O (1999a) Estado Actual de las Presas en México. In: Arreguín F, Herrera C, Marengo H, Paz GA (eds) El Desarrollo de las Presas en México. Asociación Mexicana de Hidráulica, México, pp 29-32

Ramos C (1999b) Perspectiva de la Infraestructura Hidroagrícola en el Próximo Milenio. In: Fideicomiso de Infraestructura Rural y Riego Compartido (ed) 4to Simposium Internacional de Ferti-irrigación, FIRCO, Jalisco, México, pp 15-25

Rangel J (1998) Operación de Ocho Circuitos de Control en la Red de Distribución de Agua Potable del Distrito Federal. In: Arreguín FI, Donath E, Escalante C. Fernández A, Fuentes O, García N, Gutiérrez E, Hernández E, Luna H, Martín A, Rodríguez JA (eds) XV Congreso Nacional de Hidráulica – Memorias . Asociación Mexicana de Hidráulica – Instituto Mexicano de Tecnología del Agua, México, vol. 3, pp 655-662

Russo T (1994) Making hydropower sustainable. Hydropower & Dams, 1(6):126-131

Saade L (1997) Toward more efficient urban water management in Mexico. Water International, 22(3):153-167

Sánchez J (1999) Impulsaría a Ingenios el uso de alcohol en vehículos. El Universal press note 9/Augost, México

Sandoval A (1999) En cinco años la depreciación cambiaria aumentó 60% el pago de intereses de las empresas. El Financiero press note 3/November, México

SARH/CNA (1994) Informe 1989-1994. Secretaría de Agricultura y Recursos Hidráulicos – Comisión Nacional del Agua, México

Schettino M (no date) Economía Informal, Comercio Exterior. El Universal press note, México

Schettino M (1999) Economía Informal, Crecimiento Industrial. El Universal press note 19/May, México

Seabra F (1993) Hydropower the forgotten renewable. In: International Energy Agency (ed) Hydropower, Energy and the Environment: Conference Proceedings. OCDE, Sweden, pp 45-48

SECOFI (1995) Crecimiento de la Población Trabajadora 1985-1994. Dirección general de Promoción de la Micro, Pequeña y Mediana Empresas y de Desarrollo Regional, México

Secretaría de Energía (1999) Propuesta de Cambio Estructural de la Industria Eléctrica en México. Secretaría de Energía, México

SEGOB (1998) Cuarto Informe de Gobierno 1998. Secretaria de Gobernación, México

SEMARNAP (1995) Programa de Pesca y Acuacultura 1995-2000. Secretaría del Medio Ambiente, Recursos Naturales y Pesca, México

SEMARNAP (1997) Sistema Integrado de Regulación y Gestión Ambiental de la Industria. Secretaría del Medio Ambiente, Recursos Naturales y Pesca, México

SEMARNAP (1999) Informe de Actividades 1998-1999. Secretaría del Medio Ambiente, Recursos Naturales y Pesca, México

SEMARNAP/CNA (1996) Programa Hidráulico 1995-2000. Poder Ejecutivo Federal, Estados Unidos Mexicanos, México

SEMARNAP/CNA (1997) Sistema Cutzamala. Secretaría del Medio Ambiente, Recursos Naturales y Pesca - Comisión Nacional del Agua, México

SEMARNAP/DGA (1999) Desarrollo de la Acuacultura en México. Secretaría del Medio Ambiente, Recursos Naturales y Pesca – Dirección General de Acuacultura, México

SEMARNAP/PROFEPA (1995) Mortandad de Aves Acuáticas en la Presa de Silva, Guanajuato. Secretaría del Medio Ambiente, Recursos Naturales y Pesca - Procuraduría Federal de Protección al Ambiente, México

SEMARNAT (2002) Recursos Hidráulicos, en http://www.semarnat.gob.mx/16080/sniarn/ recursos hidráulicos.html.

Solís B (1999) Por qué sigue la inflación en México. El Financiero press note 1/June, México

Solís M, Arenas R (1998) Uso Eficiente del Agua de Riego en Guanajuato. In: Universidad de Chapingo (ed) I Seminario del Uso Integral del Agua. Universidad de Chapingo, México, vol 3, pp: 2_3_1 – 2_3_3

SSA (1999) Enfermedades Gastrointestinales por Estado 1997, en http://www.ssa.gob.mx/reportes/estado/gastrointestinales 1997.html

Takeda I (1998) La Alianza para el Campo y la Sustentabilidad de la Agricultura en Riego. In: Universidad de Chapingo (ed) I Seminario del Uso Integral del Agua. Universidad de Chapingo, México, vol 2, pp: 2_7_1 – 2_7_11

Tortajada C (1999a) Approaches to Environmental Sustainability for Water Resources Management: A Case Study of Mexico. Licentiate thesis, Royal Institute of Technology

Tortajada C (1999b) Environmental Sustainability of Water Management in Mexico. Third World Centre for Water Management, México

UAS (1994) Temas sobre la Administración de Recursos Pesqueros en México. Universidad Autónoma de Sinaloa, México

UNAM (1997) Environmental Issues: The Mexico City Metropolitan Area. Programa Universitario de Medio Ambiente, Departamento del Distrito Federal, Gobierno del Estado de México, Secretaría de Medio Ambiente, Recursos Naturales y Pesca, México

Urquídi VL (1996) México en la Globalización. Fondo de Cultura Económica, México

Vázquez E, Castillo E, Méndez R (1995) Calidad del Agua de Consumo en Yucatán. In: Restrepo I (ed) Agua, Salud y Derechos Humanos. Comisión Nacional de Derechos Humanos, México

Vega M (1999) Estado Actual de las Presas, México 1999. In: Arreguín F, Herrera C, Marengo H, Paz GA (eds) El Desarrollo de las Presas en México. Asociación Mexicana de Hidráulica, México, pp 39-62

Villasuso M, Granel E, Morris B (1996) *Evaluación / Contaminación del Acuífero de la Ciudad de Mérida*. In: Asociacion Latinoamericana de Hidrología Subterránea para el Desarrollo (ed) 3er. Congreso Latinoamericano de Hidrología Subterránea. ALHSD, San Luis Potosí, México, vol 1, pp 53-65

Villegas C (1999) Fuerte caída del crédito bancario a particulares. El Financiero press note 2/Augost, México

WHO (1997) Cholera in 1996. Weekly Epidemiological Record, 72(31):229-236

WHO (1998) Cholera in 1997 Weekly Epidemiological Record, 73(27): 201-208

World Bank (1996) Staff Appraisal Report, Mexico: Water Resources Management Project. World Bank, Washington, D.C.

Printing: Mercedes-Druck, Berlin
Binding: Stein+Lehmann, Berlin